Guardians
of the Parks

Guardians of the Parks

A History of the National Parks and Conservation Association

John C. Miles

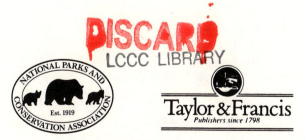

Published by Taylor & Francis
in cooperation with National Parks
and Conservation Association

Washington, D.C., 1995

USA	Publishing Office:	Taylor & Francis 1101 Vermont Ave., N.W., Suite 200 Washington, DC 20005 Tel: (202) 289-2174 Fax: (202) 289-3665
	Distribution Center:	Taylor & Francis 1900 Frost Road, Suite 101 Bristol, PA 19007-1598 Tel: (215) 785-5800 Fax: (215) 785-5515
UK		Taylor & Francis, Ltd. 4 John Street London WC1N 2ET Tel: 071 405 2237 Fax: 071 831 2035

GUARDIANS OF THE PARKS: A History of the National Parks and Conservation Association

1 2 3 4 5 6 7 8 9 0 EBEB 0 9 8 7 6 5

This book was set in Garamond by National Parks and Conservation Association. Cover and text design by Ice House Graphics. Technical development by Bernadette Capelle and production assistance by Bonny Gaston. Printing and binding by Edwards Brothers, Inc.

A CIP catalog record for this book is available from the British Library.

∞ The paper in this publication meets the requirements of the ANSI Standard Z39.48-1984 (Permanence of Paper). ♲ Printed on recycled stock.

Library of Congress Cataloging-in-Publication Data

Miles, John C.
    Guardians of the parks: a history of the NPCA/by John C. Miles.
       p. cm.
    Includes bibliographical references.

    1. National Parks and Conservation Association—History.
I. Title.
SB482.A4N37435   1995   95-134
333.78'316'06073—dc20   CIP

ISBN 1-56032-446-5

Contents

Preface

ounded in 1919, the National Parks and Conservation Association celebrated its seventy-fifth anniversary in 1994. To commission an anniversary history is not unusual; for an anniversary history, particularly of an advocacy group, to be rich, thorough, and frank is quite unusual. The association, and author John C. Miles, have had the courage to tell the story of this organization, the only one focused solely on the National Park System of the United States, honestly. All idealistic bodies suffer from inner tensions, but outsiders may not know of them, because it is a truism of historical writing that the victor writes the history. Here those tensions stand healthily and persuasively revealed.

The National Parks Association, as it was initially called, was born only three years after the creation of the National Park Service. There was then no park system as such, for the national parks and monuments administered by the new service were a mixture of the grand, the glorious, and the relatively insignificant. It was the association that first saw the need to outline a true system, by which appropriate landscapes of superlative quality might be acquired for administration by the service. Thus in time, as the service itself moved to an understanding of the need for systematic standards in the selection of new units, the two bodies—one a government bureau, the other a group of private citizens—worked together (and at times in opposition) to create what has become the finest, most elegant, and indeed most systematic collection of national parks in the world.

A history of the National Parks and Conservation Association is, thus, also a history of the American national parks. Here we see the two bodies, Park Service and parks association, struggle to define standards, to make good on the promise that the national parks would be nature's universities. Often the association was seen as too "purist," too insistent upon the highest standards of selection and protection; often the service was seen as too ready to make political compromises, too willing to damage the very resource it was committed to protect in order to ensure that the visitor had a comfortable bed to sleep in at night and a paved highway from which to view nature's (and increasingly history's) wonders. It will come as a surprise to some readers to discover that many of the presumed heroes of the national park movement in the United States, not excluding the much-lauded Horace Albright and Newton Drury, often were prepared to compromise with political and commercial pressures while NPCA (often alone) spoke out against the degradation of park standards.

To be sure, NPCA, like the National Park Service, also made mistakes. Indeed, this history is filled with facinating, bullheaded, egotistical maen (and some women), all certain they knew exactly what a national park was and must be. It was not the National Park Service so much as specific, powerful individuals within the service who wished to develop Olympic National Park or proposed putting a tramway up one of Yosemite's most splendid cathedrals of stone. It was specific individuals within NPCA—Robert Sterling Yard, Sigurd Olson, and others—who made the association what it became. Miles' history is rich in anecdote, in chapter and verse, in names and numbers, and reading it should tell us how important the quality of leadership is.

The national Parks and Conservation Association is unique among the powerful band of environmentally focused organization, for it alone takes the national parks—368 units in 1995—as its province. To have one organization that, save for moments of lapse, has kept its eye unremittingly on a single aspect of the battle to protect our nation's heritage has made possible many victories. If eternal vigilance is the price of liberty, a clear focus on a particularity of purpose is essential to success. Dr. Miles' history helps to make this clear. It is an account from which everyone interested in our national parks will benefit.

Robin W. Winks

Introduction

The National Park System ranks among the most popular of American institutions. Designate a national park and the people come—today in such numbers that the values parks were established to protect and the experiences they were to provide are threatened.

The story of the National Park System is a familiar one. Historians have examined the history of national park policy and of the government agency in charge, the National Park Service. Biographies have been written of such central figures as John Muir, Stephen Mather, and Horace Albright. Yet the role of one player in national park history remains relatively unexplored—the National Parks and Conservation Association (NPCA).

Although the National Parks Association (NPA), as NPCA was called until 1970, seems to be a persistent presence in accounts of national park debates, its role is obscure, and it is seldom credited with influencing the course of national park history. What has its role been? Where does it lie in the constellation of conservation organizations with which it has shared the conservation stage?

I discovered that if I wanted answers to these questions, I would have to excavate the details of the NPCA story myself, and that is what I have attempted to do.

No NPCA organizational history has been compiled beyond a few articles in the association's own *National Parks* magazine. My first task was to trace how the purposes, activities, and organization of the association have evolved. When Michael Cohen wrote the history of the Sierra Club, he called his approach an "inside narrative," a term borrowed from Herman Melville. It meant that his history focused on the views "of those on board the ship." Cohen's account relied heavily on records of the board of directors and other leaders of the group.

Sequoia National Park, established in 1890 as the nation's second national park, was the subject of National Parks Association concern in the 1920s and 1930s.

I, too, have written an "inside narrative," for I found most information in meeting minutes, in association publications, in correspondence of NPA leaders, and in interviews with more recent players. I have pieced together a complex organizational history primarily from these sources, consulting materials from the National Park Service and the secondary literature where I could find them. Thorough study of the association as viewed by people outside the organization might yield a different picture. I have attempted merely to describe the important ideas, events, and people in the history of NPCA, and to follow several threads through the fabric of that history.

The subject of this book is the evolution of a specialized nongovernmental organization dedicated to the idea of a system of national parks. Its major task is to describe how NPCA appeared on the conservation scene, how it defined and pursued its goals, and how it played its role in the history of national parks and conservation.

The national park idea was well formed, but had not hardened, before the National Parks Association appeared in 1919. Composed of educators, scientists, idealists, and amateurs, NPA quickly became the principal advocate of the ideal national park and a "rational" park system. In its early years, it sought to define these ideals and to fight for them in the face of pork-barrel politics, economic opportunism, and the National Park Service's expansionism.

NPA posed two related sets of questions in this struggle: What qualities should define a national park, and once a park was created, for what purpose should it be managed, and what criteria should guide that management? From its beginning, NPA took the position that only the most outstanding examples of America's heritage should become national parks, but that stand led to two other questions: Who would define what was best, and who would decide where the park boundary lines should be drawn? Struggles over these questions have been at the center of the organization's activity throughout its history.

In addressing the related question about purpose, the central dilemma for everyone, inside or outside of government, has been where to draw the line between preservation and use. NPA favored protection because its core belief was that national parks were virtually the only places where humans could seek and find answers to questions about nature and their relation to it. Outside the parks, human activity was rapidly transforming the natural world. Robert Sterling Yard, one of NPA's founders, called national parks "supreme examples" where nature "is still creating the earth upon a scale so vast and so plain that even the dull and frivolous cannot fail to see and comprehend." John Merriam, a trustee in the 1920s and 1930s, observed that "When one looks upon nature...there arises inevitably the inquiry: 'What does this mean to me? What is man's place in the world of nature?'" National parks, remnants of undisturbed nature, should help provide the answers to these questions. This core belief, which evolved over decades, shaped by both external and internal forces, has guided and inspired generations of NPCA leaders.

Four themes can be traced through the history of NPCA, and through all chapters of this book. One theme involves its constant struggle for organiza-

The automobile influenced the history of national parks, shaping the National Park System in many ways. Here, stage coach meets motor stage in an early national park encounter.

tional identity. The association claims that its distinction in the field of conservation and environmental action lies in its single-minded pursuit of a park-oriented agenda. This seems a valid claim, yet the NPCA story reveals that the association struggled constantly to define and hold its place in this field. Just as its mission and goals have evolved continuously, so have its relationships with other conservation organizations and the National Park Service.

A second theme relates to standards. During the association's first thirty years, standards guided everything. Proposals for new parks were supported or opposed on the basis of whether they met the standards. This concern about standards propelled NPA into conflicts with the Park Service and even fellow conservationists amid charges that it was "purist" and "anti-park." Over the decades, social values and the composition of the American community changed, and the conservation movement was forced to change with them. All of this required a constant reevaluation by NPCA of what should define a national park, most recently in its 1988 *National Park System Plan*.

A third theme involves the shifting nature of NPCA's leadership. The association's story is dominated by strong personalities—such as Robert Sterling Yard, Devereaux Butcher, Sigurd Olson, Anthony Wayne Smith, and Paul Pritchard—and their efforts to guide the organization in the direction they thought it should go. I have attempted to describe the influence of these leaders and to understand their motives, to explore how these powerful individuals have affected both NPCA's history and the course of national park history.

A final theme is the organization's role as watchdog of the National Park Service. This is an important aspect of its history, for NPCA has been more closely tied to the fate of a government agency than any other conservation group. Throughout the decades, the relationship between the government agency and nongovernmental watchdog has vacillated. Although both have worked to create a national park *system*, they have not always agreed on what that system should comprise. Some of these differences lie in the Park Service's contradictory mandate to preserve and protect national parks while providing public access to them. One consistent area of disagreement, for example, surfaces in debates about the place of recreation in national parks. Stephen Mather, founding director of the National Park Service, and his successor Horace Albright accepted the centrality of resource protection in the national park mission, but believed also that national parks should meet the growing need for outdoor recreation. Their support for recreation was in part motivated by their desire to garner growing public support for national parks and their agency. From its beginning, Yard's NPA placed recreation well down the list of national park priorities. Only recently has NPCA softened its stance on recreation, and only for some types of national parks. I have tried to show how the descendants of these founders have at times argued over recreation in the national park mission.

As the writing of this work progressed, I was faced with a decision: Should the National Parks Association be described as a "conservation" organization or something else? Its members called their mission "conservation" and wrote of themselves as conservationists; the word was added to the name in 1970. In view of this, I have referred to NPA as a conservation group until the emergence of "environmentalism" in the 1970s when, in its own eyes—despite its new name—it became an "environmental" organization.

Historians have made a distinction between "conservation" and "preservation," placing protection of parks and wilderness under the latter rubric. The distinction is that conservation was a utilitarian concept linked to the productive potential of a resource, often a place, which required that the resource be used wisely with the aim of making it last as long as possible. The underlying assumption was that the resource would be used, often consumed, and sometimes depleted. Preservation, on the other hand, involved protection from use. Although some use might occur, it would not involve consumption in the same sense as the "conserved" natural resource. The assumption was that the "preserved" resource would not be depleted.

The record leaves little doubt that NPCA's priorities have been protection and preservation. All of this is a caveat to explain why I refer to NPCA throughout its history as a "conservation" organization, despite its clearly preservationist mission.

NPCA has attempted to be a vehicle for people who care about national parks to work on that interest. Its special interest has been and is the National Park System. One might expect the story of such a narrowly focused organization to be straightforward, yet it is not. Its history reveals uncertainties, false starts, mistakes, uprisings, and minor intrigues. Although NPCA has weak-

Stephen Mather, third from left, Robert Sterling Yard to Mather's left, and Horace Albright on the far right of this photo were central players in the early years of the National Parks Association.

ened and nearly died several times, it has revived to be, in the 1990s, healthy and vigorous.

NPCA has been a group of private citizens who have believed in their cause and in their ability to advance that cause. Its roots lie in the nineteenth-century ethos of progressivism; its crown presses toward a twenty-first-century "ecologism," to coin a word—a resource-oriented conservation evolved into an ecosystem-oriented environmentalism, which is still emerging. The national park as an institution changes with this evolving view of how humans should treat nature, and why they should treat it differently than they have in the past. NPCA, in turn, as a primary advocate of national parks, struggles to understand how it should respond to these changing ideas of nature and parks. That struggle is its story.

No one with a deep concern for the environment in general and national parks in particular can do a project like this in a totally unbiased way. I have long enjoyed the national parks—North Cascades National Park is virtually my backyard, and I know it well—so I cannot claim to study the history of national parks dispassionately. I care about them and am thankful that they have had their champions. The work of NPCA, the National Park Service, and others in national park affairs has made possible the National Park System we enjoy today. My description of NPCA history reveals my admiration for all of these people and organizations.

xviii *Guardians of the Parks*

I

The Stage Is Set

L ate in the nineteenth century, several movements converged to allow the National Parks Association to appear. One was the push to create national parks, which resulted in a hodgepodge of areas inconsistently managed and inadequately protected. This situation gave rise to a movement to improve their care, which resulted in creation of the National Park Service. Both the parks and the Park Service were possible in part because citizens believed they should join together for civic improvement. Some of the civic organizations that appeared took on the cause of nature conservation and preservation, and a "conservation movement" arose.

The nation's capital at the time was a maelstrom of political activity, but it was also a thriving intellectual community dominated by men of science. As these scientists sorted through the masses of biological, geological, and ethnographic material collected by a series of government surveys, they voiced support for the movement to create national parks. They strove to protect places where lessons of natural history might be learned and where people outside the scientific circle might be educated and inspired. When national park advocates asked for help in forming a civic organization dedicated to the educational and scientific use of national parks, many of these men responded. The National Parks Association was the result.

On May 19, 1919, a small group of men met at the Old Cosmos Club across from the White House in Washington, D.C. That day they signed articles of incorporation forming the National Parks Association. One signatory was Charles D. Walcott, a 69-year-old geologist and paleontologist who had succeeded John Wesley Powell as director of the United States Geological Survey, had served as secretary of the Smithsonian Institution, and was then president of the National Academy of Sciences. Another was J. Walter Fewkes, also 69, a distinguished anthropologist who had conducted pioneering archaeological and ethnographic work in the American Southwest. Henry B. F. Macfarland, a

Roads for access were essential to early promotion of the national park idea.

J. Horace McFarland and his American Civic Association fought long and hard for an act to create a national park agency.

58-year-old Washington attorney and civic leader, placed his name on the articles, as did W. H. Holmes, 73, another archaeologist and artist. Holmes, who succeeded Powell as chief of the Bureau of American Ethnology from 1902 to 1909, was curator of the Smithsonian's Department of Anthropology in 1919 and would soon be named director of the National Gallery of Art. H. K. Bush-Brown, a 62-year-old sculptor of national reputation, also signed, as did Robert Sterling Yard. A former reporter, editor, and publisher, Yard was retiring as chief of the Educational Division of the National Park Service. He had been virtually a one-man "division." A 58-year-old of great energy, he had labored intensely to promote the national parks for the newly formed agency.

These men launched the National Parks Association. They were part of a thriving progressive intellectual community in the nation's capital, a group of artists, scientists, writers, educators, and civic leaders who believed that collectively, they could improve the lot of people in their community, their nation—even the world. They were joiners, nearly all listed in the Washington *Who's Who* and credited with long lists of prestigious affiliations. All were male and of advanced years.

Theirs was a different world than that of today. Women did not have the vote. Radio and film were in their infancy. Automobiles and airplanes had not made much impact on the American lifestyle. The First World War had ended a few months earlier; Americans were elated and hopeful, yet numbed by the realization that thirteen million men had died in battle. President Wilson was in Europe negotiating for the League of Nations. The Treaty of Versailles would soon be signed, and Americans looked optimistically toward a world of peace and prosperity.

Robert Sterling Yard, veteran writer and editor, was to lead NPA for more than two decades.

When these men put their pens to the association's articles of incorporation, there were forty-three national parks and monuments in the United States. The National Park Service, less than two years old, was struggling to knit these diverse units into some sort of conceptual and administrative system. Stephen T. Mather, a retired businessman of prodigious energy and ideas, was directing this task, assisted by a bright young man named Horace Albright. The U.S. Army had only recently relinquished management of Yellowstone National Park. The war over, people were vacationing in the national parks in increasing numbers, though roads and other "improvements" were, for the most part, primitive. By 1919, national parks were becoming a valued asset to American society.

What led Walcott and the others to form an organization devoted to the study and development of a system of U.S. national parks and monuments? What forces made creation of the National Parks Association possible—and even necessary? A century of events laid the groundwork for the emergence of the National Parks Association.

Stephen T. Mather was founding director of the National Park Service and an early benefactor of the National Parks Association.

PRESERVING THE NATION'S HERITAGE
The Creation of Yosemite Park

The year was 1832. Nearly thirty years had passed since Lewis and Clark made their historic journey up the Missouri River and across the Rocky Mountains to the Pacific Ocean. The fur trade was flourishing, and the Missouri was the principal route into the vast western American empire. Most travelers on the great river sought furs, wealth, or conquest, but one that year had other goals in mind. George Catlin hoped to describe and paint the Indian people who lived beyond the Mississippi. As Catlin studied Indians on the Great Plains in South Dakota, it struck him that this entire scene with its Indians, bison, elk, and myriad other creatures would change as people like him moved into this new country. Most travelers saw the West as an endless, hostile wilderness, a source of inexhaustible resources, and a land of economic opportunity. Catlin saw something else. He saw "a magnificent park, where the world could see for ages to come, the native Indian in his classic attire, galloping his wild horse…amid the fleeting herds of elks and buffaloes." This could be "A *nation's Park* containing man and beast, in all the wild and freshness of their nature's beauty!"[1] Catlin was ahead of his time with this suggestion, but he planted the seed of the national park idea in the American mind. Thirty-two years later this idea bore fruit, and a movement began to create an American national park system.

When Catlin floated his idea in the 1830s, no one paid much attention, but events in the late 1850s began moving toward his vision. In 1864, Congress ceded Yosemite Valley and the Mariposa Big Tree area to the state of California for "public use, resort and recreation."[2] The significance of this action was

John Muir, passionate and eloquent park advocate, was the spiritual force behind the early national park movement.

missed by most at the time; the nation was in the midst of the Civil War. Historian Alfred Runte notes that "In retrospect…the United States Congress had done nothing less than approve a model piece of legislation leading to the eventual establishment of national parks."[3] Yosemite Valley was to be protected "inalienable for all time." It was to be reserved from land claims. The Mariposa Grove could not be logged. The prohibition against land claims in the valley was challenged in 1868 but upheld by Congress. Catlin's idea had moved forward significantly, and an association of citizens to promote and protect parks like Yosemite became a possibility.

Yellowstone and the National Parks

Next came the creation of Yellowstone National Park. Although many consider this the first national park, Yosemite had provided the model.[4] President Ulysses S. Grant signed the Yellowstone Park Act on March 1, 1872. The act reserved more than two million acres from "settlement, occupancy or sale" and dedicated the land "as a public park or pleasuring ground for the benefit and enjoyment of the people."[5] The scale of this park was well beyond Yosemite, and unlike Yosemite at this time, it was to be administered by the United States rather than by a state.

Administration of Yellowstone was a problem: Congress created the park but appropriated no funds for its protection. The U.S. Army became its protector from 1886 to 1918. When Congress made small appropriations for the park, Army engineers built roads and other improvements. Not until 1894 did Congress pass legislation protecting wildlife in the park and providing means to enforce the regulations. The Army years in Yellowstone were spent fighting to offer minimal protection to the park's resources and to develop access to them, principally in the form of roads.

Yellowstone set a powerful example of an unquestionably *national* park. The early Yellowstone experience also pointed to problems that would plague nearly all parks from then to now: incompetent and political concessions (private entrepreneurs providing services to visitors for a fee); threats of inappropriate development; boundaries inadequate to protect resources, especially wildlife; and an inadequate budget to do the job mandated by the act creating the park. A national park was reality on paper but still far from achieving Catlin's vision.

Yosemite reappears in the story in 1890. California's management of Yosemite Valley had been fraught with difficulty. Many thought the valley should be returned to the federal government, which might do a better job of protecting its magnificence (although the federal record then in Yellowstone was not unblemished either). Only the valley had protection, and Yosemite advocates such as John Muir argued that the surrounding high country also needed protection from timber interests and other economic exploiters. The great sequoia groves were falling to axe, saw, and dynamite. After a campaign in which Muir played a prominent role, Sequoia National Park was established September 25, 1890. Again, Congress appropriated no funds to manage the park, and the military moved in to do what it could to protect the resource. Muir knew the work to protect Yosemite and other Sierra areas had barely begun and organized a citizen group that would become the Sierra Club. A new ingredient—organized citizen groups focusing on national parks— entered the scene.

The next national park established, Mount Rainier, resulted from the work of such civic-minded groups.[6]Parks such as Mount Rainier might be established for various good reasons, ranging from resource protection to scientific interest to economic boosterism, but creating a paper park was only the first step toward establishing a properly administered and protected area. For several years, Mount Rainier National Park had no appropriation and no rangers. The Washington state forest supervisor was persuaded to take charge in 1902, adding the park to his duties with no additional funding.[7] As the Sierra Club was finding at Yosemite, there was plenty of work for citizens interested in supporting national parks. Marking the boundaries was only the beginning.

Congress established Crater Lake National Park in 1902. This remarkable lake was a place clearly appropriate as a national park and worthy of Catlin's vision, but three other new parks raised a new issue. In the view of many park advocates, Wind Cave, Sullys Hill, and Platt were simply not of sufficient scenic beauty or unique natural character to merit identification as national parks. Wind Cave was described by Congressman John B. Lacey of Iowa as "substantially what the Yellowstone would be if the geysers should die," which was "a very considerable exaggeration" according to historian John Ise. In Ise's opinion, Sullys Hill was "the most unworthy park ever created, if it was created."[8] There was confusion about congressional intent in the park's 1904 designation, but Sullys Hill was considered a national park until it became a national game preserve in 1931. Congress created Platt National Park to honor a recently deceased senator from Connecticut, Orville Platt, and its principal feature was a group of springs. These springs were polluted by the inadequate sewage system of the nearby town of Sulphur, Oklahoma, and the "park" was the butt of jokes in Congress. When appropriations were sought, one jokester was moved to wonder how so much money could be spent on a 900-acre plot where there was nothing.[9]

Such parks raised a question that has been central to National Park System debates ever since: What criteria should be used for a site's inclusion or exclusion? Parks have always made good political pork, offering various benefits to congressional constituents back home. These marginal parks started a debate about the nature of national parks that raised yet more questions for park advocates and protectors. They would have more to say about Platt National Park later, and the issue of national park standards would occupy the National Parks Association throughout its history.

The Antiquities Act and National Monuments

Although national park status protected some of the giant sequoias near Yosemite, geysers and bison in Yellowstone, and even polluted springs in Oklahoma, the Indian ruins of the Southwest were being looted of their priceless archaeological treasures. The need to protect these ruins had been recognized since late in the nineteenth century, but nothing was done until 1906, when Congress passed the Antiquities Act. This act established penalties for disturbing or destroying any historic or prehistoric ruin on federal land. More important, it authorized the president to set aside as "national monuments," by proclamation, historic places, landmarks, and structures. Theodore Roosevelt used his new power to create Devils Tower, Petrified Forest, Montezuma Castle, and El Morro national monuments in 1906 and Grand Canyon in 1908, among others. Presidents William H. Taft and Woodrow Wilson proclaimed more monuments, notably Mt. Olympus, Rainbow Bridge, and Dinosaur. These proclamations protected public domain lands from land claims but provided little physical protection. Because Congress provided no money for protection, the looting of archaeological treasures continued for many years.

The Antiquities Act brought significant expansion to the emerging system of national parks. These new areas were called "monuments," not "parks," and the distinction was confusing; a new element had been added to the nascent system of dedicated public lands. Up to this point national parks had been areas of exceptional grandeur—extraordinary landscapes with "grand, monumental scenery."[10] The Antiquities Act was a public policy decision to preserve parts of America's natural and cultural heritage. Forts, deserts, cacti, church missions, and homesteads might now be given at least as much protection as the grand, scenic national parks—and these were not the only reservations established during these years. Congress had also designated national military parks, battlefield sites, and historic sites and memorials. Chickamauga and Chattanooga National Military Park and Antietam Battlefield Site had been established in 1890, Shiloh National Military Park in

1894, and Gettysburg National Military Park in 1895. With Mesa Verde National Park added in 1906, these diverse elements began to form a body of areas reserved primarily for their cultural and natural values. A systematic approach to their administration was needed. It was the first decade of the twentieth century; agitation for a park bureau or service began.

The Conservation Agenda

After the Civil War, Americans became sufficiently concerned about destruction and degradation of natural beauty, wildlife, and forests to organize for their protection. This movement began with a focus on forests (although the first formal conservation organization was the American Fisheries Society, founded in 1870). The destruction of forests, with disastrous consequences such as giant forest fires in the upper American Midwest, led to the creation of the American Forestry Association. This association's principal interests were tree culture and planting, but its formation and the emergence of state forestry associations, beginning with Minnesota in 1876, gave momentum to a rising concern for forest conservation. The American Forestry Congress was organized in 1882, aiming to develop the broad national policy of forest conservation that began with the Forest Reserve Act of 1891.

Other groups of sportsmen, scientists, reformers, and citizen activists soon organized to pursue specific conservation agendas. Gentleman big game hunters, led by Theodore Roosevelt and George Bird Grinnell (a future president of the National Parks Association), formed the Boone and Crockett Club in 1888. The Sierra Club was founded in 1892. Concern about wildlife, stimulated particularly by the slaughter of birds for the millinery trade, led Grinnell in 1886 to propose the creation of a society to protect wild birds and their eggs. State Audubon societies were soon established and organized to become the National Association of Audubon Societies in 1905. People were organizing on many fronts expressly for conservation goals.

Other organizations, for which conservation was a secondary concern, became involved in efforts to protect natural areas of particular interest to them. The Appalachian Mountain Club was established in 1876 to build camps and trails and protect mountain areas in the East. The Mazamas in Portland, Oregon, organized in 1894 and spun off the Mountaineers in Seattle in 1906. The Association for Protection of the Adirondacks was formed in 1901, and the Colorado Mountain Club in 1912. All of these groups were involved in the conservation cause, some of them specifically striving to protect their favorite areas as national parks.

Some groups focused on causes other than natural resource protection and secondarily developed interests in conservation and national parks. The garden club movement, for instance, began in the early 1900s. These groups embraced conservation, promoted it in their literature, and worked to create city parks and protect greenbelts. The American Park and Outdoor Art Society was established in 1897, followed by an American League for Civic Improvement in 1901. They merged in 1904 into the American Civic Association, the purposes of which were "the cultivation of the higher ideals of civic life and beauty in America, the promotion of city, town and neighborhood improvement, the preservation and development of landscape, and the advancement of outdoor art."[11]

One leader of the American Civic Association was J. Horace McFarland, who served as its first president from 1904 to 1923. McFarland, a businessman from Harrisburg, Pennsylvania, became involved in civic action when Harrisburg suffered a typhoid epidemic caused by the polluted water of the Susquehanna River. He crusaded for better sewage disposal and became "badly infected with the improvement bacillus."[12] McFarland fought for many years against diversion and development of Niagara Falls. He and his association allied themselves with John Muir against San Francisco's proposal to build a dam and flood Hetch Hetchy Valley in Yosemite National Park, and he became a major advocate of a national park service. A United States Forest Service had been created in 1906 under the leadership of Gifford Pinchot; McFarland envisioned a similar bureau to unify the management of the national parks, which was scattered among three cabinet departments.

A Split in the Movement

Historians have showed that the conservation movement branched in two directions early in its history.[13] One branch, led by Pinchot, aimed to "conserve" natural resources toward utilitarian ends. The other branch, led by Muir, strove to "preserve" special resources, such as inspiring scenery and remarkable works of nature. The Hetch Hetchy battle established the philosophical separateness of these two branches, with Pinchot supporting the dam and Muir fighting it. Muir lost the fight and the dam was built, but new battle lines were drawn in the national park story. Pinchot, the Forest Service, and other utilitarian conservationists would often oppose national parks, while the preservationists concentrated on creating new parks and improving the administration of existing ones. This early schism in the conservation movement was another factor that set the stage for the National Parks Association. A need emerged for a powerful voice raised specifically for national parks.

When agitation began for creation of a bureau to oversee the growing number of national parks (and, after 1906, national monuments), the conservation movement was well under way. Conservationists had many interests and goals and disagreed on particulars, but were united in their belief that the tide of "progress" was washing over nature and destroying some of its most important values. Politically, the conservationists, especially Pinchot and other utilitarians, were progressives dedicated to reforming the American political system. Stephen Fox notes that "conservation mirrored broader trends in progressivism: a shift toward expertise, 'scientific' management, planning, the ideals of economy, efficiency, stability."[14] Led by Muir, whom Fox describes as the quintessential "radical amateur" of conservation, the preservationists might agree with utilitarian conservationists on some things, but they could not embrace the idea that everything on Earth was a natural resource made for human consumption. Some "resources" should simply be left alone. They should be preserved in parks.

Fundamental Questions

Seven national parks existed in 1900. During the ensuing decade, five more were created by Congress, and the passage of the Antiquities Act resulted in immediate presidential proclamation of seventeen national monuments. The movement toward preservation was gaining momentum, yet many problems existed. Designating a park or monument did not necessarily result in desired protection. In many cases, no one was physically protecting and managing the resources. Regulations varied from park to park, as did administrative organization and responsibility. In Yellowstone, the superintendent was an Army officer appointed by the secretary of War, but "exclusive control" over the park supposedly rested with the secretary of the Interior.[15] Troops were also in charge in Yosemite, whereas Crater Lake and Mount Rainier had civilian superintendents. There seemed a clear need for coordinated and systematic administration of the rapidly growing "system" of national parks and monuments.

Another problem was the lack of definition of what qualified for national park status. Areas of doubtful park quality were proposed as parks, and some were created. No administrative authority was responsible for assessing the quality of proposed parks. John Ise describes this problem:

Some locality with an area of very modest scenic values, or perhaps nothing of value at all, with an eye to Congressional appropriations and profitable tourist traffic, might steam up a campaign to have it made a national park, and if it had an influential delegation in Congress might succeed in it."[16]

Places of dubious park quality, such as Mackinac, Platt, and Sullys Hill, thus took their place alongside Yellowstone, Yosemite, and Mount Rainier in the array of national parks.

Alfred Runte describes another problem when he notes that during the first decade of the new century, national parks were the "step-children of federal conservation policy."[17] "Conservation" was emerging as a national priority, but it was a utilitarian conservation driven by economic rationales. Its leaders, Gifford Pinchot foremost among them, thought that scientific efficiency in natural resource use should be the goal and that this could be achieved by actively managing timber, grasslands, irrigation, and hydroelectric dam sites. Pinchot created an agency dedicated to controlled forest management, and his pragmatic approach to resource questions was embraced by Theodore Roosevelt.

Runte notes that preservationists like park advocate John Muir "could not escape the certainty of head-on confrontations with advocates of utilitarian conservation. The promise of immediate returns to the national economy, as opposed to what the national parks *might* contribute to the gross national product, demanded instant rebuttal."[18] The confrontation came in Yosemite's

The first automobile enters Yellowstone, beginning the flood of auto-borne visitors that has grown ever since.

Hetch Hetchy Valley and revealed the need for what Runte called "institutionalization of the national park idea within the political and legal framework of the federal government."[19]

Such problems pointed to a need for a centralized administrative system; the drive for a national park bureau or service began. Congressman Lacey of Iowa opened the campaign when he introduced a bill in 1900. Although two bills followed in 1902, congressional support was not strong enough to allow passage. J. Horace McFarland brought the American Civic Association behind the drive to support a national park agency. Alliances formed among McFarland's group, the Sierra Club, the Appalachian Mountain Club, and a Society for the Preservation of National Parks, formed with John Muir as its president. McFarland and others took the campaign to the media. Support began to grow.

Secretary of the Interior Franklin K. Lane strongly supported the push for a national park bureau.

The National Park Service Act

Secretary of the Interior Richard Ballinger called for a national park bureau in his 1910 report and asked McFarland to confer with him about the idea. Ballinger's successor at Interior, Walter L. Fisher, continued to support the proposal, as did his successor, Franklin K. Lane. McFarland found allies among Western railroad executives, who saw the prospective park bureau as being in their interest.

In February 1912, President Taft urged Congress to establish a Bureau of National Parks. McFarland and his allies doggedly pursued their campaign against powerful opposition. The Forest Service opposed the bureau, seeing it as a competitor for resources and suspecting that an expanded national park system would be carved from lands under its jurisdiction. Some members of Congress opposed the idea of another federal bureaucracy. And Pinchot, even after leaving the Forest Service, remained ideologically and actively opposed. As Runte notes, Pinchot "spoke out against my attempt to coordinate scenic protection unless the program were handled 'efficiently, economically, and satisfactorily by the Forest Service.'"[20] Despite opposition, on August 25, 1916, President Wilson signed the National Park Service Act.

The act was not all that McFarland and his allies had sought, but direction was set for administration of a national park system.[21] The Forest Service had fought hard to retain its national monuments and had, at least temporarily, succeeded. The War Department retained its monuments. The language in the act proved to be open to considerable interpretation. Still, McFarland's camp had much cause for celebration. "The defense of the parks had been elevated from the throes of indifferent management to the full responsibility of the federal government," writes Runte.[22] Responding to priorities that emerged from the White House Governor's Conference on Conservation in 1908, McFarland wrote, "Are we to so proceed with the conservation of all our God-given resources but the beauty which has created our love of country...?"[23] Through hard work and careful alliances, he had answered his own question.

Coordinating the Effort

While McFarland and his allies worked to create the National Park Service, the Department of the Interior moved to provide some measure of coordination for the national parks. W. B. Acker, an Interior attorney, was assigned to work part time on park affairs in 1911. When Franklin K. Lane became secretary of the Interior, he brought Adolph C. Miller, a University of California economist, to Washington as assistant to the secretary and gave him the task of unifying the administration of the national parks. Miller, in turn, recruited a recent University of California graduate and law student, Horace Albright, as his assistant. Albright arrived in May 1913 and rapidly became deeply involved in park affairs. In June 1914, Mark Daniels, a landscape architect, was named general superintendent and landscape engineer for all national parks, with an office in San Francisco.

Late in 1914 Secretary Lane called Albright into his office and introduced him to Stephen T. Mather, a Chicago businessman. Miller had been appointed to the Federal Reserve Board, and Lane had invited Mather to replace him. Mather, also a University of California graduate (as was Lane), had made a modest fortune in the borax business. His biographer describes him as being, in 1914, "at the pinnacle of success, forty-seven years old, an enormously personable, energetic, and hard-driving man, who was a trifle restless and on the lookout for new worlds to conquer."[24] Mather was reluctant to come to Washington, but as a park enthusiast, occasional mountaineer, active member of the Sierra Club, and an addict to challenge, he agreed to join Lane for a year. He and Albright became a team in January 1915.

Mather's first inclination was to launch a publicity campaign. For that, he needed a writer and public relations man. While Mather had been a reporter at the *New York Sun*, he had become friends with fellow reporter Robert Sterling Yard. When Mather married in 1893, Yard was his best man, and the two had maintained their friendship while Yard pursued a career in writing, editing, and publishing. In 1915, Yard was Sunday editor of the *New York Herald*; Mather brought him to Washington as an employee of the United States Geological Survey. The Geological Survey detailed him to Mather, who personally provided his $5,000 annual salary. Yard set up his editorial offices in space provided by the Bureau of Mines and was soon producing a blizzard of promotional articles that appeared in Eastern newspapers and magazines.

When Mather called all of the park superintendents together in March 1915, he discovered that many of them were political appointees who "lacked both the ability and zeal for effective national park work."[25] Their professionalism would have to be improved dramatically if the park movement were to progress, and he set out to achieve this in any way he could. Next, to raise the profile of the park cause among influential people, he staged a camping trip through the High Sierra to show a distinguished group of congressmen, scientists, businessmen, and media moguls the nature of a great national park. The trip, in the summer of 1915, was highly successful and is credited with garnering support for the National Park Service Act from men like Gilbert Grosvenor of *National Geographic*, Henry Fairfield Osborn of the American Museum of Natural History, and eminent writer Emerson Hough.

Mather and Albright had agreed to take these jobs for only a year, but that year passed quickly. They had enjoyed the work and worked well together; progress toward creating a coordinated system of national parks in the Department of the Interior had been steady. Late in November 1915 they assessed their situation and decided to persevere for another year, with the expectation of seeing a National Park Service Act and a national parks bureau.[26] Daniels resigned as general superintendent, and in December 1915 Robert B. Marshall was reassigned from the U.S. Geological Survey and given the title of superintendent of national parks. This team, along with McFarland, his allies, and supportive congressmen, pushed the National Park Service Act to its signing by President Wilson.

The act created a National Park Service and gave it a general mandate:

> The service thus established shall promote and regulate the use of federal areas known as national parks, monuments and reservations hereinafter specified by such means and measures as conform to the fundamental purpose of said parks, monuments and reservations, which purpose is to conserve the scenery and the natural and historic objects and wildlife therein, and to provide for the enjoyment of the same in such manner and by such means as will leave them unimpaired for the enjoyment of future generations.[27]

Creating an Organization

Mather's challenge was to create an organization to accomplish this broad mandate. Albright, Marshall, and Yard were already at work. Although the Park Service was authorized in 1916, no appropriation would be made until 1917. Then the secretary of the Interior could hire a director, assistant director, chief clerk and any other help needed, up to a ceiling of $19,500. Meanwhile, the "interim service" under Robert B. Marshall would have to carry on.[28]

Mather was not happy with Marshall's work, however, and late in the year decided to relieve him as general superintendent. This was a blow to Mather's plans, for he intended Marshall to be the new agency's first director.[29] An angry Marshall returned to the Geological Survey, and Mather forged ahead without an interim director.

Mather decided to hold a national conference on national parks in Washington in January 1917. The aim was to organize to win an adequate congressional appropriation for the Park Service—money to administer and develop the parks. The stress of several years of furious effort, confrontations with the deposed Marshall, and other problems caused Mather to have a severe nervous breakdown during the conference; he would be out of action for more than a year. Despite this shocking development, the conference was a great success. Many supporters spoke in favor of the Park Service. Prospects for the appropriation were improved.

Albright and Yard

Horace Albright, barely twenty-seven years old, stepped in behind Mather. One of his first challenges was to establish his authority and, as Albright's biographer recounts, part of this challenge was Robert Sterling Yard. Yard appears to have concluded that he should be Mather's logical replacement. Albright and Mrs. Mather thought differently.

> Thoroughly horrified at the prospect of having the impetuous Yard giving orders, Mrs. Mather immediately wrote to Albright imploring him to "hold Mr. Yard in." She made it absolutely clear that Mather wanted Albright to be in charge of the Washington operation, including Yard's office, and that she herself would be "perfectly agreeable" to whatever Albright cared to do as long as he kept a tight rein on things.[30]

With this strong backing, Albright "reached an understanding" with Yard, who agreed to carry on his educational and promotional work, clearing any releases from his office through Albright. That did not end the trouble with Yard, who clashed with Mather's business manager, Oliver Mitchell. Albright and Mitchell considered firing Yard, but Albright decided against such a drastic move. "You know," he explained to Mitchell, Mather "is very close to Mr. Yard, and it seems to me that no matter what we individually think about retaining him or about the work he is doing, we would be taking a chance [of displeasing Mather] if we laid him off." Albright informed Mitchell that "we have simply to let this matter run along until we get our word from the chief."[31]

This Yard episode is significant for several reasons. Yard was not entirely happy in his role in the "interim" park service. Although he was an effective publicist, he alienated some of his co-workers. Also, the strength of Yard's ties to Mather are evident. Though Albright and Mitchell thought he should be replaced, they did not feel that they could fire him without Mather's support, and Mather was ill and was not to be bothered by matters related to parks. This incident also reveals that the relationship between Albright and Yard was strained, and this would soon be significant when Albright led the growing National Park Service and Yard headed the National Parks Association. Finally, the tiff with Mitchell involved lantern slides that Yard wished to purchase for one of his educational schemes. The fact that Mitchell and Albright thought this idea unworthy, even frivolous, indicates there was little support among Park Service leaders for Yard's educational initiatives.

This clash between Yard and other members of the infant National Park Service reveals "Bob" Yard as a complicated man. A small, intense, urbane, and opinionated person, often soft-spoken, he could be aggressive and even abrasive. His convictions were strong, and he was self-confident to the point of arrogance. He occasionally encountered others of similar confidence and arrogance—Albright and J. Horace McFarland among them—with whom he clashed. Nonetheless, as a prodigiously hard worker, gifted writer, and devotee to the national park cause, he was a valuable member of the Park Service team, as his friend Mather knew.

Why would Mrs. Mather have supported Albright rather than Yard as Mather's temporary replacement? One can only speculate, but she had perhaps heard her husband's views on the matter. Mather had suffered breakdowns before, and he might have indicated that should he again be out of commission, Albright was his choice. Yard's nature was too uncompromising. He was not Albright's equal as a political strategist and did not know the Washington scene as well. The new agency needed savvy, patient leadership, and Mather may well have decided that Albright simply fit those needs better than Yard. Whatever the reasons, Albright assumed the agency's leadership at this critical point.

Albright went through bruising congressional hearings on appropriations for the National Park Service and achieved a reasonable allotment for the new agency. A supplementary appropriation was necessary so that the Park Service could be promptly organized, and this was approved on April 17, 1917. Shortly thereafter, Albright was named assistant director of the National Park Service, and Lane promptly appointed him "acting director" until Mather could return to service. Albright set up his small Washington staff, and Yard continued as "chief, Educational Division," although there was no appropriation for that position. Yard's salary continued to be paid personally by Mather.

The National Parks Educational Committee

The stage was now set for the National Parks Association. In his own account, Yard says the move toward the emergence of the association began in mid-1917. By his own request, he had been named chief of the educational division of the Park Service, the "division" consisting only of himself and a secretary. He hoped to stimulate interest in the educational use of national parks, in addition to promoting the parks through articles, press releases, and other publications. The parks could be used to popularize study of natural science by schools and universities, he thought—the germ of what would later emerge as interpretation and outdoor education. To his frustration, Yard found minimal understanding of his educational ideas and little demand for educational services within the ranks of the National Park Service.

Yard's "inability to arouse official interest and cooperation" led him after a year "to the creation of an educational committee outside of government."32 He had received support for his ideas about educational uses of the parks from several eminent scientists, including Charles D. Walcott, secretary of the Smithsonian. Yard approached Walcott about forming a committee, and Walcott sent him to Henry B. F. Macfarland, whom Yard describes as "Washington's leading public-minded citizen." Macfarland protested that he was hardly the man to lead a national park-oriented organization. Yard quotes him: "To me

William Kent, photographed here in the redwoods with NPS Director Mather, strongly supported national parks while in Congress and was an early NPA trustee.

out-of-doors is what you pass through to the capitol and the court house—with a walk now and then in Rock Creek Park."[32] But Macfarland was excited by Yard's ideas and agreed to join Yard and Walcott in forming a National Parks Educational Committee.

Committee organizers met in June 1918 in Dr. Walcott's office at the Smithsonian. Walcott became chairman. William Kent, formerly a California congressman, was vice chairman; Macfarland was to chair the executive committee; and Yard was named executive secretary. Others in the founding group were Wallace W. Atwood of Harvard University, the explorers Henry G. Bryant and Charles Sheldon, New York Commissioner of Education John Huston Finley, William B. Greeley of the Camp Fire Club, and renowned zoologist and conservationist George Bird Grinnell. The group agreed to a set of objectives. One was to "educate the public in respect to the nature and quality of the national parks." The group sought "to further the view of the national parks as classrooms and museums of Nature" and to use "existing publicity and educational systems so as to produce a wide result...." The National Parks Educational Committee set political objectives as well. It aimed "To combine in one interest the sympathy and activity of schools, colleges and citizen organizations in all parts of the country" so as to accomplish the committee's objectives and "To make every endeavor to keep political influence out of national parks." Finally, it aimed "To study the history and science of each national park and collect data for future use."[32]

At this stage in its history, the committee's agenda was different from that of the National Park Service. One awkwardly worded objective indicates this. The committee wished "To define. . .those functions which have for their general objectives the uses of the parks by the people, in distinction from the government function of the physical development and administration of the parks for the people...." Let the Park Service develop and administer. The committee would find ways to help people derive the knowledge and inspiration that national parks were specially equipped to provide. It would "correlate" this educational and scientific work "with government functions so as to bring about an effective working partnership of common purposes."[32]

The National Parks Educational Committee had twenty-five founding members, and Yard worked furiously for the next eleven months to recruit more. The executive committee met on April 9, 1919, and could count seventy-two members, all of them distinguished leaders in science and education. The purpose of the April meeting was to decide "whether or not to recommend to the Educational Committee that it organize at once a National Parks Association to carry out the objectives set forth in the report by the Secretary." Yard had been informally surveying committee members about what such an association might do. He had ideas of his own and brought his views together with those of other committee members in "A Brief Statement of the Proposed Activities of the National Parks Association."[33]

Support from Mather

Chairman Macfarland had discussed the idea for the association with Stephen Mather and read a letter from Mather to the assembled committee.

The House Appropriations Committee tours Yellowstone in 1920 with Superintendent Horace Albright as guide.

Since my various talks with you I have been considering very fully the proposed National Parks Association, and I cannot urge too strongly the importance of getting it under way at once. Mr. Yard first proposed this organization in 1916 and gathered around him a strong body of men largely representative of public-spirited organizations and universities. The growing war cloud, and later the war, compelled its postponement; otherwise it would have come into existence early in 1917. Meantime your Educational Committee, which he organized in June 1917, to investigate the opportunities which such an Association might have for using the national parks to popularize natural science, has greatly enlarged its outlook for usefulness.

Mr. Yard, who joined me almost at the beginning of my national parks work, has labored earnestly and very effectively for its success. His study of the parks from the educational point of view...and his recently developed plans for organizing the cooperation of schools and universities of the country should be continued under freer and more permanent auspices than the government offers. It is my hope that the Association will stand earnestly behind him, and employ him in an important executive capacity.[33]

Park administrators sometimes went to great lengths to display the wonders of their parks, as here at Wawona tree in Sequoia National Park.

Mather went on to suggest how he thought the proposed association might be financed and pledged $5,000 "to pay the Secretary's salary during the formative stage and provide a fund for necessary organization expenses." Mather was willing to provide this support because of his "deep interest in the national parks and my belief that the Association, once it is successfully launched, will play an important part in their success."[33]

Mather had likely arranged with Macfarland that Yard be given this position. He had enticed Yard to Washington to help him create a National Park Service, and despite Yard's inability to get along with some members of the team, he had done a fine job at the tasks he had been given. Mather could no longer legally pay Yard's government salary, because in July 1918 Congress passed a law prohibiting the support of any government work by private funds.

Mather gave Yard severance pay and sought another position for him. He approached J. Horace McFarland of the American Civic Association, offering to donate $3,600 a year to Yard's support if he would hire him, but McFarland rejected the offer.[34] Knowing that Yard was attempting to establish an organization to pursue his educational agenda for the parks, Mather undoubtedly saw an opportunity to provide a place for him. He could help Yard get started, but after that, Yard and his organization would be on their own. As his letter indicates, he arranged this with Macfarland.

After reading Mather's letter, Macfarland led the group through a discussion of the prospects for a National Parks Association. They reviewed Yard's proposed activities. The prospects for continuing funding were examined and seemed promising, especially with Mather's financial support. The group instructed Macfarland to write a letter to the full membership of the Educational Committee informing them of the executive committee's recommendation that a National Parks Association be organized immediately, that Yard be its executive, and requesting their views on this proposal. If those views were positive, Walcott and Macfarland would call a meeting to establish the National Parks Association.

By 1919, the seed of the national park idea planted by George Catlin had sprouted into a healthy, growing plant. Many ingredients had stimulated its growth. The world had changed since Catlin had gazed on millions of bison and studied the Indians of the Great Plains. The bison and the Indians were virtually gone, only remnants surviving here and there. Country where they roamed freely had been surveyed into sections and divided among railroads, ranchers, and homesteaders. Much land that, in Catlin's day, had been public domain was now privately owned. The vast forests east of the plains had fallen under Paul Bunyan's axe, and resources of all kinds were being developed across the American continent.

A movement to conserve resources, especially forests, wildlife, and water, had emerged. Science had grown as an intellectual force in the latter part of the nineteenth century. It provided insights into nature and tools for managing forests and other resources; it also provided motivation to preserve portions of the landscape. Susan Schrepfer has argued that many geneticists, paleontologists, and biologists believed that evolution was perfectionist and "carried the belief that the universe was ordered into their preservationist efforts. They saw parks as a means of preserving the evidence of cosmic design and sought to save the most spectacular results of the culminant processes of evolution. . . ."[35] Many scientists led powerful institutions such as the Smithsonian, the American Museum of Natural History, and universities, and they brought the power of their constituencies to the national park movement.

Finally, citizen activism emerged in the late nineteenth century as a potent force for reforming government, industry, and land use. Progressivism appeared in American politics, and clubs and associations, focused on diverse reforms, appeared across the American political landscape. Virtually every cause had its citizen advocacy group; each park had its defenders.

Yet even as there was a need for a centralized government institution to tend to all parks and monuments, so was there a need for an advocate in the nation's capital for all the parks. A government agency, the National Park Service, was not sufficient to achieve the potential of the national parks. Yard saw this and, in June 1918, could write, "The big fruits of the national park movement, which…has its roots in the heart and mind of every American, can…be cultivated only by an organization of the people outside of government, and unhampered by politics and routine."[36] That organization was to be the National Parks Association.

II

The Founding of the National Parks Association

A ll the elements were present for the founding of the National Parks Association. A committed leader had appeared in Robert Sterling Yard. He had a clear vision of an organization that would work with the National Park Service in the building of a National Park System and protecting the finest examples of the natural and cultural heritage of America.

Yard had recruited a group of distinguished educators and scientists; he now faced the challenge of defining the organization's goals and creating its identity. National parks would be its special province, and education would be central to its mission. Initially NPA would address the matter of standards for a national park or monument, for this issue was central to the creation of a defensible park system.

The first order of business was to define the mission, but a statement of high ideals alone would not suffice—action would be necessary to ensure successful passage through the founding period. Yard was determined to demonstrate that the organization could act on its ideals.

Henry B. F. Macfarland, as chairman of the National Parks Educational Committee's executive committee, wrote to all committee members on April 9, 1919. He conveyed the executive committee's view that the time had come to create a National Parks Association. The response was positive, and a meeting to create the organization was scheduled in Washington, D.C. Six members

Henry B.F. Macfarland was the first president of the National Parks Association, serving from 1919-21.

of the National Parks Educational Committee signed the articles of incorporation on May 19, 1919, and the articles were filed in the Office of the Recorder
of Deeds. The National Parks Association was born.

The first board of trustees meeting of the association convened on May 29,
with Charles Walcott chairing. Consulting with Macfarland and Walcott,
Robert Sterling Yard had prepared bylaws, which were adopted. Macfarland
was unanimously elected president. Vice presidents included Nicholas
Murray Butler, president of Columbia University, and Henry Suzzalo, presi-
dent of the University of Washington; universities east and west were thus
represented. Other vice presidents were John Mason Clarke, chairman of the
Section of Geology and Paleontology of the National Academy of Sciences;
LaVerne Noyes, president of the Board of the Chicago Academy of Sciences,
who could represent Midwestern as well as scientific perspectives; and
William Kent of California, former congressman and donor of the Muir Woods
National Monument. Washington businessman Charles J. Bell was elected
treasurer; Yard was named executive secretary; and an executive committee
was elected to include Walcott, Macfarland, and Yard along with Leila
Mechlin, secretary of the American Federation of Arts, and Edmund Seymour,
president of the American Bison Society.

The slate of officers was carefully drawn from the ranks of the National
Parks Educational Committee. The officers named on May 29 brought qualifi-
cations that the founders deemed essential to the success of the new organization: national stature, representation of a range of educational and scientific
organizations, and a solid profile in the nation's capital.

A PLAN FOR ACTION
The Educational Agenda

What exactly would this new association do? Yard's goal—to educate the
public about and for the national parks—had been clearly stated during earlier discussions of the National Parks Educational Committee. He wanted people to appreciate the national parks and to learn what these places could teach
about science and history. But he also wanted more parks and monuments and
to ensure that they were managed to the proper ends. At this point, he was
careful to praise the new National Park Service and its director in all of his
writings, but even then he thought the Park Service's priorities might not be
entirely correct. An indicator is his clash with Mitchell over educational slides.
In Yard's view, the Park Service was not giving the parks' educational potential the attention they warranted. The preface to a document distributed to
NPA's trustees confirms this:

> As Congress conceives the National Parks only as concrete proper-
> ties and appropriates only for their physical protection, improvement
> and maintenance, there is no governmental provision for their study
> from any other point of view, or for their interpretation, or for prepar-
> ing the public mind for their higher enjoyment. To accomplish these

objects is the fundamental purpose of the National Parks Association.[1]

Yard recognized that Congress was setting the agenda for the National Park Service with its appropriations, especially after 1918. Mather had proffered his personal funds to bring Yard into the park business because he understood that public education about national parks was essential to public support of them. Congress would not provide any funds for park publicity and education.

Yard's ambitions had grown beyond propaganda, promotion, and publicity. Although he labored dutifully at these tasks, he tried unsuccessfully to broaden his role. He could not, however, convince Mather to fund his proposed educational activities. Had Mather not fallen ill in 1917, perhaps Yard would have been more successful in pursuing his agenda at the National Park Service. Mather's nervous breakdown and Albright's control of the new federal parks agency precluded that, and Yard was forced to turn elsewhere.

Mather supported Yard's ideas. He writes that Yard's "recently developed plans for organizing the cooperation of schools and universities of the country should be continued under freer and more permanent auspices than the Government offers."[1] Mather supported interpretation, and it soon appeared in Yosemite. In 1918 and 1919, however, organizing the new agency and creating new parks were priorities, and that is where he placed his resources. Education could come later, and if a private organization could advance the work, all the better.

The first objective of the National Parks Association was "To interpret and popularize natural science by using the conspicuous scenery and the plant and animal exhibits of the national parks, now prominent in the public eye, for examples."[1] Lantern slides of the parks could be circulated to schools, universities, and clubs. Classes could be brought to parks for their studies, photographs could be provided to schools, and lecturers could go forth to explain geology, history, and natural history. Traveling exhibits of photographs could go to public libraries across the land. Professional writers could be encouraged and assisted in preparing books and articles about the parks, and motion pictures might be made. "Scenic and travel films, now so stupidly edited, can be made intelligent and instructive, and professional producers gladly welcome improving suggestions, provided they are also popular suggestions."[1] Yard and his colleagues thought there was a strong market in the United States for such educational products.

The Search for a System

A second objective was more simply stated: "To help the development of the national parks into a complete and rational system."[1] This goal was to prove elusive; it was also to be the central concern of the National Parks Association from its inception to the present. The association elaborated this simple goal statement. The "system" was to be expanded "To the end that they...represent by consistently great examples the full range of American scenery, flora and fauna, yet [be] confined to areas of significance so extraordinary that they shall make the name national park an American trademark in the competition for the world's travel." The national monuments, distinct from national parks, would be developed "into a system illustrative of the range of prehistoric civilization and early exploration and history."[1] A theme that would govern NPA's future activities can be seen here. It is the quest for a National Park System. A system would contain diverse units subject to a common plan or, at least, serving a common purpose.

The history of the national parks to 1919 suggested no plan and little common purpose. NPA committed itself to rectifying this, in part by defining criteria for what should and should not be a *national* park or monument. Use of the words "great" and "significance" and "extraordinary" in their goal statement suggests what NPA's founders had in mind. They sought to avoid including any additional "parks" like Sullys Hill and Platt in the system.

How would they achieve such a system? First, they would study proposed new parks "to determine their fitness for admission to the system" and would inform Congress of their conclusions. They would work with state park organizations, for often an area unfit to be a national park would make a fine state park. They would cooperate with organizations sharing their objectives, such as the National Association of Audubon Societies, the American Game Protective and Propagation Association, the American Bison Society, and the Boone and Crockett Club, to promote the "natural development" in national parks and elsewhere of "native wild animals and birds." They would, with these colleagues, work for "the conservation, in an ungrazed, uncut condition, of native forest, meadow and wild flower areas."[1] NPA's founders saw their mission as extending beyond park boundaries to embrace wildlife conservation on other public, and even private, lands. They were dedicated also to what would soon be called "wilderness." They saw the parks as "destined to become the great zoological and botanical gardens of the future." Wildlife conditions were changing "with extreme rapidity," meaning that wildlife populations were declining everywhere across the United States, and they saw NPA and national parks playing an important role in wildlife conservation. This element of the NPA mission is not surprising, given that leading wildlife conservationists such as George Bird Grinnell of the Boone and Crockett Club and T. Gilbert Pearson of the National Association of Audubon Societies were among the founders.

Since the passage of the Antiquities Act in 1906, presidents had used their executive authority to create many national monuments. NPA's discussion of its second goal observes that "Our monuments, so far, have been created hit or miss, chosen without plan or purpose, upon chance suggestions."[1] Although the case may be overstated, there were no clear standards for creating monuments (or parks, for that matter) in 1919. Interior Secretary Lane, in a letter to Mather in May 1918 (written by Albright with help from Yard and others), had offered general advice on what should be included in the National Park System, but he had only begun to address this complex issue. NPA would seek to

Secretary of the Interior Albert Fall, here at Glacier Point in Yosemite with Director Mather.

create more national monuments, but it would identify criteria and work for a "rational system." This could be achieved, its founders proposed, by "a permanent committee of historical and scientific experts, whose recommendations would have weight." They would work to create such a committee.

Scientific Research and Economic Development

A third objective, related to the first, was to "thoroughly study the National Parks and make past as well as future results available for public use." NPA would conduct science in the parks, encouraging universities and learned societies to work there, and would offer the results of this work through the Library of Congress and the scientific libraries of government departments.

The final objective agreed to at the founding meeting was "To encourage travel in every practicable way." Since 1915, Yard had been trying to attract people to parks. If people saw the parks, he and Mather reasoned, they would appreciate them and support a National Park Service and an expanded National Park System.

Now Yard and his NPA colleagues saw the necessity of attracting people to parks to enjoy their unique educational opportunities. They also saw the national parks as a way of "increasing pride of country" and of "keeping as much American travel money at home as possible." Yard described the economic potential of national parks: "We possess, until recently unknown to the greater public, several Swiss Alps, a Canadian Rockies, and a greater variety of other areas of scenic sublimity than is accessible in all the rest of the world together."[2] He wrote that "No nation ever faced so great an opportunity as ours for the rapid development of a practically new income asset. But the ideal must be nurtured and a strong offensive and defensive policy outlined and persistently upheld. To this end, the value of the trademark 'National Parks of America' must be faithfully maintained lest, as its celebrity grows, local ambition and partisan expediency overload and destroy it."[2] NPA could guard against cheapening the asset.

Here again was the issue of standards, of what should bear the national park "trademark." NPA could be above politics and economic opportunism. It could encourage people to visit parks and potential parks and develop the asset while protecting the quality of the National Park System.

The founders of the National Parks Association generally supported the work of Mather and his associates in the emerging National Park Service. Their principal aim was to complement National Park Service efforts, to do work in and for the parks that the agency, because of politics and budgetary limits (which were inextricably linked), could not do. To the NPA founders, the parks were "nature's universities," places where scientists and lay persons alike could understand and appreciate natural processes unaltered by human enterprise. They saw themselves as defenders of America's natural and cultural heritage in a carefully developed system of parks and monuments. They could hold high standards and ideals without compromise; the Park Service could develop and administer, and NPA could support it.

The founders believed they could work *with* the National Park Service, at

Starving elk in Yellowstone offered NPA its first issue.

least under its initial administration. Even before the Educational Committee evolved into the National Parks Association, however, they recognized that a time might come when they would be at odds with the Park Service. In his June 1918 statement of need for the proposed association, Yard wrote, "So long as the National Park Service remains under its present management, the Association may confidently expect the liveliest sympathy, if not a great deal of definite help, there from; a very real partnership is certain." NPA must be created now, Yard believed, and "a basis for common working...established as precedent for later periods when...politics may plunge the Service so deeply into the tape basket that policies become prostituted and vision lost."[2] Looking ahead, the founders saw battles on the horizon, and hoped to establish solid cooperation with the National Park Service. During these early stages of development for both organizations and for the system of parks and monuments, the two groups must cooperate to achieve organizational strength and an improving National Park System.

THE WORK BEGINS
Finances and Membership

At the May 28 meeting, the founders got right down to business. They saw their largest initial challenge as financial and appointed a ways and means committee, chaired by Huston Thompson of the Federal Trade Commission, to raise operating funds. They agreed to organize other committees for specific tasks and to try to enlist other established organizations to the task. University and school groups, scientific institutions, "public spirited organizations of all sorts," automobile and highway associations would all be recruited. Business organizations, especially railroads, automobile manufacturers, and concessioners, would be called upon. Businesses that "will be helped by the work of the National Parks Association" would, they thought, be obvious cooperators and sources of revenue.[1] A membership campaign would be launched seeking teachers, scientists, "men and women active in public-spirited organizations of all sorts," and businessmen "of big horizons."

The founders set an ambitious agenda. They saw their organization as the heart of a national effort on behalf of national parks. The records of their early discussions suggest that they understood theirs would not be a large membership organization and would not accomplish its goals alone. Most founders represented other groups and were active in "public-spirited organizations"; they understood the need to build coalitions and what modern organizers would call "networks." They expected NPA to be a small group with a long reach.

After the founding meeting ended, Robert Sterling Yard began work as executive secretary. He developed project ideas, which the executive committee approved at its June meeting. The organization would prepare the first of a series of illustrated popular-science papers. Regular bulletins would be issued to members and allied organizations, describing issues and problems involving national parks. Yard would launch the pet educational projects that had caused problems at the National Park Service—an "extension slide service" and "library picture service." Most important, the group would mount a membership drive.

Because revenue was necessary for every association project, the first job was to build membership. Members paid $3 annual dues and sometimes gave additional donations. Yard solicited members from affiliated societies, which yielded few; the National Park Service Special List (friends of the Park Service) yielded some. Circulars went out to park concessioners, railroads, libraries, and educational conventions. Returns were below the association's expectations. The first $2,500 donation from Mather was carrying the association, and Macfarland warned the executive committee in September that NPA might not survive the year unless it could gather more funds. Mather attended the October meeting and pledged to help raise funds. Yard told the group he thought NPA would have "to show a reason for existence" before it could expect people to join. In his view, the association needed to "get into a fight for some worthy object so as to prove its spirit publicly."[3] NPA membership stood at 350 in October 1919.

Yard worked furiously. He convinced the Tourist and Publicity Bureau of the Denver Association of Commerce to do a Colorado membership drive. He toured parks and prepared pamphlets to send to members about Grand Canyon National Park and the soon-to-be Zion National Park. A trip to New York yielded several significant donations. His efforts managed to carry the association financially to Mather's second $2,500 installment on his pledge, but Yard protested the heavy workload for him and his clerical assistant.[4] He might protest, but he said, he had found his life's work. His livelihood depended on success, and he never rested. Membership had increased to 635 in January 1920.

The "Elk Opportunity"

Yard was on the lookout for a way to begin to prove the association's worth, and he soon found one. He called it "the Elk Opportunity" in Yellowstone National Park. Elk grazed along the Gallatin River, in meadows in the southern reaches of the park down into Jackson Hole. They were also common in meadows along the Lamar River, on the Black Tail Deer Plateau, and north into the meadows of the upper Yellowstone River Valley. Early Yellowstone superintendents estimated a population of 25,000 elk in the park. By 1914, their numbers had increased to a probable 35,300.[5] The summer of 1919 was dry and was followed by an early winter of heavy snows. The elk herd had grown so large that it damaged its range and this, combined with drought and deep snow, drove the animals beyond park boundaries, where they were shot in great numbers by hunters. Others faced starvation as the severe winter progressed.

The fate of the elk, particularly their slaughter by hunters, outraged wildlife and park advocates, and they called for action to save the herds. The Park Service began feeding the animals to keep them in the park and away from the hunters. A wealthy donor gave $8,000 to buy hay, and Mather worked on Congress for a supplemental appropriation to defray expenses of the elk feeding. NPA sent news bulletins to members and associates; the December 1919 *News Bulletin* was devoted entirely to "The Disaster to the Yellowstone Elk Herd." Readers learned that Yellowstone elk were in two major herds. The northern herd had ventured down into the valley of the Yellowstone River, where "Men fired by volleys into the wandering bands, scarcely taking time to aim, heedless of law or sportsmanship, carried out of their senses by greed of flesh. It recalled the dreadful last days of the buffalo. The Valley became a slaughter pen."[6] The northern herd was reduced by half. The southern herd gathered in Jackson Hole, where it was protected on a refuge but faced starvation.

The U.S. Biological Survey began feeding the southern herd, and Superintendent Horace Albright "spent his next spring's road-improvement money for additional hay" to help what was left of the northern herd. In mid-December NPA and its allies asked the governor of Montana to suspend the elk hunting season. He refused. The only way to help was feeding, so the association solicited donations for this purpose. Some were received and passed on to the National Park Service.

The April *News Bulletin* could report that the elk herds had survived this crisis principally because of emergency feeding, but the elk problem was only beginning. Elk herds would grow even as ranches and cattle on range surrounding the park would increase. Both elk and cattle "have their rights, and

Late in his tenure at Interior, Secretary Lane supported a controversial "invasion" of Yellowstone for its irrigation water.

the determination of these presently will become no small matter." NPA believed that the elk herds must be maintained, for they "constitute the noblest wild animal exhibits in North America." In NPA's view, their survival would require "ample natural range within or without the park boundaries" and control of "hideous slaughters" like that in December in the lower Yellowstone River Valley. Finally, the elk herds "must be kept within limits of size which a reasonable range will support," but slaughter by hunters would not be the proper way to achieve that goal.[7]

This elk episode reveals how NPA would approach controversies involving the parks. A position on the issue would be formulated, sometimes by the officers and executive committee, often by the executive secretary, with the ultimate approval of the trustees. Communiques would go out to members and affiliates explaining the problem, NPA's position, and recommended action. The leaders and executive would write or visit key players to try to influence their decisions and actions. NPA's effort would be a combination of education and lobbying. It would not be the only player, and often not the central player, in the effort. Its role would be to present the issue and the conservationist's positions at the seat of the U.S. government while soliciting support for that position from concerned organizations of various interests nationwide.

Did NPA accomplish much in the elk crisis? Probably not. The Biological Survey and the Park Service responded, and the principal tangible results of NPA's efforts were donations to the hay fund and a small informed cadre of concerned citizens. The crisis probably did more for NPA than NPA did for the elk. It gave the fledgling organization an opportunity to demonstrate its aims and methods, to be part of a significant national issue about which its members cared. By the May 1920 first annual meeting, membership had grown to 833. The National Parks Association had survived its first year.

"The War On the National Parks"

The spring of 1920 brought relief to the elk and the beginning of a real fight to the association, when Yellowstone National Park was again at the center of controversy. The issues were irrigation and protecting national parks from intrusion. Representative Addison T. Smith of Idaho introduced a bill into the Sixty-sixth Congress to create an irrigation reservoir in the Falls River-Bechler River basins within the southwest corner of the park. The water would flow to farmers in Idaho. Interior Secretary Lane favored the project. Mather and Yellowstone Superintendent Albright did what they could to slow necessary surveys, and NPA and other organizations opposed the legislation in congressional hearings. Secretary Lane resigned in February 1920, and without his support, the legislation lost momentum. This was only the opening skirmish of what would prove to be a battle lasting for decades.

Another bill approved by the Sixty-sixth Congress marked, in Yard's words, the opening salvo of "The War on the National Parks." Yard and his conservation associates knew that a bill was in preparation to create a Federal Power Commission that would have authority to permit water development on public lands, but they expected national parks and monuments to be exempted. According to Yard, on the morning he learned that Congress had adjourned and the power bill had passed, he went to the Department of the Interior to confirm this exemption. To his shock, he found that the act gave the commission authority to permit development in national parks and monuments. Immediately he sent telegrams nationwide, informing members and allies of this unexpected development. In response, a flood of telegrams came to President Woodrow Wilson protesting this intrusion on national park sanctity. Wilson stated his intention to use a pocket veto to stop the legislation.

Another wave of telegrams flowed into the Oval Office, this time from Western power interests, warning the president that "if some means were not found to sign the bill, several western states might be lost in the November elections."[8] The president reversed his intention and signed the bill. Yard's view of this was that "Congress had killed the fifty-year principle of complete conservation by conferring upon the Federal Power Commission the power to grant leases in national parks and monuments at will." Mather, with the assistance of new Interior Secretary John Barton Payne, urged the president to veto the bill unless he could exact a pledge from leading congressional water power advocates to amend the act in the next session to exempt national parks and monuments. Wilson exacted that pledge before he finally signed the bill.[9]

Yard knew that pressure was essential to ensure that this pledge was kept. He set out to develop a national network so that by the beginning of the second session of Congress, "every Senator and every Representative should receive advice from his people at home to restore the national parks to their former status and to hold them safe."[8] Working with allies such as the Appalachian Mountain Club, Sierra Club, Mazamas, and Mountaineers, he set up regional organizations in Boston, Chicago, San Francisco, Portland, and Seattle specifically to address this issue. He also enlisted the support of the new National Association of Business and Professional Women and the General Federation of Women's Clubs. Women's clubs were organized in every congressional district and pursued a legislative agenda as part of their regular activities. Their agenda for the next session was already set, but Yard was determined that they add this new element. When he wrote the federation president, Mrs. Winter, and received no response, he went to her home in Minneapolis and then to the national board meeting of the General Federation in Washington, D.C. At the board meeting, the federation passed a set of resolutions "word for word as we wrote them, and assurances were given me that the Federation would work heartily from top to bottom." NPA would provide the Women's Clubs with the information they needed and handle correspondence on the issue. "Your Executive Secretary accepted the conditions," Yard wrote in his report to the NPA Board, "with a smile as broad as if he knew where our December rent was coming from."[8]

Yard's campaign continued. The association issued its first "magazine" during the summer of 1920 titled, *The Nation's Parks*. Yard did not call this a magazine, wishing to avoid any suggestion that there would be future issues, because no funds were available for regular publication. *The Nation's Parks* was nicely produced—reflecting Yard's vast experience in journalism and publishing—and featured photographs of parks and proposed parks. The association's concern about the threat to Yellowstone by irrigation schemes and the Federal Water Power Act was stated in the lead article, titled, "Hands Off the National Parks." In it, the association explained its work to readers and solicited their support.[10] Although NPA would not be able to afford another publication of the quality of *The Nation's Parks* for many years, it was Yard's first step toward an effective vehicle to convey the message that the National Parks Association was alive and well and hard at work on its missions of education and park protection.

As promised, Representatives John J. Esch of Wisconsin and John J. Rogers of Massachusetts and Senator Wesley L. Jones of Washington introduced bills in the next session of Congress to exempt parks and monuments. The exemption legislation became known as the Jones-Esch bill. Yard meanwhile continued his membership recruiting and fund raising for the association, in the course of which he solicited A. W. Harris, president of the Harris Trust Company of Chicago. Harris agreed to become a member of NPA's "Group of Fifteen," a group of large donors who pledged $1,000 each year for three years. As work progressed on the Jones-Esch bill, Harris wrote to Yard protesting NPA advocacy of the legislation. An exchange of letters ensued, with Harris saying that if he had known the association would stand in the way of the country's development he would not have become a member. Yard tried to resolve what he thought was a misunderstanding. "Please understand," he wrote, "that all we are trying to do is restore the exact national parks situation which existed at the time you made your contribution to the National Parks Association."[11]

Hubert Work reassured national park supporters when he replaced Albert Fall as Secretary of the Interior.

Harris replied he was glad to learn that the "National Parks Association will be satisfied to have the amendment to the Water Power Bill apply only to the national parks as at present constituted."[11] NPA would be satisfied with nothing of the sort. Harris went on to inform Yard that a Mr. Pierce would be coming to Washington to "look after the interests of the water power people." He would offer an amendment to the Jones-Esch bill confining the exemption to existing parks and monuments and opening future parks to permitting by the Federal Power Commission for dams, powerline rights-of-way, and other developments. Harris would appreciate any help Yard could give in smoothing the way for Pierce's amendment.

Yard was dismayed. Henry J. Pierce, he learned, was president of the Washington Irrigation and Development Company of Seattle and represented Pacific Coast power interests. These interests had plans for water projects in Western valleys being considered for future parks, particularly the Tehipite Valley and Kings River Canyon in the proposed enlargement of Sequoia National Park. These power interests were willing to relinquish claims on existing parks and support Jones-Esch if future parks were exempted.

Harris, Yard believed, was a plant. His company was one of the so-called insider banks used by water power interests to sell their securities. Harris was part of a scheme to maneuver NPA into the position of being faced with supporting the amendment or losing the Jones-Esch bill altogether, thus leaving all present and future parks vulnerable to water power development. NPA ultimately supported the bill with Pierce's amendment. Both houses of Congress approved the measure, and President Wilson signed it before leaving office in March 1921.

This incident was a lesson to Yard and the association. Yard was a journalist, not a politician. He learned that he and his NPA colleagues would have to play hardball politics to achieve some of their goals. They were creating a political network of considerable force, but it would do them little good if they were outmaneuvered, as in the Jones-Esch case. They would study their political lessons. The struggle over the Federal Water Power Act allowed NPA to build broad alliances, especially with influential women's groups, and play a lead role on the national stage in park protection. But the value of the effort to enhance NPA's prestige, and thereby its membership and resources, was diminished by its defeat in the larger effort of protecting present *and* future parks.

Internal Strife

NPA suffered other growing pains. The attempt to broaden the NPA effort had led, in August 1920, to the creation of a National Parks Committee that would be advisory to NPA. George Bird Grinnell chaired the committee, with Yard and J. Horace McFarland as vice chairmen. Just before the NPA executive committee meeting in March 1921, Yard was summoned to a meeting of this advisory group and called to task by McFarland, who contended that Yard's aggressive campaign against irrigation interests was hurting the national park cause, particularly in the West. He thought that all NPA publications coming out of Yard's office should be reviewed by the committee before release, and a resolution to that effect was passed.

Yard was outraged and said so when he met with the NPA executive committee shortly after his encounter with McFarland. He angrily told the committee that he thought McFarland suffered from a severe case of "association jealousy" and thought NPA was competing with his own American Civic Association. Yard claimed McFarland had openly belittled NPA on more than one occasion. Walcott, chairing the meeting, asked Yard what authority the National Parks Committee had over NPA, and Yard said it had "not a whit." Walcott ruled that the whole business was of no concern to the executive committee and the group resolved that "It is the desire of this Committee that any change in policy must be referred to the Executive Committee for consideration or action."[12] They were telling Yard to forget the incident and, in future, avoid such pitfalls by working within the structure of the National Parks Association. March of 1921 was a bruising month for Robert Sterling Yard.

Yard's confrontation with McFarland reveals another layer of politics NPA encountered in its emergent period. McFarland's American Civic Association had been a major force in creating the National Park Service. McFarland could be justly proud of long and effective service for national parks, but he did not like Yard. The antipathy between the two men may have set their organizations, which should have been fully allied, at odds.

Charles D. Walcott served as NPA president from 1921-24.

The Yellowstone Defense Continues

Even as Yard and the association were occupied by work on the Federal Water Power Act and internal squabbles, Representative Smith of Idaho and his ally, Senator Thomas J. Walsh of Montana, were pressing their invasion of Yellowstone National Park. Their first attempt to push a Falls River-Bechler River basins reclamation project through Congress had been defeated. A congressional hearing on the bill on May 27, 1920, drew opposition from a coalition of outdoor and conservation groups, National Park Service Director Mather, and Yellowstone Superintendent Albright. New Interior Secretary Payne made his opposition clear in a letter, and the bill was dead for the session.

Senator Walsh introduced a new bill on December 7, 1920, which would permit the state of Montana to dam the Yellowstone River near the outlet of Yellowstone Lake. Behind the proposal was the Yellowstone Irrigation Association, a group of promoters from Livingston, Montana. Walsh held surprise hearings when the Livingston backers could be present and the opponents, not knowing of the hearing, would be absent. Senator Charles L. McNary called for opposition hearings in the next session, but Walsh insisted they be held immediately. The second hearing convened on February 28, with yet another in March. Twenty-six organizations, including NPA, were represented "and made such a shambles of the arguments of the promoters that the Walsh bill was not reported."[13]

The Sixty-seventh Congress convened, and in April, Walsh introduced two bills that were reworded to make them more palatable. He issued a circular to every member of Congress restating the arguments for the Yellowstone Dam. At hearings Albert Fall, newly appointed secretary of the Interior, opposed the bills and stated that if dams should ever be built in national parks, the federal government should build and control them, not private or state interests as the Walsh bills proposed. NPA and other conservationists were pleased with Fall's opposition to the proposed legislation but said he "indirectly committed himself against National Parks conservation" by suggesting that there might be "cases where it is necessary and advisable in the public interest to develop power and irrigation possibilities in National Parks...."[14] Fall did not seem to understand what NPA saw as an inviolable principle governing national parks: that they cannot be dedicated to any commercial interest. The next two years would bring clashes between Fall and NPA on this and other fronts.

TAKING STOCK

The second annual meeting of the association gathered in the garden of the Cosmos Club on May 20, 1920. Forty of its 1,300 members attended. Before lunch and their meeting, they walked to the White House for a reception with President Warren G. Harding. Leaders of the association had despaired that it would live to its first birthday, yet here it was, celebrating its second with a reception by the president of the United States. The reception was arranged by Henry Macfarland, the association's president, and testifies to his power and connections. Still, the fact that the president would receive NPA indicates that it had become a respected force in Washington conservation politics. Its membership consisted of well-connected people, and it had made an impression with its work on the Federal Water Power Act and Yellowstone Dam issues.

After lunch, Robert Sterling Yard gave his annual report as executive secretary. The Jones-Esch bill and defeat of Walsh's latest proposals were counted as victories, although in his view, the "war on our national parks" had only begun. The amendment to the Jones-Esch bill meant that each new park proposal would have to exclude the proposed park from the Water Power Act. "As soon as one new park is created under the Water Power Act, the precedent will call for water power entering all the parks.... It will become the duty of the conservationist to rectify the illogical and impossible condition of two classes of national parks and monuments as soon as possible."[15] The next battle over water power was already looming, he told them: the city of Los Angeles had applied to the Federal Power Commission for dam sites and water power rights in the Tehipite Valley and Kings River Canyon in California. These valleys were in the proposed Roosevelt-Sequoia National Park. He expected this battle to be a long one.

Yard's report indicated that the "defense campaign" involving the water power and irrigation issues had resulted in creation of a large, effective network of cooperating groups, and he hoped this network could be sustained. "This Association's principal contribution toward these victories...was the spreading of the news of danger. Its part was the part of Paul Revere. But we not only called to arms; we gave historical settings and explained principles. The campaign west of Pennsylvania was made on a one-page statement entitled Essential Facts."[15]

The campaign's sheer bulk of publication and mailing had been costly. Yard estimated that 54,000 pieces of mail had gone out, including 17,000 copies of "Essential Facts," 1,800 newspaper reprints, 7,000 pamphlets, and 20,300 *Bulletins*. The association's finances were in the black, but just barely. No financial reserve was available, but the executive secretary, as always, was optimistic that the means to continue the work would be found. His main regret was that he had not yet found the means to establish a magazine. Although various schemes to do so had fallen through, he remained dedicated to a magazine as the keystone of the educational and political work of the association.

LOOKING FORWARD

The second annual meeting marked the closing of the initial chapter in the story of the National Parks Association. The organization had survived its infancy under the careful direction of its distinguished president, Henry B. F. Macfarland. This would be Macfarland's last meeting; he died unexpectedly that October. He and Charles Walcott had been the driving agents in helping Yard launch the association. Macfarland personified the association's founders—a man of power and influence in the nation's capital, a civic activist who gave his time to many causes. For him, the national parks were another civic cause. He lent the cause his executive skills and judgment, but he was not a wealthy man, and he did not enjoy asking for money. He once offered to resign as president so that a man of more wealth and fund-raising ability could lead the organization; his offer was not accepted. He and Yard worked well together, Macfarland tempering Yard's ambition and occasional spells of abrasiveness and overzealousness.

NPA established its approach to park preservation during these first two years. Yard and his scientist and educator colleagues were dedicated to NPA's educational mission, but they found early on that financial and time limitations prevented the launching of many of their favorite educational projects. In 1920 Yard proposed, and the board approved, production of a *National Parks Album* "To do for the Association what the *Portfolio* did for the Service."[16] The *National Parks Portfolio* had been designed by Mather and Yard and published in 1917 as a luxury picture book, which they hoped would stimulate park travel and support.[17] Yard had done a good job of writing, selecting photographs, and laying out the book, and it was well received. Mather and seventeen railroad companies underwrote the $48,000 cost of publication. Although Yard's proposed NPA *Album* was then financially beyond the struggling association, the idea would be realized many years later by another NPA executive, Devereux Butcher.

Yard also proposed publishing an extensive series of interpretive pamphlets on subjects such as "Birds of the Rockies" and "Our Native Wild Animals." A few were published, but the means to establish a publication program lay far in the future. When funds were limited, as they had been throughout the history of NPA, the organization decided to give priority to the park protection work, which meant publishing politically useful materials such as the "Essential Facts" bulletin issued during the water power and irrigation battles. Yard hoped to publish a magazine; he really wished to be an editor rather than an executive, and said so. He knew that this would be a useful vehicle to achieve both educational and political objectives, but it was simply beyond financial reach.

NPA found it could be "Paul Revere," watching Congress closely from its Washington base and alerting people across the nation when problems were discovered. It could go to Capitol Hill and testify at congressional hearings, educate Congress and civic activists in the field. Although it might not lead many of its fights, NPA could offer its national network and Washington influence to groups far afield. Through its evolving educational programs, it could dispense "essential facts" about issues across the country. Its function in the conservation field was established: it served as the Washington watchdog and information clearinghouse for the national parks protection effort.

The founders thought there was national park work that the National Park Service could not (and perhaps would not) do. The agency would be subject to the will of Congress and the executive branch of government. To some degree, politics would dictate its priorities. The founders were idealists who thought they could transcend these politics, could work for the "big fruits of the national parks movement. . .unhampered by politics and routine."[2] There is no evidence that they disagreed with the National Park Service during this founding period, nor that they contributed to park protection in a way closed to the Park Service. They were, however, preparing for the time when NPA would stand firm for its national park ideals against powerful political forces within the government and even the National Park Service.

Robert Sterling Yard emerged as the pivotal figure in the new association. Although not a young man, Yard threw himself with seemingly boundless energy into the task of creating the National Parks Association. The record shows that Yard was, in every sense, the founder of NPA. There is little doubt that NPA was his idea; it certainly became his passion as well as his livelihood. Yard recruited the key founding members. He raised money to keep the association solvent (the Ways and Means Committee provided little of either, its chairman, Huston Thompson, too busy to lead a fund-raising effort). Yard wrote the press releases and bulletins, testified in Congress, and recruited allied associations. In short, Robert Sterling Yard *was* the National Parks Association during this period. He may have made enemies, such as J. Horace

McFarland, and mistakes, as in the Harris incident and the Jones-Esch bill, but he must be given credit for launching the association.

What role did Stephen Mather play in founding NPA? He gave money and moral support. His donation allowed Yard to draw a salary during the association's first year. He must be given credit for luring Yard into the national park field. Yard had a reputation for abrasiveness, and Mather made Yard's appointment as executive secretary a condition of his donation to the fledgling organization. Some of the founding group may have been reluctant to hire Yard, perhaps because of McFarland's antipathy toward him (though nothing in the record confirms this). After attending the October 1919 board meeting, he disappears from the association's records except for the withdrawal of his financial support and occasional references to his role as director of the National Park Service. He helped start the organization, then let it be. Yard built it into a significant force in the effort to protect the national parks.

III

Coming of Age in the 1920s

The National Parks Association's founders thought they were organizing to advance the causes of education and scientific research in national parks. They quickly found that the work needed was less scientific and educational than political. No sooner did they begin work than their mission changed in response to public policy debates that required action they had not anticipated. Yard learned a few hard lessons during the first two years, but despite his stumbles, the association survived its fragile infancy and grew.

NPA became fully engaged in the politics of national parks—politics that strained relations between Yard and Mather. The association remained a supporter of parks and of the Park Service. Increasingly, however, it found itself not only defending specific parks from commercial invasion but defending national park ideals from what it saw as compromise and retreat. During this period, the issue of the purpose of national parks grew in importance as public demand for recreation grew. Were parks merely outdoor pleasuring grounds, or were they much more? NPA's answer was that they were more: they were "museums of undisturbed nature" and the "national gallery of scenic masterpieces." NPA became the leading advocate of this view in a debate about national outdoor recreation policy.

DEVELOPMENT IN THE PARKS
Power for Los Angeles

Representative Henry Ellsworth Barbour of Fresno, California, introduced a bill in Congress on June 29, 1921, to create a Roosevelt-Sequoia National

Car camping emerged in the 1920s as a new approach to outdoor recreation in the parks.

Park. Barbour's bill was one in a series of proposals to enlarge Sequoia National Park. The original legislation for the park, a sweeping proposal to protect most of what is now included in Sequoia-Kings Canyon National Park, had been introduced by California Senator John F. Miller in 1881. Miller's proposal was too large for the time. Timber, grazing, and irrigation interests would not allow such a large area to be set aside. Yet logging threats to remaining stands of giant sequoias were so great that park proponents could achieve part of their goal by focusing on these trees. Thus, Sequoia and General Grant national parks were created in 1890. Efforts to expand these rather small parks began almost immediately, spurred by active hydroelectric development around and even within them. Bills failed in Congress in 1911, 1917, and 1918. Stephen Mather made enlarging Sequoia National Park an early priority in his administration.

Mather and other park backers sought a national park five times the size of those created in 1890. Included would be the Kings River drainage and potential commercial timber resources along its western and southern edges. A bill for such a park, introduced by Senator James D. Phelan of California (infamous in conservation circles for his dogged pursuit of a dam in Yosemite's Hetch Hetchy Valley while he was mayor of San Francisco), passed the Senate in 1919 but was opposed in the House. Phelan reintroduced his bill in 1920, and it drew even more opposition, this time from the Forest Service, which did not want to lose land under its jurisdiction, and the city of Los Angeles, which coveted hydroelectric sites in the Kings River Valley. Mather negotiated with Forest Service chiefs Henry Graves and William Greeley and agreed to trade the existing park's three southern townships and the J.O. Pass area of the proposed park for the larger park he sought.[1] Barbour then introduced his compromise bill.

The Park Service Opposes NPA

The National Parks Association entered the picture at this point, strongly opposed to the Barbour bill. Yard expected the power interests to take the opportunity provided by the Pierce amendment to the Jones-Esch Act, which would allow power development in new national parks at the discretion of the Federal Power Commission. Barbour's bill contained no clause exempting the proposed park from the Pierce amendment, and Yard knew Los Angeles had already applied to construct an elaborate water control and power generation system on the Kings River. This was the initiative from power interests he expected and feared. If it succeeded, future national parks would enjoy far less protection than those already created.

Opposition to Barbour's legislation came from many quarters and once again squashed park expansion. The association devoted four issues of the *National Parks Bulletin* to the problem. The Sierra Club led a successful fight to amend the bill, only to have the amended legislation fall victim to the city of Los Angeles and its politicians. Mather's dream of an expanded Sequoia park was frustrated again, and his relations with Yard and NPA were damaged by the politics of the issue.

An editorial cartoonist depicted the battle over the national parks and what might happen if park defenders were defeated.

NATIONAL PARK
S THE PEOPLE INHERITED IT—

THE LOGICAL FINISH
IF WE LET DOWN THE BARS

Yard explained in his April 1922 report to his board that the association had lost two major contributors, one of whom was Stephen Mather. The director's financial support had been essential to launching the association, and he had pledged ongoing support of $1,000 per year. When his contribution was not received on schedule in July 1921, Yard thought he might have forgotten, but he also suspected that Mather was not pleased with NPA's stance on the Barbour bill. When Yard had traveled to Yosemite to inspect the Sequoia situation, he had met with Mather who, by Yard's account, did not wish to discuss his position on the Sequoia legislation. Yard was aggressively pressing for an amendment to the Barbour bill. In a meeting with Francis Farquhar of the Sierra Club, he learned that Mather and the club had agreed to back the bill as written. Barbour's district strongly supported power development, and Mather's approach, as Farquhar explained, "was to back the bill as it stood long enough to get it reported, and then turn on it and have it amended on the floor of the House" or, failing that, to support better legislation in the Senate and resolve the problem in conference committee.

Yard thought Mather's approach too risky and refused his support. On principle, NPA would oppose any legislation that allowed development in parks. When Yard later discussed the matter of contributions with Mather, he was told that "his subscriptions to us put him in a false position and he had stopped them." Mather's final contribution of $250 came a few weeks later.[2]

At its founding, NPA knew that one day it would be at odds with the National Park Service, but Yard and his colleagues did not expect that moment to come so soon. They knew there was a need to create an organization beyond political influence; perhaps Yard thought Mather's compromising was the result of such influence. Albert Fall was then secretary of the Interior. Although conservationists were generally pleased with his opposition to the Walsh bills and his appointments in Interior, they remained suspicious of his politics and were particularly concerned that he did not oppose *all* development in national parks.

Was Mather yielding to Fall's pressure? Perhaps, but whatever Mather's motive, Yard could not join him. The two old friends parted ways in the cause in which both so deeply believed. After this, neither would have much to say about the other. Ironically, the Barbour bill was amended, as both men knew it must be, and then died from opposition by development interests.

Robert Sterling Yard was emerging as an uncompromising fighter for the complete protection of national parks. In his pragmatic way, Mather might attempt to work with business and compromise to achieve his goals, but Yard simply would not. Yard's stance was a strategy of the association—hold the line on national park ideals more resolutely than any politician or government official could.

This stubbornness was also a quality of Yard's character. As friend and foe sometimes said, he was a "purist" in his view of national parks: that nature was yielding everywhere to the onslaught of human enterprise, and the remaining few places where nature could be studied and contemplated in its

An editorial cartoonist depicted the battle over the national parks and what might happen if park defenders were defeated.

original state should be inviolable. Better to have no park, he thought, than a substandard park that might, in the long run, compromise the integrity of the national park ideal. "If there is any part of any proposed national park which is more valuable for commercial uses than for park uses, it should be cut out of the proposed boundaries and left to carry out its mission of prosperity in the National Forest where it belongs."[3] In Yard's view, the Kings River in the proposed Roosevelt-Sequoia National Park was unquestionably more valuable for park uses. He and his association would fight for those values, but if the area could not be dedicated to park values, it should be dropped from the national park.

Albert Fall and Expanding Commercialism

Mather was in a difficult political situation in the Roosevelt National Park matter. This was not new to him, but it may explain why he thought it necessary to disassociate from the National Parks Association. His boss was Albert Fall, with whom he was enjoying a surprisingly good relationship. When Warren G. Harding had been elected president, Mather and his associates expected to be replaced, because they had all been appointed by Democrats during the Wilson years. Horace Albright described their expectations:

> When he announced as his nominee his old Senate crony Albert Fall, gloom settled over all of us in the National Park Service. Not only was Fall one of the committee members who had opposed us on the [Yellowstone] dam issue, he was known to have personal interests in mining, stock-raising, and ranching and, as far as we knew, he had no leanings toward protection of national parks. It looked like very bad news indeed.[4]

To everyone's surprise, Fall kept Mather, Albright, and the rest of Mather's administration in place and was generally supportive. Mather proceeded with his initiatives but knew there were "certain conspicuous holes in Fall's philosophy of land use."[5] Fall had opposed Walsh's Yellowstone Lake dam scheme, but not all dams in national parks. One night around a Yellowstone campfire, he attacked the policy of oil land withdrawals, and he and Albright exchanged opposing views on the need for conservation. Fall expressed his faith in progress and reputedly said, "I stand for opening up every resource."[6]

Mather had to move cautiously with such a man, and was doing so in the Roosevelt National Park case. He very much wanted the expanded park, was a pragmatic and political bureaucrat, and was willing to play politics to achieve his goal while keeping his workable relationship with Fall. Mather knew that open opposition to Fall would necessitate his resignation and likely thought the Barbour bill was not important enough to require sacrifice of his position and all of his other projects. Yard, on the other hand, did not need to appease Fall. He and NPA could and did take the hard preservationist line. Mather could not have it both ways; thus, he was forced to cut his ties to the National Parks Association. Yard alluded to this situation in his annual report in June 1922:

Horace Albright, from his post as Superintendent of Yellowstone, was a powerful Park Service leader during the 1920s.

We have readjusted our relation with the National Park Service, reaching an understanding that, working to an ideal in common, each shall establish its own policy independent of the other and pursue it in its own way, with the best effort and good will on both sides toward making our activities cooperative.[7]

A few days earlier Yard had reported to the executive committee that NPA must press for repeal of the Pierce amendment. Fighting water power interests in each new park proposal would be overwhelming, as the Roosevelt National Park issue was demonstrating. Yard said he had not yet pursued this because he was certain that both Secretary Fall and Mather would oppose him. Mather would take this stance "because of his confidence that the water power companies could be persuaded by diplomacy to back complete conservation."[8] Even Yard did not want to challenge Mather too aggressively; the consequences might be severe. He also did not want to risk sinking the entire Roosevelt-Sequoia National Park initiative and alienating NPA's California allies, principally the Sierra Club, which were pushing hard for the park. As he reasoned to the trustees, "We do not dare risk any break in the splendid spirit of cooperation in the growing national organization of [national park] defense."[8]

Perhaps unwittingly, Albert Fall thus drove a wedge between NPA and the National Park Service. Mather eventually achieved part of his dream of an expanded Sequoia Park long after Fall's tenure, although he compromised again and removed the Kings River section from the proposal. This, ironically, is what Yard begrudgingly said must happen if the Kings River dams could not be avoided. Although the Senate dropped the Roosevelt name and approved an expanded park in July 1926, the fight over the Kings River went on. Congress would not repeal the Pierce amendment until 1935, and NPA and others would fight the water power issue each time a new park was proposed.

The Battle Line Is Drawn

Even as Yard was reporting the NPA split with the Park Service to his board, Secretary Fall was hatching a scheme that would put Mather under even greater pressure and reunite the park interests. As a New Mexico senator, Fall had twice tried to create a national park in the Mescalero Indian Reservation, first in 1913 and again in 1916. Objections to the previous bills had included the lack of suitability of the reservation for national park status, the loss of Indian rights that would result from park designation, and the personal gain that would accrue to Fall, whose ranch was adjacent to the proposed park. Cynics were certain that Fall's only interests in the park were his personal political and financial gains.

The *National Parks Bulletin* summarized Fall's new proposal:

This proposed national park consists of a number of little wooded spots, miles apart in the valley bottoms in the Indian reservation, plus a bit of bad lands 40 miles away, plus a sample of gypsum desert 38 miles away, plus a Reclamation reservoir 90 miles away, all these across deserts of heavy sand. [9]

Fall's ranch would be in the midst of these widely dispersed and unconnected units, which Horace Albright described as looking like "a kind of strung-out horseshoe."[4] The park, referred to in the legislation as the "Mescalero Indian Reservation Act" but called the "All-Year National Park" bill by Fall and others (presumably because it could be used all year for recreation), was backed by the "Southwestern All-Year National Park Association." This association consisted principally of boosters from chambers of commerce in the small southern New Mexico cities of Tularosa, Las Cruces, and Alamagordo, and from El Paso, Texas. The Texans were interested mainly in the Elephant Butte Reservoir on the Rio Grande, which would be circled by roads connected to a popular El Paso resort south of the reservoir if development plans proceeded.

Economic boosterism associated with national park proposals was nothing new, but the all-year park went too far. As Fall drafted the bill, the park would be open to water power development, irrigation, hunting, mining, grazing, leasing of commercial privileges, and timber cutting. Yard wrote that with this proposal, "Secretary Fall at last came out in his true colors as champion of interests destructive of National Parks Conservation."[10]

Fall was a wily operator. Drawing on his years of experience in the Senate, he engineered quick Senate approval of the legislation. He made the bill appear as a measure to improve the lot of the Mescalero Apache, whose reservation had been created by executive order, giving them no vested rights to their property.

All national park supporters were united in their opposition to Fall's bill. They agreed with NPA's view that "The safety of the entire National Parks System from commercialism depends on its defeat."[11] With Yard and J. Horace McFarland leading the charge—these old rivals often worked together despite their personal antipathies—they prepared to protect park standards by defeating the bill in the House of Representatives. Yard mobilized his "national organization of defense" to lobby congressmen, and presented NPA's strong opposition to the All-Year National Park in the *National Parks Bulletin*.

When this *Bulletin* came to the notice of Secretary Fall, he was furious and invited Yard to visit him to discuss the matter. Yard describes the encounter:

> For nearly an hour, Mr. Fall sought to win me to his plans, which he explained with all his big enthusiasm. When, putting his question, he found me still in opposition, he suddenly flew into a rage, called me a meddler, denouncing and defying all conservationists. His voice raised to a roar, he gesticulated, repeatedly leaned over to glare in my face, laughed scornfully at my replies, and wound up by demanding in tremendous tones whether I knew that I was talking to the Secretary of the Interior and was interfering with his sworn duty.[7]

Yard would not be intimidated. He came away from the meeting more determined than when he went in. Fall not only had tried to change his views on the all-year park proposal but had told him that he had a plan "to transform the National Parks System into a wide-open system of merely recreation areas...."[7] Fall, who seemed committed to finding a way to change the nature of the national parks, knew that many localities in addition to southern New Mexico desired to cash in on what they thought was a national park bonanza. Fall probably did not know he was touching a vital nerve in Yard and NPA—the issue of park standards. His bullying only outraged and energized them.

The *Bulletin* went out to the network, and opposition to Fall's proposal grew. Yard had strong contacts in New Mexico's Women's Clubs, as the following excerpt from a pro-all-year park newspaper in southern New Mexico reveals:

> An unfair and widespread attack on the proposed creation of the Yearly [sic] National Park is being circulated especially in the northern part of the state. It has been so effective the women of the Women's Club of Albuquerque have been so misled as to pass resolutions opposing it. These attacks emanate from what is known as the National Parks Association, one of the many lobbying institutions in Washington, D.C., run by a person by the name of Robert Sterling Yard who calls himself "executive secretary." One of the principal

objections advocated is that in the Act creating this park provision is made for power development at Elephant Butte Dam.

We respectfully suggest to the Women's Club of Albuquerque that before they further continue any activities based upon any propaganda from Yard or his association, that they discover who Yard is and what his association is. Has Yard, who pretends to be the executive of the Association and who is advising people of the United States as to the creation and conduct of national parks, ever himself been in a national park?[12]

Sentiment in New Mexico was divided, with southern interests supporting the park and northern New Mexicans opposed. The *Albuquerque Herald* editorialized against it and defended NPA:

The editor of the *Las Cruces Citizen* does not give us credit for much intelligence. Far from being "misled" by that association, we earnestly IMPLORED its aid to prevent New Mexico's being made a laughing stock of the nation, and her representation in the system the "lame duckling" among national parks, as the editor of the *New York Times* dubbed it; also that New Mexico's real material for a national park shall be ignored and forever barred from entrance into the system.[13]

Fall had stirred up a hornet's nest in his home state, and NPA played a big role in the agitation.

The Decline of Fall

The issue reached a climax in 1923. Fall's bill languished in the House and did not receive a hearing until Congress convened for a new session in January. The bill went to the Committee on Indian Affairs, which scheduled a hearing on January 11. In Albright's words, "The hearings opened with the big guns of the conservation movement loaded for action."[4] Yard, J. Horace McFarland, and George Bird Grinnell spoke in opposition. McFarland told the committee that in his view, Congress was in danger of making a national joke instead of a national park. Proponents made their case; Fall testified for three hours, being forced by tough questioning to abandon many elements of his proposal. According to Albright, the committee seemed ready to compromise and approve a recreation area under the Bureau of Indian Affairs, but Fall would accept nothing but a national park, and the bill was not reported out of the committee.

The end of Fall's assault on the national parks came with his resignation in late February 1923. He allied himself with two corrupt oil men, Edward Doheny and Harry Sinclair, giving them control of immensely valuable naval oil reserves. The Elk Hill reserve in California was leased to Doheny's company, the Teapot Dome reserve in Wyoming to Sinclair's. Fall received at least $100,000 from Doheny and $300,000 from Sinclair and eventually was

Herbert Hoover helped pull NPA from its doldrums during his year as association president in 1924-25.

convicted and imprisoned for these transgressions.[14] Ironically, Fall's resignation was the result of dogged investigations by the conservationists' antagonist Senator Thomas Walsh of Montana, who still had his eye on Yellowstone Lake.

A CRISIS OF LEADERSHIP
Herbert Hoover to the Rescue

Fully engaged as it was in defending the national parks, NPA was still struggling as an organization. President Macfarland's death in the fall of 1921 created a crisis of leadership. Macfarland had led the board and provided support and much-needed counsel to Yard. Stepping in as acting president was Charles Walcott, but he was very busy, quite elderly, and not really interested in being president. Yard made a plea to the board at the June 1922 annual meeting for assistance in raising funds; the work of park protection was restricted by his constant need to raise money. The board dithered over the matter for the next year, and Yard continued to carry the burden. A Committee on Permanent Organization was formed and pressed the search for a president. Finally, after several candidates declined, a president was elected at the February 1924 board meeting—Secretary of Commerce and future president of the United States, Herbert Hoover.

Why would Hoover accept the NPA presidency? A man of his prestige and rising political influence would have been in high demand at the time, and there is no evidence he had been active in NPA. Albright notes that Hoover and Mather were friends and that Hoover was an avid fisherman.

Venerable conservationist George Bird Grinnell succeeded Hoover as NPA president in 1925 and served until 1929.

Albright's talks with him revealed "that he truly loved the parks."[4] Although national parks were not a central concern, he seemed to appreciate their value, as indicated in a letter to Walcott accepting the association's presidency:

> The defense and preservation of our national parks is a most worthy effort. Their stimulative, educational, and recreational values are, all of them, of vital importance to all of our citizens. Recreation grounds and natural museums are as necessary to our advancing civilization as are wheat fields and factories. Indeed, I should like to see the Association not alone devote itself to defense of the areas that have been set aside by our Government for perpetual use in these purposes, but to expand its activities in the promotion of other forms of recreational areas.[15]

As this letter suggests, Hoover was interested not only in national parks but in the broader problem of providing outdoor recreation areas to a growing number of users. He brought this up repeatedly in discussions during his year as association president.

Outdoor recreation policy was a concern of the Coolidge administration at this time. A rising standard of living and increasing ownership of automobiles allowed more people to visit national parks and forests and other outdoor recreation areas in their leisure time. Theodore Roosevelt, Jr., undersecretary of the Navy in Coolidge's administration, was pressing for a national policy on recreation uses of federal lands, and Hoover was aware of Coolidge's interest. While Hoover presided over NPA, Coolidge called for this national policy and convened a National Conference on Outdoor Recreation, in which NPA played an important part. When Hoover resigned the NPA presidency in November 1925, he expressed his belief that a strong, centralized governmental authority was needed for conservation and use of natural resources. "If anything is certain," he said, "it is that the Government should have a continuous, definite, and consistent policy directed to intelligent conservation and use of natural resources. But it can have no such policy so long as responsibility is split up among half a dozen different departments." He thought an "Under Secretary of Conservation" was needed, someone with responsibility for conservation and natural resource policy "with the spotlight of public opinion continuously focused upon him."[16] Yard told the retiring president that he hoped he would take this idea to the public. A "Department of Conservation," which Hoover seemed to advocate, was to be a controversial proposal in the years ahead.

Grinnell to the Helm

Hoover was replaced as NPA president by a distinguished and accomplished man of an entirely different sort, George Bird Grinnell. The seventy-six-year-old Grinnell was, in every respect, a veteran of the conservation struggle of the past half- century. He was graduated from college in 1870, before the first national park had been created. He was a member of one of the early exploratory expeditions to Yellowstone in 1875, and his reports on the mammals and birds of the region were the standard references for years. He visited the Black Hills with George Armstrong Custer's first expedition to that area in 1874, and earned a Ph.D. in zoology in 1880. Editor and publisher of *Forest and Stream* magazine for thirty-one years, he had become a leader of the conservation community through his writing and activism. With Theodore Roosevelt and others, he founded the Boone and Crockett Club, of which he was president at the time he became NPA president. An authority on the Plains Indians, he was sent by presidents Grover Cleveland and Theodore Roosevelt to negotiate with Indian tribes in two crises and successfully resolved the disputes. Grinnell was one of the leaders of the successful effort to ban the killing of game in national parks (a ban was passed by Congress in 1894), and he was the driving force behind the creation of Glacier National Park in 1910. In 1925, President Coolidge awarded him the gold medal of the Roosevelt Memorial Association "for distinguished service in the promotion of the outdoor life." Grinnell would serve as NPA president until 1929.

The fact that Hoover and Grinnell would volunteer to serve as president is testimony to the status of the National Parks Association in the 1920s. The association numbered only about 2,000 dues-paying members when Hoover assumed its leadership, but they included leaders of science, education, and business. The founding group was aging, and some could not continue in the mid-1920s. Low attendance at meetings, Yard's continued complaints about the burdens of fund raising, and the inability to find a president for several years suggest that the association went through a crisis of morale after Macfarland's death. Yard worked hard, but the internal affairs of the organization bogged down. Hoover, with his stature and commitment, energized the association; he regularly presided over board meetings during his tenure as president despite his packed schedule as secretary of Commerce. Attendance at meetings revived, and the association forged ahead.

Support from Interior

This revival was assisted by Harding's appointment of Hubert Work to replace Fall as secretary of the Interior. Work, formerly postmaster general, was a welcome relief from the duplicitous Fall. He pledged that "in the conduct of the Department of the Interior there shall be no submerged or camouflaged policies, no issues tucked away behind smoke screens, but an open and frank exposition of all actions deemed essential to the public interest."[17] Work was charged by President Harding to repair Fall's damage to the Department of the Interior's credibility, but he also proved to be a strong park supporter in his own right. Once again, Yard was invited to call upon a secretary of the Interior, but this time his reception was cordial, even warm. Work assured him of his sympathy with the principles and purposes of NPA. Then he asked Yard to brief him on the history of national parks and provide suggestions for administrative policy. Yard obliged.

Yard was immensely pleased to report to his board in June 1923 that times had changed at Interior. He quoted a speech just delivered at Yellowstone by the administration's representative, Dr. John Wesley Hill:

> Regardless of all facts and figures…any plan however meritorious on its face for the commercial exploitation of parks must by the very nature of its aims and purposes be immediately doomed to failure….for it is at last the established policy of Government that our national parks must and shall forever be maintained in absolute, unimpaired form, not only for the present, but for all time to come, a policy which has the unqualified support of the great American now in the White House….[18]

This was music to the ears of all NPA stalwarts and led Yard to an optimistic assessment of prospects. "With Fall whipped and out, Walsh nearly through, Addison Smith waiting for a turn of the tide which will never come, and a sympathetic Secretary of the Interior, we may confidently assume that never again will battle require all our effort, and that a few years will see it disappear from our activities."[19] Battle was not to end, but Yard could be permitted his optimism, and the immediate future could be devoted to something other than fighting commercial invasion of the national parks.

NATIONAL OUTDOOR RECREATION POLICY

High-Level Interest

One consequence of Hoover's leadership was NPA involvement in an effort to define national outdoor recreation policy. President Coolidge convened a National Conference on Outdoor Recreation in May 1924. More than 300 delegates representing 128 national organizations, ranging from conservation groups to athletic clubs, gathered in Washington, D.C. The conference was to advise the newly appointed President's Committee on Outdoor Recreation, which comprised five cabinet members, including Hoover, and was chaired by Theodore Roosevelt, Jr. This committee was charged with developing a national policy that would make outdoor recreation available to all, and with resolving difficulties that blocked access to this public resource. The ultimate outcome would be a National Plan for Outdoor Recreation.

Parks and Forests

This high-level interest in outdoor recreation policy came from several directions, the most important of which was rapidly increasing outdoor recreation activity by the public. The growing demand for outdoor recreation services raised a flock of policy questions for the National Park Service and Forest Service as well as natural resource decision makers at all levels of government. Late in 1925, Yard described this in the *National Parks Bulletin*:

> The era of outdoor recreation has rushed upon us with the speed of the automobile, which brought it, and there can be no denying its pace, its power and its permanence. It swept from east to west... engulfed our National Parks, swept down our coasts, overran valleys and mountains, and now is invading desert and wilderness. Ungoverned and ungovernable, at least it can be directed. The General Conference on Outdoor Recreation is...born of civilization's instinct for organization and control.[20]

Additionally, the drive to create new national parks and expand existing parks was generating conflict between the Park Service and the Forest Service. At a hearing on the Barbour bill in late 1921, Chief William B. Greeley of the Forest Service had raised the issue of what should be a national park and national forest. NPA had devoted an entire issue of the *Bulletin* to Greeley's remarks. He argued for clear distinction between parks and forests, stating that as "a matter of national policy, we should have one class of reservations which exist primarily for commercial use—that is, the national forests. We should have a second class of reservations established for recreation and the specific forms of development that contribute to recreation. I do not think we should mix these two purposes in our national parks."[21]

From the perspective of the Forest Service, each new or expanded park was at its expense—land was taken out of a national forest and placed in a national park. This was the case with the Barbour bill, which would add national forest land to the Roosevelt-Sequoia National Park. The Forest Service did not wish to lose any domain. Greeley thought that if an area had commercial value, it should remain in a national forest. He used his own version of a purist approach in defending national forests against Mather's expansionism.

Yet even as Greeley was testifying about this issue, some members of his organization were proposing the allocation of portions of the national forests exclusively for recreation and preservation. The national forest wilderness idea was emerging from the work of Forest Service landscape architect Arthur Carhart and forester Aldo Leopold. By 1924, a movement was afoot in the Forest Service, led by Leopold, to create formal wilderness reservations. The first—the Gila Wilderness Area in New Mexico—was approved on June 3, 1924. The clear distinction that Greeley was attempting to draw between national parks and national forests was blurring, and this fact raised new questions about what should and should not be a national park, and even about the nature and purpose of the National Park System.

Another impetus for the outdoor recreation conference was a growing concern about development in national parks. How much development was too much? One especially pressing question was where roads should be built. Mather, always promoting national parks to increase appropriations to the National Park Service and gain public support for new and expanded parks, was thought by some to be too development oriented. He had long and vigorously courted the support of automobile associations and had worked to improve access to and accommodation in the national parks. If people could not comfortably reach the national parks and enjoy their stay once there, he reasoned, he could not expect them to help him in his struggles to build the system. Some thought he erred too far toward one side of the dual mandate of the National Park Service Act, emphasizing enjoyment at the expense of conservation.

The National Conference on Outdoor Recreation would also address the role of the federal government in relation to state government in providing outdoor recreation. When should an area be a state park rather than a national park? What should be the policy on the vast unroaded parts of the public domain, regardless of who administered them? Roads were rapidly being built across these lands, and, as Aldo Leopold would point out to the conference, opportunities to retain wilderness and naturalness were declining with equal speed. Finally, concern was increasing about the fate of wildlife, particularly on public lands. What should be done to protect wild creatures?

NPA and the Presidential Conference

The conference gathered on May 22 at the New National Museum, with Colonel Theodore Roosevelt presiding. Speeches were made, resolutions adopted, committees appointed, and reports prepared. NPA was well represented. Yard was the official NPA delegate, and six members of the NPA board and executive committee served on conference committees. Yard gave a wide-ranging address to the conference titled, "Scenic Resources of the United States." He started with a recommendation: "The scenic resources of the United States ought to become the object of an immediate survey under the auspices of the Government." Only in this way "shall we secure to this generation and to posterity the enjoyment, health, education, and mental and spiritual inspiration which are the Nation's due from its heritage of wilderness." He extolled the scenic resources of America and said a land policy should be the task of the conference. Principles must be established "concerning the uncommercial uses of the public domain, outlining the governmental systems through which its highest use may be accomplished in practice, and protecting each by definition.[22] He concluded with a call for a new national system of recreational lands that would meet the growing demand for outdoor recreation. This system would lie between the extremes of the preservation Yard thought should be provided national parks and the economic exploitation he deemed appropriate in national forests.

Yard's call for a survey was echoed in the report of the Committee on Federal Land Policy. The committee also called for a clear definition of the primary functions of the two major agencies "touching the field of recreation," the Forest Service and National Park Service. The Park Service should preside over parks, "protecting inviolate those wonderful or unique areas of our country" which must be "protected completely from all economic use." Areas administered by the Park Service should "represent features of national importance as distinguished from those of sectional or local significance." The committee agreed that national forests "are areas set aside to protect and maintain, in a permanently productive or useful condition, lands…capable of yielding timber or other public benefits." The resources of these lands, including recreation, "should be developed to the greatest possible extent consistent with permanent productivity in such a way as to insure the highest use of all parts of the area involved."[23] The committee was certain that with such a definition and with reliable information about outdoor recreation resources, the recreation needs of the American people could be met by an improved system of public outdoor recreation areas. The "highest use" of areas could be identified, and they could be transferred to the jurisdiction of the appropriate agency.

Recreation Survey of the Federal Lands

The conference agreed that a survey should be undertaken and assigned the task to a joint committee of NPA and the American Forestry Association. The survey would serve as the basis for national planning. The committee was officially named the Joint Committee on Recreational Survey of the Federal Lands. William P. Wharton, an NPA member destined to be its longest serving president, was named chairman; Yard was secretary, and Ovid M. Butler of the American Forestry Association was treasurer. The Laura Spelman Rockefeller Memorial Fund provided a grant for the committee's work.

The committee was to complete the survey in two years but it proved a huge undertaking. A second general conference was convened in January 1926, but the survey was far from complete. Of the many talks given at the second conference, two are important to the NPA story. NPA trustee John C. Merriam spoke at great length about the responsibility of federal and state governments to provide for recreation. The national parks were his major interest. Although such parks are important recreational resources, he argued, their principal value is inspiration. "To me the parks are not mere places to rest and exercise and learn," he said. "They are regions where one looks through the veil to meet the realities of nature and of the unfathomable power behind it." These places must be managed for "complete conservation," which is "protection with all natural features unimpaired." He recognized the difficulty of doing this when there were also great economic resource values—as in the California redwood groves he was working to save. But protection can be justified when "higher educational and spiritual values...offer the greatest and most noble uses to which any possession can be put."[24]

Merriam introduced two themes that would occupy him throughout the next decade with NPA. First, national parks were the last places where "unmodified primitive life of the world" could be preserved and examined; parks should preserve the "primitive." The second theme, related to the first, was that national parks should be primarily places for education and inspiration, not merely for recreation.

The second important speaker was Aldo Leopold. He, too, spoke to the value of the primitive, but not with a focus on national parks. Wilderness conservation was his topic. In the two years since the first meeting, the Forest Service had continued to debate the idea of reserving wilderness, but there was no national wilderness policy, and Leopold saw an urgent need. Wilderness, he said with typical simplicity, "is the fundamental recreational resource. It is the food, and all the other things are merely the salt and spices which give it savor and variety." Preserving wilderness is very difficult because it is not a "crop," and "economic laws work relentlessly and irrevocably toward its progressive and complete destruction."

> In short, I am asserting that those who love the wilderness should not be wholly deprived of it, that while the reduction of the wilderness has been a good thing, its extermination would be a very bad one, and that the conservation of wilderness is the most urgent and difficult of all the tasks that confront us, because there are no economic laws to help and many to hinder its accomplishment.[25]

Leopold did not agree with Yard that a new kind of federal reservation was necessary. Portions of the national parks and forests could be dedicated to wilderness, "the national parks devoted to the gunless type of wilderness trip, another kind in the national forests devoted to all types of wilderness trips including hunting." Wilderness should be "a specialized form of land use within our existing or prospective forests and parks."[25]

Attractions such as the bear show at the Yellowstone dump raised questions in the 1920s about the nature and purpose of national parks.

Merriam and Leopold were singing a duet. During the ensuing years, both men would develop their ideas on wilderness. A decade later, Leopold would join Bob Marshall, Benton MacKaye, Robert Sterling Yard, and others in founding the Wilderness Society. Merriam would become a leading voice in NPA's advocacy of a wilderness policy in national parks. Wilderness would emerge as a central issue in park battles and in the rivalry between the National Park Service and the Forest Service.

Critical Findings

The Joint Committee on Recreational Survey of the Federal Lands finally published its report in 1928. The report identified problems and recommended elements that must be included in a federal outdoor recreation policy. The National Park Service received a measure of criticism. Stephen Mather had aimed to popularize parks to advance their cause, bringing Yard in to do the job, and had succeeded too well. The publicity campaign, coupled with the increased use of the automobile, resulted in an enormous rise in park visitors and the growth of the "people's playground" view of national parks. "The playground feature overshadowed the primary purpose of the parks and tended to increase the pressure for ill-advised additions to the parks which were simply pleasant outdoor playground areas and nothing more."[26] Also, the report continued, pressures for development to accommodate visitors had increased, and "one finds developments and activities not entirely compatible with the primary park purposes." Camp and hotel accommodations should be concentrated in one or two places in each park, and roads should be limited. Development locations should be considered carefully, and "it may well be questioned whether these areas of concentration need be the gems of the national park system—as, for example, the floor of Yosemite Valley."[27] The survey identified a trend toward development in national parks that the committee thought should be curtailed.

Turning its attention to national forests, the committee found the Forest Service responsive to the growing demand for outdoor recreation. More should be done, however, to protect national forest wilderness:

National parks will always have large regions including forested areas that will be retained in primitive state, but these will naturally and properly represent features outstanding for their scenic beauty and impressive representation of educational and inspirational values. It is on certain of the national forests, however, that the greatest possible opportunities for preservation of untouched regions favorable for wilderness recreation are found.[28]

The report identified 12.5 million acres suitable for wilderness preservation and suggested that wilderness might take three forms: "wilderness areas" would be the least developed, but grazing and timber operation not requiring roads would be allowed; "semi-wilderness areas," in which "tame" forms of recreational development (presumably summer homes, hotels, restaurants) would not be allowed, but some commercial use and development would be permitted; and "natural preserves," rather small areas, dedicated exclusively to research. The committee was not ready to exclude *all* commercial activity from any portion of national forest land. It kept the historic distinction of "conservation" versus "preservation" in its analysis, even while liberally quoting Leopold and echoing his concern that the rapidly disappearing wilderness "can not be re-created and therefore wilderness recreation is the one form of outdoor life which can not be restored or built to order. Whatever opportunities exist to-day [sic] must be preserved, or they are gone for all time."[29] The Forest Service was rapidly building roads and would largely eliminate wilderness unless measures were taken for its protection.

This report was confusing.[30] It reflected a persistent lack of clear definition for wilderness areas, probably reflecting a disagreement among committee members. The term "wilderness" applied to areas on which uses would vary considerably. The wilderness described was far from that conceived by Aldo Leopold, who would have no development whatever in his wilderness. At this time, the Forest Service was trying to use wilderness as a land classification to foil National Park Service proposals for new parks. It was suggesting that "wilderness" would give greater protection than would national parks, which were being developed. Yet the Forest Service could not entirely erase from management of any portion of its domain its central principles of utilitarian and multiple use. The strength of its commitment to these principles reached into the joint committee, where it had friends among the American Forestry Association (AFA) members and influenced the report.

The report concluded with a set of general recommendations that the committee thought should govern formulation of federal outdoor recreation policy and public land management. Several were important to national parks. More "interbureau cooperation of regional studies and planning" was essential to "most completely realize potential, educational, scientific, inspirational, and recreational values of national parks and forests." The "objects and standards" of the National Park System should be better established by law to "differentiate this system clearly from other land systems, Federal and State, and (b) provide a definite basis for development of recreation in the parks in coordination with recreation in national forests and other permanent Federal reservations." New areas "fully meeting the national park standards" should be added to the system, and units not meeting the standards should be eliminated. The secretary of Agriculture should delimit wilderness areas in the national forests.

As NPA's history reveals, central issues in outdoor recreation and land policy in the years after this national conference involved "standards"—what should or should not be a national park, and why; what should be proper uses of national parks; and how much wilderness should be preserved, and by whom. The Forest Service and National Park Service would argue repeatedly over these and other questions. New parks on the horizon, such as Shenandoah and Great Smoky Mountains, would bring these issues into sharp focus. The National Conference on Outdoor Recreation provided an assessment of the condition of recreation and federal land resources in the mid- to late-1920s and was an agenda-setting event.

The role of the National Parks Association in the conference was important. The NPA-AFA joint committee report was perhaps the most tangible and influential outcome of the conference, although national events conspired to limit the impact of the conference's work. Herbert Hoover, who participated in the conference and was an advocate of a stronger federal outdoor recreation policy, became president of the United States, but a grim economic situation and the onset of the Great Depression greatly reduced government response to the conference's recommendations. Still, the debates of the conference continued and profoundly affected national park and forest policy.

CONSERVATION POLITICS
Yellowstone Once More

While the Barbour and Fall initiatives and the national conference posed new challenges, NPA and other Yellowstone defenders faced continuous skirmishes. Senator Walsh submitted his third bill for damming Yellowstone Lake in 1924. In April 1926, Representative Smith renewed his efforts to draw irrigation water from Yellowstone's Bechler Basin for his Idaho constituents, this time proposing to cut the basin out of the park altogether. These threats were defeated once again in the Sixty-ninth Congress, only to reappear in the Seventieth. The cagey Smith attached his proposal to delete Bechler Basin to an Interior Department bill to revise Yellowstone's boundaries. Yard and others thoroughly denounced the proposal in a hearing, and Interior Secretary Work opposed it. Smith, knowing that the National Park Service wanted the boundaries changed, held the boundary revision hostage to his Bechler Basin proposal, and neither was approved.

NPA continued to publish the *National Parks Bulletin* through the 1920s, its format changing as dictated by the precarious budget. Yard still pursued his dream of a regular NPA publication of magazine quality, and decided to publish relatively few high-quality *Bulletins*—two or three per year—and supplement them with regular news releases. These monthly releases, which came from the "National Parks News Service," stated NPA's positions on issues and provided detailed information. Yard strove to maintain his national network of park defenders, and the news service was an important vehicle for doing so.

Prospects for Confederation

During the 1920s, there were efforts to consolidate conservation organizations with closely related missions. Chauncey J. Hamlin, chairman of the National Conference on Outdoor Recreation, proposed to Yard in early 1925 that NPA affiliate with the American Civic Association, the American Park Society, the Park Executives Institute, and the State Parks Conference. These groups would become the American Federation of Parks and Planning. Separately they competed for income and members; together they could eliminate the competition and cut their overhead. Proponents of the idea thought the causes of park and recreation planning, promotion, and protection would be better served by such a merger.

NPA and later U.S. President Herbert Hoover, posing here with Albright and a string of Yellowstone trout, was an avid fisherman with an interest in outdoor recreation policy.

The idea went through various revisions, emerging as a federation that would allow the separate organizations to maintain autonomy while they pooled administrative resources and coordinated activities. NPA President Hoover favored the idea, but the majority of trustees did not. NPA declined to join the Associated Societies. Although J. Horace McFarland had retired as president of the American Civic Association and now chaired its National Parks Committee, his rivalry with Yard continued. There were many good reasons for NPA to maintain its separate identity, but undoubtedly one of the most compelling was the prospect of Yard losing any of his control over the affairs of the association. The memory of McFarland's earlier attacks was strong, and Yard wanted no organizational connection to his old rival.

Another merger was considered in 1928, this time with the American Forestry Association. This idea found strong support within NPA. William P.

National parks often served as refuges for wildlife species, such as bison in Yellowstone.

Wharton, a trustee of the forestry association, was chair of the joint committee conducting the recreational survey and a vice president of NPA. He proposed a close partnership in which the organizations would share offices, operations, and publicity. NPA would benefit from a stronger financial base; the forestry association would strengthen its editorial and political resources. Ovid M. Butler, Yard's executive counterpart at the forestry association, went further and proposed that the two merge into the "American Forest, Park and Wildlife Association." Butler would be the executive, and Yard, editor of publications. Yard told his board he would enjoy this arrangement, for he still aspired more to editorial than executive duties.

George D. Pratt, American Forestry Association president, was concerned that the proposed arrangement would not be supported by Stephen Mather and asked whether he would approve. According to Yard's account, Pratt would not support the move unless Mather did. Mather approved, "provided that precautions shall be taken that the national park cause shall be able fully to preserve its autonomy so that present functions of defense, promotion and education upbuilding shall operate unimpaired."[31] NPA approved the merger in principle at the annual meeting in May 1928, but it came to nothing.

The fact that the American Forestry Association and NPA seriously considered a merger indicates changes in conservation politics at the time. The strictly utilitarian approach to national forests advocated by Gifford Pinchot and imbued in Forest Service philosophy was cracking. The growing demand for outdoor recreation was pushing foresters and the Forest Service in new directions. The idea of national forest wilderness was beginning to blur the clear difference between national parks and national forests. Some conservationists, Yard among them, were moving to the position that recreation should be minimal in national parks. Historian Susan Schrepfer summarizes this view: "Within an ordered, hierarchical public land structure, the national parks were...to fulfill a unique and crowning function. The Park Service was to permit only minimal conveniences and such mild outdoor activities as hiking and camping, so that the parks could serve educational, scientific, and even spiritual purposes."[32] One way to take recreation pressure off national parks was to promote it in national forests; thus, Yard and others who had focused exclusively on national parks began to think about national forest policy. The National Conference on Outdoor Recreation facilitated this shift.

Preservation-oriented conservationists were also entertaining the possibility that creating a national park was not the only, or even the best, way to preserve an area's natural values. Some criticized Park Service development of national parks and the emphasis on recreation in their management. "Purists" like Yard, Merriam, and William Colby of the Sierra Club were becoming critical of the "boosters" in the late 1920s. Shrepfer notes that "boosters" thought they had to promote park development. "Mather and Horace Albright believed the parks needed to be 'sold,' promoted in exchange for congressional appropriations and popular support. With this emphasis on numbers, park administrators minimized questions of proper use."[32] Perhaps portions of the national forests could be better protected than many of the parks. Although the Forest Service was far from committed to complete preservation of parts of the forests, perhaps this offered preservationists a new arena for activism. Thinking about the NPA-AFA merger, Yard, Wharton, and others may have considered this prospect.

NPA's Expanded Mission

Although park and forest defenders were drawn together by these forces, no merger occurred, and NPA focused on its mission as redefined in 1925. In light of the changing scene, the board changed NPA's statement of purpose in a revealing way.[33] The association's mission broadened. The original emphasis on education remained, as did the aim of enlisting a broad base of support for the national parks. Nature conservation remained the overall goal. The original objective of extending and developing the national parks into a "complete and rational system" was replaced by the broader aim of developing a "national recreation policy." The association no longer aimed to encourage travel; protecting parks and park standards became central.

These changes are understandable considering events since 1919. Threats to parks had been continuous. NPA had not been able to pursue its educational agenda as originally conceived, but it continued to try. The founders thought the major challenge was promoting new parks, but they learned that protecting existing parks and the *ideal* of a national park was an even greater task. Their idealism had been tempered in the heat of political struggle. They had not lost those ideals but had come to see that hard, unrelenting political work was necessary to maintain the qualities of the "museum system of undisturbed nature" that was the national parks.

Perhaps most significant in the association's new objectives was its pledge to uphold the standards of the National Park System. The question of what qualities should define a national park had been around for decades, but events in the 1920s brought it to the forefront. Fall's All-Year National Park was not up to the standard. If development was allowed in a national park, was the area still worthy of being called a national *park*?

New proposals for national parks popped up everywhere (thirteen were proposed to the Sixty-eighth Congress). Yard saw a "localist invasion" of the National Park System, with boosters attempting to help their local economies through the attraction of a national park. He knew NPA must fight this invasion on every front. When Hubert Work succeeded Fall as secretary of the Interior, he asked Yard, among others, to brief him on the history and problems of the National Park System. Yard did so, and was elated when Work wrote a letter to Senator Duncan U. Fletcher of Florida affirming a policy of maintaining high standards. Yard featured the letter on the front page of the January 21, 1924, issue of the *National Parks Bulletin*:

> The national parks…must not be lowered in standard, dignity, and prestige by the inclusion of areas which express in less than the highest terms the particular class or type of exhibit which they represent.[34]

Work's commitment to this statement was soon to be tested, for he authorized a survey of the southern Appalachian region "for the purpose of selecting an area that will be typical of the scenery, plant, and animal life of this range for a National Park."[35] The battle over standards would be joined.

IV

What Is–or Should Be–a National Park?

THE NEED FOR STANDARDS

Early in 1927, the trustees of the National Parks Association passed a resolution that there should be a single standard for selection of national parks, and that standard should be the "historic" standard. National parks must be unmodified and natural. They must be the finest example of their kind of scenery in the country. They must be of national significance, and their boundaries should be drawn so that the park values might be protected and effectively administered. NPA did not offer specific criteria, but suggested that these broad concepts and dedication to the educational and inspirational mission of the National Park System would allow park professionals to make the decisions necessary to preserve the quality of the system. They expressed confidence that park professionals could and would exercise good professional judgment in doing this.

Such a resolution was necessary, because they believed the quality of the system was seriously threatened. The demand for outdoor recreation had raised questions about the mission of national parks, and the push was on to create many new parks that were not, in the opinion of many, up to the standard of most existing parks.

The issue of standards had always been of concern to NPA, but the late 1920s brought it to the top of the association's agenda. A review of the standards problem and of how NPA engaged that issue in the 1920s and 1930s reveals much about the nature of the organization at this early stage of its development.

The Olympic Mountains of Washington were a battleground for NPA in its defense of "standards."

What Is a Park?

When Congress created Yellowstone National Park in 1872, boundaries were drawn entirely on paper and rather arbitrarily. These boundaries were not marked, or even known, on the ground. They could be drawn easily to enclose the geysers, falls, and lakes that early explorers found unique because few people other than Indians lived on the public domain lands in the vicinity. When the next park was created in 1890, however, the situation had changed—human settlement and exploration reached everywhere, and fights over park boundaries were a feature of every new park proposal. Someone had designs on nearly every part of the American landscape for its logs, minerals, water, or other resources. The issue of what should be in a park and what should not became important and contentious. Designing a national park was assigning one resource value rather than another—scenery, for instance, rather than timber. A clear standard was necessary to resolve these struggles over priorities. Such a standard was difficult to achieve and became even more so as the population of the United States and accompanying development pressures increased, as outdoor recreation became a popular pastime, and as the movement to create new parks grew.

By the 1920s, the question had become not only what standard should be used to decide what should be a national park but what should not. The "All-Year National Park" brought this issue into clear focus. A large part of the challenge was to maintain a high standard for national parks despite an onslaught of park proposals by chambers of commerce and other boosters desiring the economic benefits of a park.

Yard, delighted that Fall had resigned and that the All-Year National Park idea was dead, warned his board in June 1923 that the "ambition of localities to possess parks of national prestige" posed a serious threat. If places such as those Fall had proposed for his friends in New Mexico (and El Paso) were admitted to the National Park System, they "would utterly destroy its conservation or its museum value."[1] At the same meeting, Yard spoke to the growing demand for outdoor recreation, which he saw as another threat to national parks. NPA, he argued, should propose to Congress a new system of public reservations that might be called "Federal Recreation Reserves." Such a system would deflect some of the recreation demand from the national parks. In his view, national parks would be damaged, even desecrated, by too much recreation use and development.

Only a few years before, Yard had been a major promoter of park use and new parks. What had changed? Why was he now so worried about "standards"? The park world was changing, and the automobile was the principal agent. The earliest parks were accessible primarily by railroad, which limited the visitation rate. As people bought cars and roads were extended across the landscape, more people could reach even remote parks. Once there, they demanded accommodations, which meant development. Furthermore, boosters seeking economic benefits from park visitation could realistically expect a road to "their" park, when only a short time before the prospect of a railroad for access had been slight to none. This led to proposals that would not have emerged, because they were not worth the trouble, in the age of railroads. Yard and other conservationists were aware of the economic value of parks and were not averse to taking advantage of that value; he had unabashedly recruited the assistance of Colorado boosters and concessioners in NPA's first- year recruiting effort. But now he feared that too many people and too much commercialism would overwhelm the higher purposes of the National Park System.

The Problem of Management

The standards problem was viewed as having two parts: What criteria should determine whether or not an area should be a national park and, once it was a park, what values should guide its management.

"Standards" had been a concern of the National Parks Association from the beginning. At its founding, the association dedicated itself to developing the national parks into a "complete and rational system." Doing this would require the parks to be increased in number "to represent by consistently great examples the full range of American scenery, flora and fauna, yet confined to areas of significance so extraordinary that they shall make the name national park an American trademark in the competition for world travel...."[2]

NPA was even then identifying criteria for defining a national park. For instance, parks should be outstanding examples of unique American landscapes. The greatness and uniqueness might be defined by scenery or by natural features. Parks would be "confined" to areas of "extraordinary" significance. No definition of "extraordinary" was provided; the founders knew what they meant and assumed that others would, too. The NPA statement also suggests that the system they were seeking would be complete only when the "full range" of American landscapes was represented among the parks. The association would soon face dilemmas posed by this vague goal.

The original NPA statement of objectives did not use the terms "primitive," "primeval," or "wild." The founders did allude to a goal of "conservation, in an ungrazed, uncut condition, of native forest, meadow and wild flower areas." Very soon the concepts of wild and "primeval" parks would complicate matters. Perhaps the founders saw no need in 1919 to include these qualities among the criteria, because there was still much wild land across America; new parks would come from this vast pool of pristine nature. Road building on a grand scale was only beginning. Yet they knew that a multitude of threats faced these wild places, and that one reason for national parks and for their association was to counter these threats with conservation and preservation. Whatever the reason for overlooking the quality of the "primitive" in defining national park standards, NPA and others soon corrected this oversight and, in doing so, brought the issue of standards to the forefront of the national park debate.

Robert Sterling Yard wrote often of the origin of national park standards. He describes how Mather and his staff puzzled over the matter as they created the National Park Service:

At the very beginning arose among us the question: What are National Parks anyway? Everyone knew generally and no one knew specifically. Albright, the lawyer, searched law books and records in vain for a definition. Mather and I asked officials, members of Congress, park-makers in the West, seers generally wherever found. A dozen offered definitions differed radically.[3]

Finding no legal definition, they decided "the parks themselves must furnish the definition." Excluding the few obvious misfits like Sullys Hill and Platt, they concluded that national parks "were areas of unmodified natural conditions, each the finest of its type in the country, preserved forever as a system from all industrial uses...."[4] But even they did not write the definition down, says Yard, "probably because this definition was to us self-evident and no one in or out of Congress raised the least objection to it."

Mather, Yard, Albright, and others, notably Frederick Law Olmsted, Jr., wrote and lobbied for a bill to establish the National Park Service, the language of which did not set criteria or standards for national parks but specified their purpose and thus indirectly suggested the criteria. In the National Park Service Act of August 25, 1916, Olmsted stated that the fundamental purpose of the parks "is to conserve the scenery and the natural and historic objects and the wild life therein and to provide for the enjoyment of the same in such manner and by such means as will leave them unimpaired for the enjoyment of future generations."[5] This statement of purpose suggests that national parks will have scenery and natural qualities worthy of the attention of generations of Americans. Again, perhaps, the drafters of the act assumed that national park qualities would be clear, or at least that those defining new parks (who would include the experts in the park agency created by the act) would know how to make choices in expanding and managing the National Park System.

The National Park Service was set up, and Albright had the job of writing policy objectives for the new agency. He drafted them, sent them to reviewers such as Sierra Club leaders, J. Horace McFarland, and Yard, a Park Service employee at the time. Mather approved the draft, and Albright sent it to Interior Secretary Lane, who issued the policy objectives as a letter to Mather on May 13, 1918.[6] The letter made twenty-three policy statements for the National Park Service and touched on the problem of standards.

> In studying new park projects you should seek to find scenery of supreme and distinctive quality or some natural feature so extraordinary or unique as to be of national interest and importance. You should seek distinguished examples of typical forms of world architecture, such, for instance, as the Grand Canyon, as exemplifying the highest accomplishment of stream erosion, and the high, rugged portion of Mount Desert Island as exemplifying the oldest rock forms in America and the luxuriance of deciduous forests.
>
> The national park system as now constituted should not be lowered in standard, dignity, and prestige by the inclusion of areas which express in less than the highest terms the particular class or kind of exhibit which they represent.[7]

NPA argued that Eastern national parks, like the proposed Great Smoky Mountains park, must be held to the same standard as the Western parks.

According to Lane, the National Park Service was to hold park proposals to a high standard of national significance, but the specifics of these standards remained ill defined.

Criteria for New Parks

The early 1920s brought proposals to use existing parks for commercial purposes, and proposals for new parks that all guardians of standards, both inside and outside the National Park Service, agreed were a threat to the quality and integrity of the system. Secretary Fall proposed his pet park, but also expressed his intention to "transform the National Park System into a wide-open system of merely recreation areas, big and little, involving the destruction of its conservation and its quality of scenic magnificence."[1] This "merely recreation" approach to national parks horrified Yard and NPA. When Fall resigned in disgrace, their relief was evident.

New Secretary of the Interior Work reassured the conservation community in a letter in January 1924. He restated and slightly reworded the part of Lane's 1918 policy letter that alluded to standards. When Work's letter was published in the *National Parks Bulletin*, the headline above it read: "Secretary Work Defines National Parks Policy: Confirming the Practice of Fifty-two Years of Government, He Declares for Highest Scenic Standards and Complete Conservation."[8]

When Herbert Hoover became NPA president, he advocated a response to growing recreation pressure. NPA began exploring the idea of a new system to meet this need, which might be called the "National System of Recreation Reservations." This system would complement national parks and be distinct from them. The *Bulletin* declared that NPA had shown how citizens could fight invasions of national parks. The real challenge "lies not so much in invasion as in keeping out of this system park units which already contain industrial works, or which include natural resources manifestly necessary for industrial utilization at some future time." The number of such proposals for parks "below the required standard of beauty, is increasing with great rapidity...." Action in defense of standards was imperative.[9]

The councils of the National Conference on Outdoor Recreation worked on the problems of standards. Yard exchanged letters on the matter with Secretary Work. The association also addressed the problem in its review of new park proposals. (The *National Parks Bulletin* of January 21, 1924, lists proposals for thirteen new parks.) The issue came to a head with the proposed parks in the southern Appalachians. Early in 1924, Secretary Work announced that the administration wished to establish a national park in the southern Appalachian mountains. He appointed a committee to select an area that would conform to national park standards. William C. Gregg of NPA was one of the committee's five members. This area, Work charged the committee, should be typical of the plant and animal life and scenery of the region. The park should be of majesty and size.[10] Because the federal government did not own land in the area, creating this park would require the acquisition of land. Another challenge would be locating pristine land equal to the highest quality national parks.

The Southern Appalachians

Since its origin, the National Parks Association had advocated a national park in the Appalachians, but it was chagrined at the course of events set in motion by Secretary Work's initiative. The Southern Appalachian Committee first settled on the Great Smoky Mountains between Tennessee and North Carolina as the most suitable area for a national park. Then—under political pressure, in Yard's opinion—it shifted in favor of the Shenandoah region. Secretary Work sent the report nominating the Shenandoah to Congress on December 12, 1924, along with a bill for appropriations to survey the proposed park. When the legislation, called the first Temple bill, was introduced in the House, the Tennessee and North Carolina delegations demanded that the Great Smoky Mountain area be included, and this was done.[10]

NPA had concerns about a Shenandoah National Park, but thought the Great Smoky Mountain region met the highest national park standard. Next the Kentucky delegation demanded similar consideration—Mammoth Cave should be included in the bill. Although the committee had not even examined that area, it was written in as a location to study. Still other southern areas tried to include their favored parks in the bill but failed.

The advisory committee, which had become the Southern Appalachian Commission, visited Mammoth Cave but made no report on it. Matters proceeded with Shenandoah and Great Smoky; the supporters of each park would raise subscriptions to purchase parklands, and a bill would be introduced to create them. They raised the funds, and on April 14, 1926, the second Temple bill creating the two parks (when certain specified areas were purchased and presented to the government) was submitted to Congress by Secretary Work.

Most observers thought the Mammoth Cave proposal was dead; they were wrong. The Kentucky park proposal reappeared in the commission's recommendation, which was enclosed in Secretary Work's letter transmitting this second Temple bill to Congress. Work had not acted on the commission's recommendation and wrote, "I express no opinion and make no recommendation at this time as to the desirability of the inclusion of the Mammoth Cave area within a national park."[11] When the Temple bill, which did not include Mammoth Cave, was introduced, yet another was offered in both houses, identical to the Temple bill except for inclusion of Mammoth Cave National Park. When the Public Lands committees sent this new bill to Work for a report, he replied, obviously stalling, that the National Park Service would have to prepare the report, and it had not studied the site.

The result was that Congress passed a bill authorizing a Mammoth Cave National Park without a report from the secretary of the Interior or examination of the merits of the proposed park by National Park Service experts. Congress circumvented the high standards espoused by Secretary Work and Director Mather. "Politics," wrote Yard, "controlled the situation."

> With the entire south aflame for National Parks, and the Congressional primaries at hand...the fitness of any of the areas for admission to the system had ceased to count.[12]

Yard saw three causes for this compromising of national park standards. One was "ignorance of the nature, purpose and national destiny" of the system. A second was "a passion for national recreational expansion." And the third, and most serious, was "the fatal belief that different standards can be maintained in the same system without the destruction of all standards."

The National Parks Association and its allies were working to address this critical threat to the National Park System. For years they had been advocating a *system* and standards to define it. They had gathered under the auspices of the Camp Fire Club in 1923 and drafted a declaration of policy on standards based on the work of Albright and Mather. They were already working in the various councils of the National Conference on Outdoor Recreation to address the threats and opportunities posed by burgeoning demand for outdoor recreation resources. The debates over Roosevelt-Sequoia and All-Year national parks had been about maintaining the national park ideal. The crisis involving the southern Appalachians was merely the newest challenge in the campaign for an ideal National Park System, and NPA decided to intensify its focus on the problem of standards.

The proposed Shenandoah National Park offered a particularly difficult dilemma. NPA had initially approved of a park in the Shenandoah area. It

The proposed Shenandoah National Park was not, in NPA's view, up to the standard.

believed the area met appropriate standards, but when Yard learned that far less of the area was "primitive forest" than had been originally thought, he recommended that the association oppose the park. At the 1926 annual meeting in May, trustee Charles Sheldon summarized NPA's stance:

> Areas better fitted for national parks than Shenandoah have been turned down. It can be duplicated anywhere in the Appalachians. At first it was thought that Shenandoah preserved a great area of primitive forest in the Appalachians. Then it was discovered that this was not true. Looking into the future, the Shenandoah Park is going to raise an avalanche of bills in Congress to produce national park areas that are really state parks-commercial propositions. A different standard for the East is going to appear. Western men are going to use this loose Eastern standard to invade the Western parks.[13]

The association went on record as embracing the national park standards adopted by the national conference in May of 1924, and planned what Yard called its "campaign of righteousness" on the issue of standards. That campaign would involve education about what standards should be, a direct attack on political manipulation of the park selection process as demonstrated in the Mammoth Cave case, and advocacy of a new approach to reviewing national park authorizations.

The *National Parks Bulletin* of April 1927 was devoted largely to the issue of standards. The articles reviewed the history of how and why parks were chosen, and prominent people presented statements about standards. NPA called for protecting park standards by law:

The protecting standards of the National Parks System are defined in the parks themselves and in the thoughts and aspirations of the American people, but they will not be wholly safe till they are defined also in law.[14]

In NPA councils, Yard pressed for strong public protest of the Shenandoah and Mammoth Cave park proposals, which were moving ahead, but the trustees counseled moderation. A proposed park in Arkansas—Ouachita National Park—provided a vehicle in the fight for standards that all NPA leaders could support, and Yard took the campaign in that direction.

Representative Otis Wingo and Senator Joe Robinson of Arkansas introduced a bill into the first session of the Sixty-eighth Congress to carve a Mena National Park out of the Mena National Forest in western Arkansas. Both the Forest Service and Park Service examined the area and agreed that it was not of national park quality, and the proposal languished. When Robinson and Wingo reintroduced the legislation in 1928, however, it passed both houses despite opposition from the Forest Service and Park Service, the secretaries of Agriculture and Interior, a majority of the House Public Lands Committee, and NPA and other conservation groups. The area was now called Ouachita National Park.

This, in Yard's view, was even worse than the Mammoth Cave case because Congress was creating a substandard national park against the judgment and even the active protest of the national park experts in the Department of the Interior. NPA sent out a series of informational releases about the Ouachita situation through its National Parks News Service. Hearing from Yard and many others about the threat to standards in this proposed park, Representative Don B. Colton, chairman of the House Public Lands Committee, wrote a letter to President Coolidge, urging him to veto the bill. Coolidge did. Colton told the president that if the Ouachita bill was signed, the committee would have to report twelve other bills proposing inferior parks.[15]

Local Opportunism

Specific park bills were lined up on the committee's agenda behind the Ouachita park proposal. Then a Wisconsin representative introduced a bill that would establish a national park or forest *in every state*. As Yard wrote repeatedly in this campaign, in most cases, these were not areas of national significance. They involved local opportunism rather than national interest.

With help from allied associations, particularly the American Forestry Association, NPA was able to lead a successful campaign in this round against compromising park standards. Ouachita legislation, however, would be reintroduced in the Seventy-first, Seventy-second, and Seventy-third Congresses, to be defeated each time.

Yard pressed the cause on another front. He addressed a letter to Chairman Colton urging the Public Lands Committee to continue "to submit every national park proposal to the Interior Department for expert study of its standards" and to "give preferential consideration to its advice as architect and builder of the system." Further, he urged public hearings on all park proposals. Yard's letter was prompted by Senate Public Lands Committee approval, without Interior Department review, of the proposed "Teton" national park in South Dakota. Yard closed with an explanation of NPA's larger concern:

> Please understand that we register no protest against creation of public parks, only against forcing into a single classification, namely, the National Parks System, areas which do not conform to its standards. We believe profoundly in parks.[16]

Yard urged Colton and other members of the House and Senate public lands committees to study the Report of the Joint Committee on Recreational Use of Federal Lands to the National Conference on Outdoor Recreation. All members of Congress had received a copy of the report.

As Yard's letter indicates, the NPA campaign was moving on several fronts. Yard was active on Capitol Hill, communicating with key congressmen, sending regular information circulars to all members, and testifying in hearings. He sent NPA circulars to allies across the country, and they responded. The Colorado Mountain Club, for instance, let the House Public Lands Committee know of its opposition to the Ouachita proposal. When its letter on the subject was read in the hearing, a Colorado member of the committee stated that the club was being used by Robert Sterling Yard for "propaganda" purposes. This elicited a strong response from the club. George Harvey, Jr., chairman of the Mountain Club's National Parks Committee, wrote declaring that club members had learned of the ill-advised nature of the park proposal from an NPA bulletin and taken it upon themselves to make their views known. "Assuming this information to be correct," wrote Harvey, "we needed neither Mr. Yard nor anyone else to tell us what to do or how to do it."[17] This is an example of how NPA was reaching out to allies. The network Yard had built in the early 1920s was still functioning.[18]

A CHANGING SCENE
New Faces and the Great Depression

The campaign to defend standards continued into the late 1920s, when there were several significant changes of players. Late in the Coolidge administration, Interior Secretary Work, with whom NPA had enjoyed a good relationship, was replaced by Roy West. When the election of 1928 brought Herbert Hoover to the White House, he appointed Ray Lyman Wilbur as Interior secretary. Wilbur promised to be an advocate of parks, though he proved to have less interest in and understanding of them than Work. A former medical doctor, Wilbur was president of Stanford University, had served on the California State Park Commission, and was an advocate of wilderness parks. In November 1928, Stephen Mather suffered a stroke and resigned as director of the National Park Service in January. Horace Albright, who had been superintendent of Yellowstone National Park for a decade, was sworn in as Park Service director on January 15, 1929. Albright was reluctant to take the job, but many urged him to accept the challenge, including Yard.

This wave of change reached to the National Parks Association. At the annual meeting in 1929, George Bird Grinnell stepped down as NPA president; the "burden of his years" required that he curtail his activities. Dr. Wallace W. Atwood was installed as his successor. Atwood had long been an NPA trustee and was president of Clark University in Worcester, Massachusetts. A geographer, he had pursued a distinguished career with the U.S. Geological Survey and had recently been appointed a member of the Committee to Advise the National Park Service on Education and Development of National Parks.

The economic environment also changed. The 1920s had been a decade of unprecedented economic growth and prosperity. As secretary of Commerce, President Hoover had been architect of many policies that had fed the economic prosperity of the decade, and his election to the presidency inspired confidence that led to a booming stock market. Employment, construction, loans, and other business indicators swung upward until, on October 29, 1929, the stock market crashed. The economy slid rapidly into the Great Depression that would last for thirteen years and change every aspect of American society, including conservation and the work of the National Parks Association.

New Functions for NPA

Despite this tumultuous social and economic environment and changing of the guard, NPA kept up its campaign to maintain and define national park standards. In 1930 it took the work in a new direction. A special meeting of the trustees was called on December 5 to change the organization and redefine its purpose. The board was enlarged, and classes of trustees were created. They decided to address the issue of standards on two fronts: how to "maintain proper balance between the protection of primitive features in the parks and development of the parks for the purpose of making them accessible to the people," and how the National Park System should grow. This latter issue would be addressed in terms of the use and function of national parks. How do national parks differ from city and state parks, state, and national forests?[19] These questions had been addressed in the presidential conferences, but more work was necessary. Two committees were appointed. The group examining "protection of the primitive," called Committee Number One, was chaired by Wallace Atwood; Committee Number Two was chaired by trustee Henry Baldwin Ward.

A Comprehensive Set of Principles

Both committees offered progress reports at the annual meeting in June. Committee Number One had concluded that visitors in national parks should be directed to specific "observation stations" where they could experience the park's attractions and be educated about them. Development in parks should be strictly limited and subjected to stringent criteria. Buildings and roads should be carefully sited, always with protection of the "principal defining features of the parks" a primary consideration. Committee Number Two concluded that expansion of the National Park System would be limited "because little is left that is truly unique." National parks should be "primitive"; recreation should occur in other areas, whereas national parks should "emphasize educational and spiritual values." Potential national parks should be surveyed by experts, and the "question of whether the government can afford to pay for an area should not determine its fitness to become part of the national park system." Finally, nothing less than "total exclusion of all commercial development is essential for the maintenance of our national parks."[20]

These preliminary conclusions comprehensively represented the views of the association in 1931 about the principles that should guide the selection, development, and management of national parks. These principles would govern NPA positions on many issues until the 1960s.

The association's deliberations show that it still thought national parks should be outstanding and nationally significant examples of unique American landscapes. They should be largely primitive. New parks should come only from a limited pool of candidate areas and only after careful study by the Department of the Interior. Once an area was identified as an outstanding example of a natural or scenic wonder and was made a national park, the top management priority should be to preserve its natural and primitive qualities. Every management and development scheme should be governed by the goal of preserving the natural values that had made the area worthy of national park status. People should be encouraged to come to the parks only for the inspiration and education to be found there. Those seeking recreation or amusement should go elsewhere, to national forests and state parks. Roads and buildings in parks should be kept to a minimum and, when built, should blend into the setting and not affect primary park values. No commercial activity should be allowed, because it would inevitably be inconsistent with the educational and inspirational values of the parks.

This stance was the work of many members of the association, but particularly Robert Sterling Yard and John Merriam. Yard had emphasized the educational value of parks since his arrival on the national park scene in 1915. Merriam was a distinguished scientist, educator, and conservationist greatly interested in the meaning of nature and how that might be understood by people not trained in science. Historian Susan Schrepfer notes that as a park advocate, Merriam "interpreted the rationale of preservation. He saw nature—be it a petrified woodland, a living redwood, or the Grand Canyon—as more than a series of scientific technicalities....Merriam argued that the forces of nature held philosophical and religious truths."[21]

In the spring of 1928, Interior Secretary Work appointed Merriam to head the Committee on Study of Educational Problems in National Parks, which issued its report the following year. That report asserted that "The distinctive or essential characters of National Parks lie in the inspirational influence and educational value of the exceptional parks."[22] It established that "the primary function of National Park administration concerns the use of the parks for

NPS Director Mather was a pragmatist who did not share NPA's "purism" on the issue of national park standards.

their inspirational and educational values." The committee defined the purpose and value of national parks, but Merriam thought its statements needed to go further and explore the implications of these values for park selection and management. Thus, he argued for NPA studies that produced the suggestions and recommendations made at the May 1931 annual meeting.

THE VIEW FROM THE PARK SERVICE
The Politics of Pragmatism

Was NPA in step with the National Park Service and Horace Albright in this thinking? Yes and no. The record clearly shows that Secretary Work and directors Mather and Albright were concerned about standards and opposed to substandard areas in the National Park System. They were, however, constantly faced with the need to practice pragmatic politics in a way NPA was not. As NPA founders had anticipated, times and issues would come when politics would require the Park Service to compromise and when NPA could hold to its ideals. The late 1920s and the 1930s became such a time, and standards became the issue. Mather and Albright faced the challenge of balancing the National Park Act mandates of conserving parks "unimpaired for the enjoyment of future generations" while "providing for the enjoyment" of the American people. Both Park Service directors interpreted their responsibilities to embrace a greater element of "recreation" than NPA thought appropriate. While agreeing that the parks must be protected, Albright thought they should be developed and managed so the public could use and enjoy them. "The greatest good for the greatest number," he wrote in 1935, "has to have a small place even in national park administration."[23]

NPA, by its very nature, could define a position in contrast to the Park Service. With no contradictory legal mandate, the organization could stake out a philosophical position that might seem extreme, from the perspective of the pragmatic administrator. Led by Yard and Merriam, NPA did just that. In Albright's view, the association became overly "purist" in its stance on preserving the primitive. When Yard accused Albright of softening his attitude toward national park standards in the 1930s, Albright responded: "I have not changed my views since I entered the National Park Service way back in the days when Steve Mather first came to Washington," he told Yard. "You are the one who has altered yours."[23]

Albright, trained in the law, had always approached national park management pragmatically; as his responsibilities grew, so did his pragmatism. Yard, in his entirely different role, had become ever more idealistic. Yard's emphasis on preservation deepened. The idea of wilderness, of the centrality of the primitive qualities of parks, had gripped him. In the late 1920s and early 1930s, Yard was becoming an advocate of wilderness. By the mid-1930s, this advocacy would lead him to become a founder of the Wilderness Society, which he would serve for nearly a decade as executive and president. Albright and the National Park Service simply did not embrace the ideal of maintaining naturalness in national parks to the extent that Yard, Merriam, and NPA thought they should.

Albright thought Yard strident, undiplomatic, and self-righteous, which he could be. He had thought Yard impulsive in 1917 when Yard wished to step in for the ill Mather, and he still thought him so. An exchange of letters in 1930 reveals that Albright and others believed Yard could, at times, be a liability rather than an asset. Frederick Law Olmsted, Jr., the distinguished landscape architect and NPA trustee, chided Yard in response to his request for Olmsted's counsel on a Yellowstone matter:

> An appearance of losing one's temper or of being ready to think ill of or to discredit the motives of those who differ with one is, on the lowest plane, apt to be a tactical blunder; and is too often significant of a state of mind in which one's judgment about more *vital* essentials is undependable.[24]

Olmsted shared his letter with Albright, who replied to Olmsted:

> I just want to thank you for writing the letter you did, under date of February 11, to Robert Sterling Yard. I hope it will do him some good. I have about reached the point where I think that his Association is not of very much value to the National Park Service and that we will have to rely on the American Civic Association for our outside support. Poor Bob Yard antagonizes far more people than he is successful in making friends with. He has been particularly unfortunate in his dealings with Congress.[24]

NPA trustee Frederick Law Olmsted, Jr., chided Robert Sterling Yard for his abrasive way of doing NPA business.

Apparently relations with Albright and the National Park Service were strained not only by the positions on issues Yard and NPA might take but also by Yard's style. If Olmsted thought it necessary to write as he did, he must have believed Yard was transgressing enough to affect NPA's influence.

Historic Preservation

Another note of discord between NPA and the Albright administration sounded over the matter of preserving sites of national historic significance. No national parks and monuments under jurisdiction of the National Park Service had been created primarily to preserve historical values. National monuments created under the Antiquities Act protected Indian sites; national battlefields were administered by the War Department. Albright and others had long believed sites of national historical significance, along with Indian sites, should be administered by the National Park Service. NPA did not disagree and proposed to the National Conference on Outdoor Recreation in 1926 that the military parks be transferred to the National Park Service as National Historical Parks. When Albright became director, he set out to effect this transfer and bring several other historic sites under Park Service control. His first success was with Wakefield, the birthplace of George Washington.

Seeing this movement toward historic preservation, NPA created the Committee on Preservation of Historic Sites and charged it with recommending a course of action. The committee, under John Merriam's leadership, reported late in 1932 that historical preservation was very important and proposed the establishment of a system of historic sites representing periods of American history. Committee members, however, urged NPA to move slowly on the matter. They were concerned again about standards. What should be a national historic site under National Park Service control rather than state or local administration? They noted that many historic sites were admirably protected and administered by local government and private entities; they counseled that local control, where successful, should be encouraged. The federal government should be involved only in sites of national significance that represented important historical periods, and only when someone else could not or would not do the job well.

NPA's concern about this issue was a familiar one, revolving around standards. It feared that hundreds of sites would vie for national status, dilute the quality of the system, siphon limited funds away from truly worthy parks and historic sites, and damage the whole National Park System. Even so, it applauded the creation of the Morristown National Historic Park, the first of what it hoped might be a new and *separate* national system. Morristown and the proposed Saratoga and Colonial national monuments were, they thought, the best examples of Revolutionary War sites and would be a fine start to the new system. NPA proposed that these be called "national historical *sites*" rather than parks, to differentiate them from scenic and natural parks. It wanted nothing that might cloud the definition of *national park*.

Albright pursued his goal of bringing historic sites into the National Park System. In April 1933, he told President Franklin Roosevelt "how wasteful and inefficient it was to have several different organizations handling parks, and why the Interior Department and the National Park Service should have control over them all."[25] To Albright's gratification, the president agreed.

Albright got all he wanted and more. An executive order from Roosevelt moved the War Department's parks and monuments, as well as the monuments administered by the Forest Service, to the National Park Service. The reorganization of August 10, 1933, added to the agency's domain a dozen predominantly natural areas in eight Western states and the District of Columbia, and forty-four historical areas in the District and eighteen states, thirteen of them east of the Mississippi. Albright had more than achieved his goal of adding a system of historical parks to his agency's domain.

Fallout over Expansion

The National Parks Association applauded parts of this reorganization but remained concerned. In August 1933, the *National Parks Bulletin* recorded the changes as a rather small news item—surprising, considering the extent of the expansion and its many implications. The same issue also reported Albright's resignation as director and the appointment of Arno B. Cammerer as his successor. Minutes of association meetings immediately after the reorganization make no mention of it. Albright became a member of the NPA executive committee in 1933 after leaving the Park Service, and Yard resigned as general secretary at the same meeting in which Albright was installed. Yard continued to serve as editor of publications.

Then, in 1936, the *Bulletin* discussed the reorganization under the headline, "Losing Our Primeval System in Vast Expansion." Movements were afoot in Congress to continue expanding the park system. This response offered several revealing comments:

> The time was when the National Parks Association, in its upbuilding of the primitive system and defense of standards, closely paralleled in interest the interest of the Service; but that inimitable partnership of the creative years necessarily ceased about three years ago.[26]

Three years reached back to the reorganization. The author, probably Yard, raised a litany of NPA concerns about expansion, again dealing with standards. NPA's main interest "will continue to be that of the last seventeen years, namely, the development, beneficent use and protection of the great standard National Parks System as a unique expression of primeval nature in supreme beauty." Threats were looming: there was a "muddle of many diverse ingredients...;" the "lustre" of the system was being "dimmed" by roads and by "the deadly precedent of forcing Jackson Lake Reservoir into Teton National Park." The system's identity was being lost and its "standards confused by continued official neglect to distinguish it from the National Historical Parks and others."[27]

The expansion involving historic sites was only part of the problem, but a serious one. Other difficulties had appeared since reorganization: continued growth of recreational use of national parks; road building everywhere (in part, a product of federal job programs such as the Civilian Conservation Corps, which was hard at work in many parks); congressional directives to the National Park Service to advise on development and planning for state and local parks; and the proposed inclusion of substandard areas and commercial developments in national parks. All threatened the standards.

NPA had concluded in the early 1930s that the National Park System was essentially complete—most appropriate areas had been identified and included in the system. It viewed this new activity as driven largely by the bureaucratic expansionism of the Park Service. NPA would continue its fight. It would continue "promoting education in realization of the parks' higher use...and preservation of national park standards from unnecessary roads, skyline drives and destructive forms of expansion...will justify the Association's years of laborious, patient, often misrepresented work."[27] William P. Wharton had become NPA president, and in 1937 he asked, "With all the reservations now supervised by the Park Service becoming increasingly known to the general public as just 'National Parks,' where were the real National Parks 'going to get off?' Obviously something had to be done," he went on, "to save them from submergence in this matter of miscellaneous reservations which, however desirable their purposes, are most of them about as far removed from old time National Parks as it is possible to imagine."[28]

NPA's solution to this problem was to designate a separate system for these "real" national parks, called "National Primeval Parks." These would be "those national parks which, by reason of possessing primeval wilderness of conspicuous importance and supreme scenic beauty, conform to the standards originally recognized under the title of National Parks."[29] Parks such as Yellowstone, Sequoia, Mount Rainier, Glacier, Rocky Mountain, and Grand Canyon would be part of this system. These parks should be administered as a separate system, because management of primeval parks required special training, skills, and administrators "imbued with National Primeval Park ideals." If this recommendation were adopted, the National Park Service would administer a system with four distinct parts: national primeval parks, national monuments, national historical sites, and national military parks. The special needs of each subsystem would be identified, standards and selection criteria established, managers trained, and the integrity and standards of the entire system maintained.

NPA placed official recognition of this national primeval park system at the top of its agenda and kept it there until World War II. In the *National Parks Bulletin* and News Service, the association thereafter referred to the proposed primeval parks by that title, as in "Yellowstone National Primeval Park." Perhaps it hoped that calling them primeval would keep them so. The record suggests that NPA did not pursue this goal very aggressively but became occupied with other issues.

An incident in 1938 reveals how little NPA's campaign for standards seemed to impress the administration of the National Park Service. In January the American Planning and Civic Association hosted a National Park Conference. At this event, Ovid Butler of the American Forestry Association affirmed the support of his and other organizations, including NPA, for the national park standards as described by the Camp Fire Club of America nearly a decade earlier. In his response, Park Service Director Cammerer stated his belief that, although these standards might have been appropriate at one time, they were now outmoded.

> I would much rather have a national park created that might not measure up to all that everybody thinks of it at the present time, but which, 50 or 100 years from now, with all the protection we would give it, would have attained a natural condition comparable to primitive condition.[30]

Should a bill for a park or monument come down to a little mining or no national park, continued Cammerer, he would rather have the park and allow the mining (or logging or other commercial activity). There was precedent for Congress to revoke rights for such intrusive uses after establishing a park. Former Director Albright, now an executive in the potash industry but still active in national park affairs, agreed with Cammerer in his remarks closing the conference. Though he claimed to have helped draft the Camp Fire Club statement, he thought strict adherence to it was no longer possible.

NPA was appalled. Cammerer's and Albright's remarks suggested a shift in official policy that confirmed their worst fears—standards would be compromised beyond repair. Wharton interpreted those remarks as suggesting that recreation demand had become the principal driver of national park policy.

The real impetus behind the new drive seems to originate in the recently conceived idea that the Park Service is the only federal agency fitted to administer recreation on federally owned or controlled lands. Some persons even go so far as to assert that its proper function is to stimulate and direct recreational travel throughout the country. That is a long step from preservation for inspirational and educational use of great natural masterpieces.[31]

Allowing commercial uses in national parks, argued Wharton, would make the parks little different from national forests. The final outcome might well be reversion of large parts of the National Park System to national forests.

At the annual meeting of 1938, NPA passed a resolution stating that the association "deplores the deliberate abandonment by the Director of the National Park Service of the national park standards....The national park system as we have known and cherished it cannot endure if the ideals of its founders are discarded and a purely opportunist policy of expansion and promotion is substituted."[32] This resolution would be conveyed to the director. Wharton was instructed to press Cammerer for a separate classification and system of national primeval parks, and to invite him to write an article for the *National Parks Bulletin* on his views of this proposal. Relations with Cammerer were strained.

Olympic National Park

The issue of standards also led to trouble with other conservationists. One example in the mid-1930s involved Olympic National Park. Advocates of this park had been working for decades against strong opposition from the Forest Service, the timber industry, and sometimes even the National Park Service to pass legislation creating a large wilderness park on Washington's Olympic Peninsula. Legislation had been introduced, backed by Secretary Harold Ickes, which had specifically called for a "wilderness" park. This wilderness idea was not popular with the Park Service, which argued that a wilderness clause was made unnecessary by the 1916 Park Service Act.[33] NPA agreed.

The Park Service had other reasons to oppose prohibiting development in land under its jurisdiction.[34] Cammerer expressed his view of wilderness when he wrote in 1938, "Certainly no wilderness lover could selfishly demand that the national parks be kept only for those who are physically able to travel them on foot or horseback, for they are definitely set aside for the benefit and enjoyment of all."[35] Some NPA leaders, notably Yard, did not agree with this, but even so NPA found itself aligned with the Park Service, the Forest Service, and the timber industry against the Olympic Park legislation it had once supported, and incurred the wrath of conservationist colleagues. The problem of standards led to this odd alliance.

The leaders of the long effort to create Olympic National Park, which included the forested valleys as well as the rock and ice uplands of the Olympic Mountains, were Willard Van Name, Irving Brant, and Rosalie Edge. In 1930, these three founded the Emergency Conservation Committee, a tiny

but potent group that took the Olympic Park cause as one of its major projects. Van Name was pamphleteer and financial supporter, Brant was the master strategist and political mover, and Edge was the organizer of popular support for the project. When an Olympic Park bill was finally approved, Edge wrote her coalition of scientific societies, museums, academicians, conservation organizations, and media (the same coalition, ironically, that Yard had successfully used for so long) that "only two organizations whose cooperation we sought failed to help us—the National Association of Audubon Societies and the National Parks Association.

> We do not think that the Audubon Association was actually opposed to the Olympic Park—but it was lethargic and indifferent. The National Parks Association was opposed to the park. Its President, Mr. William Wharton, even telegraphed President Roosevelt, urging him to veto the bill. It is significant that Mr. Wharton represents the American Forestry Association on the Board of the National Parks Association; Mr. Wharton is also first Vice President of the Audubon Association."[36]

Edge implied in her statement that Wharton and NPA had favored forestry, that is, the timber industry, over a national park that included disputed timber stands. Carsten Lien, in his history of Olympic National Park, makes precisely this claim. Did NPA "come out for the timber industry," as Lien claims?

In 1936, NPA passed a resolution supporting the Mount Olympus National Park advocated by the Emergency Conservation Committee. In February 1937, this position was reaffirmed in a motion made by NPA trustee Robert Marshall—the same Bob Marshall who was then creating the Wilderness Society with Yard and others. During the summer of 1937, NPA Executive Secretary James Foote made an extensive tour of Western parks, including the proposed Mount Olympus National Park. He presented a long report on his findings to the executive committee on October 14, 1937. He recommended that NPA support the Wallgren bill that had been introduced into the Seventy-fourth Congress. This bill differed from the earlier Emergency Conservation Committee proposal. The boundaries of the proposed park were considerably reduced, and it lacked a wilderness clause that was in the earlier version.

> To me, the present position of our Association [backing the Emergency Conservation Committee proposal] in regard to the proposed Mt. Olympus Park is unhealthy and decidedly untenable. I cannot give it my personal unqualified support for the reason that I consider the present boundaries of the proposed park to be entirely adequate and more to be preferred from the standpoint of national park standards than the original and larger boundaries.[37]

The giant timber stands of the Olympics were coveted by the timber industry, which used every stratagem to oppose the park.

He offered reasons supporting this position. First, the park as proposed by Wallgren complied with national park standards of size, beauty, and primeval condition. Second, the lower Bogachiel River Valley was properly omitted from the park because, although it contained a superb stand of Douglas fir, part of it was in private hands and would require condemnation. Placing such an area in the park would be dangerous policy precedent because the private holdings might be logged before the government could acquire them. The Lake Quinault section was left out of the bill because it, too, was spotted with private landholdings. The section contained nothing of scenic or natural value that was not represented in other parts of the park. "I am convinced," said Foote, "that I am not overstating when I say that there is nothing of National Park caliber in the Quinault section which was left out of the present Wallgren Bill."

Finally, Foote thought there were economic issues to consider. "The question of park making in the Olympic Peninsula is unique in that a careful consideration of the economic life of the peninsula enters into the picture." An economic crisis brought on by "ruthless logging" was forcing a transition from saw timber to a pulp industry. The original bill, which would withdraw seventeen billion board feet of timber from harvest, had attracted powerful local opposition. The new bill, withdrawing twelve billion, would not be so strongly opposed.

> I honestly believe that if we insist upon—and get—the original boundaries, we will not be able to hold them, and since the present smaller boundaries will give us a fine park, why chance a breakdown in policy and standards by insisting on the additional allotment.[37]

Foote was expressing concern about the economic welfare of the timber communities, but he believed that park quality, and perhaps the park legislation, would be threatened by asking for too much.

This was consistent with the position stated by NPA Committee Number Two in 1931. A conservative approach to extending park areas was required, it cautioned. "The incorporation into the National Park System of extended territory will reduce those areas available for commercial utilization and give rise to conflict of interests that would jeopardize the safety of the National Park System."[20] The committee feared that grasping for too much would give commercial interests an excuse to invade the parks. Its solution was to advocate quality of parklands rather than quantity.

Although Foote recommended accepting the smaller park, he objected to Wallgren's removing the wilderness clause. The chances of reinstating it were slim, he thought, and "any similar guarantee [of wilderness protection] will have to be made by the Park Service or the Secretary of the Interior; and, somehow or other, I don't trust either of them."[37] He proposed a *quid pro quo*: NPA would be willing to "sponsor" the smaller park if the Park Service provided "an absolute guarantee of wilderness preservation and non-development similar to Section 4 of the first Wallgren Bill." The executive committee accepted

Foote's recommendations and passed a resolution supporting the pending legislation and urging inclusion of a wilderness preservation clause. In the end, the Wallgren bill failed, and the Emergency Conservation Committee-backed bill for the larger park was approved.

Wharton, stung by Edge's criticism, explained his wire to President Roosevelt to the NPA board at its twentieth annual meeting on May 18, 1939. He had urged the president to veto the bill, but not, as Edge and others thought, because he was doing the business of the timber industry. As explained in an earlier chapter, the Federal Water Power Act gave license to approve water projects in national parks unless a clause strictly forbidding it was included in the legislation creating a new park. Wharton said that in read-ing the Olympic Park bill, he had seen no such clause, and so had requested that the bill be vetoed. To his embarrassment, he learned a few days later that the Federal Water Power Act had been amended in 1935 specifically to exempt all national parks from power development. Learning this, he withdrew his request. "I have no excuse to offer for my own ignorance," he told the board, "and can only say that I acted with the best intentions. I am glad to say that the record clearly shows that our association from the first favored and worked for the establishment of a properly safeguarded Olympic National Park."[37] The record does show that, Wharton's mistake aside, NPA's strong stand on standards led it into a position at odds in this Olympic case with much of the conservation community.[38]

THE MUSEUM OF AMERICAN NATURE
NPA Purism

The motto for NPA during the 1930s might have been "Save only the grandest and most sublime scenic and primeval landscapes for the highest educational and inspirational purposes." In its view, the National Park System should be a pure expression of conservation. National parks should be muse-um pieces, and just as a rare and historic cultural artifact is given unequivocal protection from alteration and commercial exploitation, so should a national park. Just as museum artifacts are the object of careful study and meditation, so are national parks. The finest museum specimens, examples of the priceless heritage of human experience, are chosen with care, and so should be the national parks. During this period, the National Parks Association saw itself as struggling for the ideal museum of American nature. Adhering to this high cause placed it at odds not only with those bent on grazing, mining, logging, and otherwise profiting from the exploitation of parkland but even with allies that believed in national parks, if not in the pure way embraced by NPA.

Critics then and since have accused NPA of "purism," characterizing it as a group of uncompromising idealists who sometimes did more harm than good for the national park cause. NPA split with Stephen Mather over the Barbour bill because it would not compromise standards to create a Roosevelt-Sequoia National Park. Horace Albright, always the pragmatic administrator, thought NPA obstructionist in its opposition to some of his plans for expansion. He was willing to have a reservoir in Grand Teton National Park, but NPA was not. Purism entered into the association's decision to support a reduced Olympic National Park bill and would lead to a split with the Sierra Club in the late 1930s over the Sequoia-Kings Canyon legislation.

There is no question that NPA was "purist" in its views of the purpose and management of national parks. Occasionally national park supporters saw NPA as less than helpful in its fight for new parks. Horace Albright, blocked at one point in his effort to expand Grand Teton National Park, fumed that "Ward [the NPA vice president] and his associates are idealists without any respect whatever for practical problems involved in maintaining national parks or other preserves. It is a pity that conservation is always being thwarted by its friends."[39] Donald Swain, in his biography of Albright, defines a purist as "one who believed that the natural features of the parks should never be disturbed under any circumstances, and that the protection of the parks from commercial utilization and spoilation by the tourists should be the primary job of the National Park Service."[40] When the Park Service failed to live up to this high standard, it was criticized by NPA, and Albright and others leaped to the agency's defense. NPA saw its role as necessarily uncompromising in the fight for the highest quality in the National Park System, and recognized that this essential role would not please everyone. "Our method of conflict is unwaveringly impersonal," said Yard in a 1931 description of the association. "We fight *for* a principle, never *against* a foe."[20] Thus it might oppose the Park Service position in one case, then support it in the next.

Historian John Ise tried to weigh the justice of NPA criticism of the Park Service. He argued that the two groups simply disagreed, and did so for understandable reasons. NPA could afford to be idealistic—it did not have any administrative responsibilities. The Park Service had to run the parks and satisfy the visitors and Congress, which appropriated its funds. NPA criticized the Park Service for paying too much attention to recreation. The agency was, Ise admits, heavily engaged in promoting programs involving recreation, but the need and demand were there, and "the Park Service was probably better qualified to do this work than any other agency."[41] Ise summarizes his assessment of purism:

> The purists and the wilderness advocates have performed a real service in stressing the dangers of over-development and the value of wilderness preservation, for most political forces push the Park Service in the direction of development, and park Directors have not always been free of the common bureaucratic ambition for growth and expansion....[42]

In Ise's view, the Park Service tried to follow a middle-of-the-road policy between recreation development and the preservation of nature. NPA and other "purists" played a useful role as advocates at one end of this continuum.

The Standard of Quality

The fight for standards did not end in the 1930s, although it occupied NPA's attention during that decade more than any other time. The association appointed a committee in July 1944 to revisit the matter. The committee produced a revision of earlier statements, titled "National Primeval Park Standards." Its description of standards closed with the statement, "These standards shall apply also to national monuments that are of similar character and purpose as the national primeval parks."[43] Then, in 1956, NPA rewrote them again. This 1956 version, reflecting changes that will be described later in this story, dropped "Primeval" from the title and defined national parks as "spacious land and water areas of nation-wide interest established as inviolable sanctuaries for the permanent preservation of scenery, wilderness, and native fauna and flora in their natural condition."[44] The association added a definition of "National Nature Monuments" to this statement. Such monuments "are established to preserve specific natural phenomena of such significance that their protection is in the national interest; they are the finest examples of their kind, and are given the same inviolate federal protection as the national parks...their primary purpose is to protect geological formations, biological features and other significant examples of nature's handiwork."[44]

The association's position on standards was evolving, adapting to changing times. But the goals remained the same—a system of national parks and historic sites of unequaled quality, a National Park Service that put its preservation mission first, and a rational organization of diverse Park Service-administered areas and sites of national significance.

Surviving the 1930s

The decade of the 1920s saw the National Parks Association define itself as advocate and defender of national parks and national park ideals. The parks needed protection from those who would build dams in them, mine them, convert them to playgrounds and recreation areas. The ideals needed protection from politicians, boosters, and entrepreneurs—even from ambitious bureaucrats who would degrade the concept by creating parks that did not meet criteria of exceptional scenery, wildness, and national significance. NPA advocated a system of national parks that would protect part of the American natural and cultural heritage by preserving examples that would otherwise disappear under a tide of development. And the principal purposes of that system must, they insisted, be education and inspiration.

Led by Robert Sterling Yard, the association had found a way to conduct its business on a limited budget. Yard had sought allies in the cause where he could find them. Using advocacy and information, he organized them in the park defense work. He sat in Washington, D.C., watching Congress and the Park Service, corresponding with allies around the country, sounding the alarm. Small though it was, the association rose rapidly to a position of prominence among conservation organizations.

Despite its prominence, NPA was not a strong organization. Its power resided principally in Robert Sterling Yard, and its influence derived from its distinguished trustees. Its membership was relatively small, and it had no financial reserves. Yard was approaching seventy. Although he was totally dedicated to the cause, he did not enjoy fund raising and was not good at it. As the 1920s ended, the collapsing American economy threatened the organization's existence.

Robert Sterling Yard was a tireless advocate of parks and wilderness, though he admitted he was a "tenderfoot" more at home in the city than the mountains.

NPA had also established itself as both critic and advocate of the National Park Service. Yard was so dominant in NPA affairs in the 1920s that his personality as well as his ideas often defined the relationship between the two organizations. His friendship with Stephen Mather had cooled, yet Yard and NPA consistently argued that the professionals of the Park Service must be central players in all matters involving national parks, especially in assessing new areas for inclusion in the system. With Mather gone and Horace Albright directing the agency, new challenges faced NPA. The decade of the 1930s promised tests of the association's viability and of its value in the national park arena.

Financial Straits

At two in the afternoon on November 25, 1929, after lunch in the Cosmos Club, the trustees of the National Parks Association held their belated tenth anniversary meeting. Originally scheduled for May 19, the meeting had been postponed to allow revision of the bylaws. The stock market had crashed in October, and massive economic troubles for the nation and the association lay ahead.

Despite the economic cataclysm, the meeting was typically optimistic. NPA had barely survived its first year, but during the 1920s it become an established player in the games of conservation. The locus of its activity was Washington, D.C. There it monitored Congress, the Department of the Interior, and the National Park Service, lobbied, and sounded the cry of alarm or opportunity to national park enthusiasts across the United States. During the 1920s, membership had crept slowly upward, approaching a high of 2,000 late in the decade.

The NPA operation was still small. Yard and a clerical assistant did virtually all of the work. A board of trustees met twice annually and an executive committee several more times to make policy and usually endorse the actions of the executive secretary. NPA expected its influence on national park policy to grow in the future.

Yard offered his usual lengthy secretary's report. After reviewing the past decade, he stated that "the destiny of this Association is clear as daylight."

> Its function, from its uninformed, unshaped, crusading beginning has been to shepherd multitudes, to interpret to them truth, to rally them in defense of standards, and use them in dissemination of knowledge. Its method also is clear. A decade of achievement proves beyond question that our most effective tool is the printed word....Our destiny is to become this movement's mouthpiece to the people.[1]

Wallace W. Atwood was elected president to replace the aging Grinnell and pledged to do all he could to improve the association's financial well-being.

Finances had always been tight. Yard had repeatedly asked the board to authorize hiring an assistant for him. The response was always the same: sympathy, a pledge to hire someone when funds allowed, followed by another pledge to raise the necessary funds, but Yard still had no assistant. Yard had also repeatedly argued for regular publication of the *National Parks Bulletin*.

Membership would grow, he thought, if members received tangible benefit from their membership, such as a regularly issued, high-quality publication. Such a publication would also be an invaluable tool in the association's work, just as occasional publications had been already. Ironically, the board had not yet fully realized what was happening to the economy; it voted at the 1929 annual meeting to publish the *Bulletin* ten times per year and to raise $40,000 to enlarge the operation. The Great Depression, followed by World War II, would reduce membership and make funds even tighter until the mid-1940s.

A reorganization of the trustees approved at this meeting separated the executive committee chairmanship from the association's presidency, a move that would cause problems later. Officers' terms were reduced to one year, and a council elected by the board to one-year terms was created. The aim was to broaden participation in association business, especially fund raising. In his report at the 1930 annual meeting, Yard alluded to what the reorganization was trying to address. "As compared with other vigorous and influential organizations of our times, our trustees have perhaps lacked group responsibility and initiative. With the growth of these, which I see coming," he added, "we should soon equal the American Forestry Association, and the American Association of Museums in prosperity and fullness of achievement."[2] The very financial instability of the association deterred successful fund raising. Wealthy prospective donors worried that their money would not be well spent on a faltering organization. In making this point, Yard quoted a conversation he had with a man of means:

Another man of wealth said to me:
"The ideals of our organization are right and its achievements are rather remarkable. Your failures of the last eleven years are mighty few considering the changing times, and you are quick on recovery. One gets a little enthusiastic just following along. I'm glad to give you a little money."

"But why not a lot of money?" I asked.

"To whom am I giving it?" he retorted. "You have many splendid men in your crowd and they seem very much in earnest. But it is a crowd not an organization. It doesn't give me a sense of security and permanence. Suppose I should give you $10,000, and you, for instance, should die; how long would the crowd stick? Or who would gather up the fragments? And what would they do with them? Get my point?"[2]

Regardless of the literal quotation, the fact that Yard said this was an admission of organizational weakness and recognition that NPA's business was riding too much on the shoulders of one man. Yard was approaching seventy. Although he seemed as vigorous as ever, donors saw an organization overly dependent on an elderly man. Yard himself seems to have thought this a serious problem that demanded action.

Affiliated Organizations

Caspar W. Hodgson, elected chairman of the executive committee six months earlier, had confidently pledged to raise large sums for the association. He had not yet succeeded and, times being difficult and his business more demanding, he offered to resign. John C. Merriam, citing his experience with the Save the Redwoods League, argued that the association should try to raise funds but not be too discouraged during hard economic times if the effort was unsuccessful. The work could proceed without a large bank account.

> Money is, unfortunately, too large a power in the United States. If you have a lot of money you can deluge people with literature and propaganda, which is not our way. The other way is by stating the case so well that, with reasonable opportunity to get it to the people, it would be acceptable. I am more concerned with the statement of the case than with its circulation; if you have the case to state, you can find means to state it.[2]

The case should be stated more clearly, said Merriam, and he proposed a way the association could expand its influence without great cost. His idea was to expand the newly created council by building its membership with representatives from scientific, educational, conservation, and other organizations that shared NPA's interest in maintaining the standards and educational uses of national parks. Organizations would be carefully examined and invited to nominate representatives to serve.

The association embraced Merriam's ideas and adopted them at a special meeting of the trustees in December 1930. Two classes of trustees were established: those representing allied organizations and those elected at large by the association. The cooperating groups would be invited to join the council, and their representatives would be approved and duly elected. Twenty-one invited organizations nominated representatives, and they were elected. The organizations ranged from the American Association for the Advancement of Science to the Association of Adult Education, the American Society of Landscape Architects, the General Federation of Women's Clubs, and the Sierra Club. This action formalized the network of cooperating organizations that Yard had developed nearly a decade earlier. Several organizational representatives had already served on the NPA board, and this reorganization of leadership led to an even greater interlocking of the conservation community.

Merriam recommended another step to strengthen NPA. More than defending parks was necessary, he thought. A broader, more clearly stated program was required. "If you go before a Committee in Congress, you set up an ideal that is better than any that anybody else sets up and see that its case is well stated. You announce that here is a thing so important that it must be protected by the nation."[2] Merriam suggested that the functions of NPA be studied and articulated, and committees one and two (see previous chapter) were established.

Association efforts moved forward under the new organization, but as the Great Depression deepened, so did NPA's financial plight. The offices could no longer be rented, and Yard's salary could not be fully paid. George Washington University provided the association office space without rent, and Yard accepted a cut in his salary in March 1931. His salary was reduced again in November, but he labored on. He was assigned to work with Committee Number One and Committee Number Two, and some of his executive duties were transferred to the new position of "director." To this point, the general secretary had been the association's executive. Now, as of 1931, the director would, "have general charge of the business of the Association."[3] In 1930, the trustees had agreed to raise funds to support an assistant to Yard for three years. This assistant was expected to succeed Yard, who would return to his "natural field of publication and related activities." Yard seems to have been easing himself out of executive duties, a goal which he appears to have nurtured since NPA's earliest days. Loren W. Barclay, a young businessman from New York with an interest in conservation, was hired as director and began to assume Yard's executive duties.

Yard Steps Down

NPA's financial crisis deepened in 1932. Membership dropped steadily, from 628 in 1931 to 480 in 1932 and 321 in 1933. When the District National Bank collapsed and its assets were placed under a conservator, $4,300 was frozen there. The association could not meet its payroll or pay its printing bills. Barclay could not be paid and left the association. At the November 1933 executive committee meeting, Yard resigned as general secretary and was elected "editor of publications," remaining both a trustee and a member of the executive committee. For lack of resources, the association would operate without an executive until 1936.

Did Yard leave his executive position under duress? Was he forced out because some members of the association leadership thought he had outlived his usefulness? Although some thought he should move on, the association's records do not suggest that he retired for reasons other than that the association could not afford his salary. If he could not be paid, he would prefer to focus his volunteer efforts on writing and editing and his new passion for wilderness. Yard may well have been planning a 1933 "retirement," because a three-year timetable to train a successor was set in 1930.

Might Yard have become too "purist" for his associates in NPA? Again, there is no evidence that he clashed with association colleagues over purism and standards. The work on standards went on aggressively even after Yard left his executive role. Albright certainly did not agree with Yard's purism. Others on the NPA board may have objected to Yard's unwillingness to compromise and wished him gone, but no evidence supports this. Yard's concern about wilderness deepened, and he became interested in wilderness preservation beyond the national parks. He became acquainted with Bob Marshall, a vigorous campaigner for wilderness in national forests, when Marshall joined the NPA board in 1933. The two discussed their interest in creating a group to

aggressively promote wilderness. Yard wrote John Merriam in March 1934, suggesting that a wilderness group be created, which Merriam might lead.

> Organization to preserve the primitive should begin without delay to gather facts, establish relationships with scientific societies and national state [sic] government bureaus, and promote interest and education among associations and leagues of associations of many kinds throughout the country. Its work should be studious, educative and promotive, taking no part in political activities....This is by no means a new idea with me. I have dreamed of it for at least four years. Coming now, it would meet the immediate need of the primitive and solve other problems besides.[4]

Merriam responded that he liked the idea but would not have time to lead such a group. Yard sent a copy of his exchange with Merriam to Bob Marshall, who responded:

> It certainly seems as if the idea of a Wilderness Society began simultaneously in many different places within the past 6 or 8 months. In date of priority, I should say you and Dr. Merriam started the ball rolling in March, then H.A. Anderson came along in July and then Benton MacKaye, Harvey Broome and I came along in August. It will be splendid when we all get together and really rally support behind what is to many of us the finest value in the country.[5]

When Marshall (along with MacKaye, Anderson, and Broome) decided in November to move on the idea, he sent an invitation to six people to join the organizing committee of the Wilderness Society. Yard accepted and went on to become secretary and president, serving as the society's executive from 1935 until his death in 1945.

Some accounts attribute Yard's retirement as NPA general secretary to his "weak" administration.[6] No evidence supports this. By 1930, Yard had ably served as the association's administrator for more than a decade. Undoubtedly he was tired of the job, and this may have affected his work. Whatever Yard's reasons for relinquishing the association's administration, he did not disappear from its councils. He continued as editor of publications until 1942 and served on association committees until his death three years later. Had Yard been forcibly removed from his executive role or stepped down because of serious disagreement with NPA leaders and direction, it seems unlikely that he would have continued his long and important service to the association. All indications are that Yard made the move on his own initiative, for his own reasons.

Why did Yard not create a "wilderness committee" within NPA rather than move on to help start a new organization? One reason was that NPA was in financial trouble and could barely sustain its current programs. Another was that a group outside NPA ranks, interested primarily in wilderness in

Havasu Falls, Grand Canyon National Park

national forests, was moving to convene this new organization, regardless of Yard's inclinations. Bob Marshall, then an NPA trustee, intended to provide financial backing for the new group. The only other NPA member strongly interested in wilderness at the time was John Merriam, and he was too busy to devote much attention to the cause. When invited to become a founder of the Wilderness Society, he declined, stating that he simply had too many other commitments. Yard seems to have been ready for a change, and circumstances conspired to provide him new challenges and opportunities.

Education and Inspiration

While NPA struggled with finances and organization, the park work continued. The defense of standards was ongoing. Yard produced releases for NPA's National Parks News Service, although the numbers declined as the budget tightened. Four issues of the *National Parks Bulletin* were published between 1930 and 1936, none in 1931 and 1932.

John Merriam's NPA Advisory Board on Educational and Inspirational Uses of National Parks, established in February 1927, went forward with its work. It defined two tasks: to identify broad principles that should guide educational and inspirational programs in the national parks, and to develop a demonstration project that would show how these principles could be applied. In June 1927, Merriam reported that the committee had decided to focus development of a model observation station on Yavapai Point, Grand Canyon. The Park Service and various cooperating agencies were working on how best to present the complex story of Grand Canyon geology; Merriam saw participation in this project as the best way to address his committee's goals. The Park Service was becoming more involved in education. At Mather's urging, in 1928 Secretary of the Interior Roy O. West appointed the Committee on Study of Educational Problems in the National Parks.[10] Four of the new committee's five members were on NPA's Educational Advisory Board, with Merriam again chairing. This Interior committee in turn recommended creation of an advisory board to counsel the director of the Park Service on educational policy and development. This became the Department of the Interior Educational Advisory Board, established in March 1929, also chaired by Merriam and stocked with NPA leaders. The Interior committee (the second in this confusing array) also recommended establishing a major educational unit within the National Park Service, to be supervised by high-level staff in Washington, D.C. In 1930, the Branch of Research and Education was established, with Harold C. Bryant in charge as assistant director of the National Park Service.[8]

The extent of NPA influence on this educational activity in the National Park Service is difficult to assess. Interest in education had grown within the agency. By the late 1920s, there were museums in Yosemite, Grand Canyon, Yellowstone, Mesa Verde, and Lassen Volcanic national parks. The American Association of Museums was strongly behind this movement, and Merriam was a prominent member of its Committee on Museums in National Parks. Interpretation programs, although small, were under way at Yosemite, Yellowstone, Rocky Mountain, Glacier, Grand Canyon, and Zion. Stephen

Landscapes, such as the Grand Canyon northwest of Lincoln Point, were principally valuable, in the NPA view, as places for learning and inspiration.

Mather had created a Park Service education division in 1925 and became a more active supporter of education in the parks. Merriam revived NPA's dedication to education, which had been superseded by the necessary defensive park fights in the early and mid-1920s. There is reason to believe that NPA was important in the movement for education in the national parks; many of the principal players were active NPA members. The time was right for NPA's advisory board. The creation of the two government committees on the heels of NPA's action, involving virtually the same people, suggests that the Department of the Interior and the Park Service took NPA's initiative seriously and carried it further than it could have gone if it had remained exclusively an NPA project.

Fresh water flows to the sea along mangrove-bordered streams in the Everglades.

NEW PARK BILLS AND CONCERN FOR STANDARDS

The Everglades

Another NPA project in the early 1930s involved Florida's Everglades. This subtropical region on the southwestern point of Florida had been abused for three-quarters of a century. The vast swamps were scenes of slaughters, their great variety of birds butchered to supply feathers for the millinery trade. Some species were threatened with extinction. From its beginning, the National Association of Audubon Societies had been fighting to defend Everglades wildlife. Several of its wardens were killed in the effort.[9]

NPA had called for an Everglades National Park as early as 1920, as had Mather in 1923. The first park legislation for the area was introduced in 1926. Not until 1929, however, did Senator Park Trammell of Florida win congressional authorization for an Interior Department investigation of the feasibility of a "Tropic Everglades National Park." A committee, including Albright and Cammerer, reported that the area was of park quality and recommended a park that would preserve the primitive character and wildlife of the region.[10] Senator Fletcher and Representative Ruth Bryan Owen of Florida introduced Everglades park bills in both houses in the Seventy-second Congress.

At this point, NPA entered the picture. Concerned about standards and presumably mistrustful of the investigating committee's judgment, NPA on October 9, 1931, appointed Frederick Law Olmsted, Jr., and William P.

Wharton to study the Everglades. They spent ten days in the field traveling by auto, motor cruiser, small boat, on foot, by blimp, and by airplane. They found a strong park advocacy group in the Tropical Everglades National Parks Association and came away supporting the proposed legislation with minor reservations. They submitted their report to a special meeting of the trustees on January 18, 1932.

Speaking directly to the Fletcher Senate bill and the Owens House bill, Olmsted and Wharton presented several conclusions. They thought the park "highly desirable, from a strictly national standpoint and with scrupulous regard to standards proper for the national park system" and judged the area to be of sufficient size and "primitive natural conditions of nationally outstanding distinction." Natural values of great importance would be preserved in the park, especially the mangrove forests and bird life. They recommended that action on the park legislation proceed as quickly as possible because of the continuing destruction of wildlife, the threat of damaging wildfires set by hunters, and the prospect of private exploitation. They addressed boundary questions at length. Although most land proposed for the park was appropriate, they expressed concern about a few substandard tracts. Several areas outside the proposed boundaries should be included, even though they were separated from the main park by the Tamiami Trail. Although detached pieces of land in a park presented problems, areas such as the last large stand of royal palms were of such significance that "important objectives of conservation should not be sacrificed through timidity about technicalities of precedent."[11]

In the discussion that followed Olmsted and Wharton's presentation, everyone agreed that the report should be widely distributed. Director Albright attended this meeting, and he suggested that Senator Fletcher be asked to make the report a Senate document. It was ordered to be printed on January 22, 1932, and received wide distribution.[12] The association agreed to support the legislation, but despite its efforts, no Everglades bill was approved in 1932.

After this failure, NPA decided to take the lead and assemble a coalition of Everglades proponents to set goals and formulate a strategy. The coalition met in 1933 and 1934. The driving force behind the effort to create Everglades National Park was the Tropical Everglades National Park Association and its tireless chairman, Ernest F. Coe. NPA contributed as part of the Washington lobby for the park and as the convener of the coalition. The conferences called by NPA were attended by representatives of the American Forestry Association, the American Game Association, the Society of American Foresters, the Izaak Walton League, the American Civic Association, the American Association for the Advancement of Science, the National Association of Audubon Societies, the Ecological Society of America, the Garden Club of America, and the Everglades National Park Association. The coordination of this formidable array of conservation groups was a significant contribution to the drive for the park.

Everglades bills were introduced into the Seventy-third Congress and were supported by President Roosevelt. Florida's senators pushed the bill through their chamber, but once again, the "Republican filibuster" held it up in the House. As the debate in the House continued, the coalition managed to amend the bill. In a hearing before the House Wildlife Committee on March 19, 1934, the coalition raised objections. It said that by creating the park before lands had been selected and giving the Interior Department responsibility for boundary selection, the bill was removing the public (meaning conservationists) from these critical decisions. Also, the bill did not specify preserving the primitive as the park's foremost objective. President Cloyd Heck Marvin and First Vice President Wharton of NPA and their American Forestry Association counterparts met with Director Cammerer and Horace Albright and agreed to an amendment that read:

> The said area or areas shall be permanently reserved as a wilderness, and no development of the project or plan for the entertainment of visitors shall be undertaken which will interfere with the preservation intact of the unique flora and fauna and the essential primitive natural conditions now prevailing in this area.[13]

Director Cammerer also agreed to appoint representatives of interested organizations to an advisory committee to the Park Service on the selection of lands to constitute the park.[13]

The Everglades Act, as finally passed on May 30, 1934, provided for a park within an area of approximately 2,000 square miles. Because the land was state and private, it would be presented to the United States by the state of Florida, which would purchase private tracts, a procedure used in earlier Eastern parks. When the jurisdiction for parkland was accepted from Florida, no further action of Congress was necessary to establish the park. The act did not create an Everglades National Park so much as authorize one. Hard work by many parties, including NPA, lay ahead before an Everglades National Park would become a reality.

As the struggle for the Everglades progressed, other new and expanded national parks occupied NPA's agenda. In the 1930s, NPA devoted attention to Shenandoah, Great Smoky Mountains, Teton, and Kings Canyon national parks as well as Olympic and Everglades. In each case, an underlying issue was maintaining standards and natural and primitive values. The association was pleased to see so much movement toward protection of more parkland, but the issue of quality remained. The 1930s saw the emergence of a strong state park movement, and NPA leaders wrestled with what should be a state park versus a national park. Although NPA was not the leader in any of these efforts, it worked on all of them from its Washington, D.C., perch, raising its voice in support or protest as its reading of the situation dictated.

A "Crown Jewel" in the Appalachians

The movement to create a national park in the Appalachians had been under way for nearly fifty years. The logging threat to the remaining virgin forests in the rugged southern Appalachians grew in the early twentieth century, generating concern for the future of both timber resources and mountain scenery. An Appalachian National Park Association was organized in 1899, and the federal government's Bureau of Forestry and the U.S. Geological Survey mapped 8,000 square miles from Virginia to Alabama. The agencies conducted various studies, among them an examination of the suitability of the area for a national park. Although there was national park-quality land in the region, particularly in the Great Smoky Mountains of North Carolina and Tennessee, virtually all of it was in private hands. The time was not yet right to attempt to create a national park out of private land. Events in Maine—the creation of Acadia National Park (originally established as Sieur de Mont National Monument in 1916, changed to Lafayette National Park in 1919, and finally to Acadia in 1929) through acquisition of private land—and the emergence of Stephen Mather on the national park scene, helped pave the way for serious consideration of a national park in the Appalachians.

Mather envisioned a large national park in the Appalachians equivalent to Yosemite or Rocky Mountain. He began promoting the idea in 1919 and included it in his 1923 National Park Service Annual Report. Several motives prompted him. He was genuinely interested in protecting the natural values of the area, particularly the shrinking virgin forests, but he had political motives as well. As Darwin Lambert, a historian of Shenandoah National Park, has noted, "He worried that a park system with 'crown jewels' only in the West might fail to hold majority support in Congress when funds were scarce. He needed citizen support where most of the people lived."[14] Automobiles and roads were rapidly improving in the 1920s. Creating a high-quality national park within driving distance of the nation's capital would not hurt Mather's promotion of the national park cause.

At Mather's urging, Interior Secretary Work appointed the Southern Appalachian National Park Committee and charged its members with finding the most appropriate area for the proposed park. This event triggered a scramble among boosters of prospective parks in Tennessee, North Carolina, and Virginia. Even Arkansas and Kentucky hoped to tie their fortunes to the Appalachian park movement. After an exhausting review of the possibilities, the committee recommended in December 1924 for a park in the Virginia Blue Ridge between Front Royal and Waynesboro, but it also hoped that a way could be found to create another park in the Great Smoky Mountains. What resulted was the "Three Park" bill: Shenandoah, Great Smoky, Mammoth Cave. It, in turn, gave new status to the Southern Appalachian National Park Committee by changing it to a commission and providing it "a few thousand dollars to hire a clerk and for members to scout where boundaries might be if any park were to be established and if anybody might donate any suitable land or money to buy some. Congress had never appropriated money to buy national parks and wasn't about to start doing so."[15] Associations and commissions were formed in Virginia, North Carolina, and Tennessee to raise money to purchase land. By the late 1920s, significant progress had been made toward acquiring land necessary to meet the

Shenandoah National Park was popular in the East, posing a dilemma for NPA.

acreage minimums established by Congress to create Shenandoah and Great Smoky Mountains national parks.

National Parks Association concerns had first been raised in 1925. Although the association "is committed fully to the Shenandoah and Great Smokies National Parks proposals," there was "a serious danger which it is our own special duty to take precaution to avert." This danger was the possibility that Congress would appropriate funds to buy private land for the parks. Should this happen, the "open gates to the United States Treasury may prove to be flood gates through which will flow innumerable demands for the purchase of all manner of National Parks. Herein will be the greatest danger which National Park Standards have ever faced."[16] Congress, however, proved reluctant to appropriate funds for land purchase. Most land purchases for Eastern national parks were made with state and private funds.

NPA generally supported the Great Smoky Mountains National Park project. The original acreage minimum of 704,000 set by Congress in 1926 was a concern because the association feared that substandard land would have to be included to achieve such size. After more thorough study of the park prospect, however, the minimum was reduced. NPA then promoted the Great Smoky Mountains park in its *National Parks Bulletin*.

The Shenandoah park, on the other hand, was a continuing concern. The executive committee resolved in 1925 "That it must oppose the admission of any National Park whose forests are not primitive and which is not scenically an outstanding example of its kind."[17] Because there had been little "primitive forest" in the proposed Shenandoah park area when the Forest Service surveyed it in 1914, Yard urged opposition. Other trustees counseled caution. Committees were appointed to recommend NPA action but quietly disappeared without reporting. The association took no public stand opposing the park, but internal debate continued right up to the establishment of Shenandoah in 1935. If Shenandoah became a national park, some argued, its precedent would lead to many "inferior" parks. The park's boosters were not interested in protecting sublime scenery and preserving the primitive, but in commercial benefits. If a new and lesser standard for Eastern parks was officially recognized, Westerners would use that standard to invade existing Western parks. Far better, thought some, to make Shenandoah a *state* park. To the end, the argument remained internal.

Why did the association, in uncharacteristic fashion, keep its concerns about Shenandoah to itself? There are several possible reasons. One is that the trustees wished to avoid the political fallout in Washington that would result from vocal opposition to a locally popular idea. Also, the park was genuinely supported by many backers of the general national park movement, as well as by some of the Washington-based NPA leaders. NPA pursued a broad agenda that might be compromised if its action on Shenandoah alienated its supporters both inside and outside of government. It is likely that all of these reasons combined to mute the association's concerns about Shenandoah National Park.

Other Eastern national parks were being boosted by local proponents and studied by the National Park Service. NPA opposed them all. The Mount Katahdin region in Maine was one. NPA Executive Secretary Foote made an extensive scouting trip to national parks and candidate areas in the summer of 1937. He reported that Katahdin "is not national park caliber and to it applies not one national park standard."[18] A Mount Mansfield National Park in Vermont's Green Mountains was similarly judged to be more appropriately a state park. Reporting on a visit to the National Park Service in

Washington, D.C., Foote summarized NPA's view of virtually all park proposals in the East.

> I had occasion to visit the National Park Service office the other day and was told by the Associate Director that of all things he accomplished in the National Park field he would get the most satisfaction in having his name linked with a project that took into the national park status a second-rate area for the purpose of conserving it that it might return within fifty years to primitive character. I personally cannot find a national park standard in a policy of this kind.... The statement made me wonder if this is the future policy of the Service. If it is, then we are bound to see more areas like Mount Katahdin being dragged into the National Park System with the attendant result to the prestige of the primeval parks.[18]

Expansion of Grand Teton

NPA also found threats to park standards in the West. In his 1937 report, Foote counseled strong opposition to proposed expansion of Grand Teton National Park. A small Grand Teton National Park had been established in 1929; that park included most of the peaks in the Teton range and Leigh and Jenny Lakes at its eastern foot. Even before creation of this limited park, Yellowstone Superintendent Horace Albright, working with John D. Rockefeller, Jr., and his Snake River Land Company, started moving toward an expanded park that would add more than 200,000 acres to the east boundary. The company was secretly buying private land, and additional parcels of federal land, much of it in the Teton National Forest, were to be part of the expansion. Jackson Lake, a natural lake that had been expanded by damming for irrigation in the first decade of the century, was to be in the enlarged park.

Bills to expand Grand Teton National Park had been introduced in 1934 and 1935 and had not progressed. NPA opposed the expansion because it regarded Jackson Lake and the large areas of sagebrush lands east of the Snake River as below national park standards. Historian Robert Righter has ably summarized the reasons for NPA opposition. Conservationists had been especially sensitive to reservoirs in national parks since the loss of Hetch Hetchy Valley in Yosemite National Park in 1913.

> Although they lost, preservationists had waged a national campaign, arguing that a reservoir was a violation of the sanctity of a national park. It now seemed incongruous to voluntarily allow inclusion of Jackson Reservoir into the National Park System. It would establish a dangerous, perhaps disastrous, precedent. Yellowstone National Park was under constant attack by Idaho and Montana irrigationists, demanding that dams be built to provide water storage for downstream users. Particularly threatened was the beautiful Bechler Valley and the immense Yellowstone Lake. If Jackson Reservoir could be "lived with" by the National Park Service, surely they could be flexi-

Jackson Lake was enlarged by a dam and was thus a "reservoir" and a bone of contention for those who opposed expansion of Grand Teton National Park.

ble to the needs of farmers dependent upon Yellowstone waters. The National Parks Association was convinced that defense of the national parks would be strengthened by the exclusion of Jackson Reservoir from the system.[19]

NPA had been fighting against water projects in national parks since 1919. It was a continuous battle in Yellowstone. Water projects had been an issue in the long effort to expand Sequoia National Park by adding the Kings River area, and NPA saw a new chapter ahead in that struggle. Robert Sterling Yard, in an NPA National Parks News Service release in January 1935, described how Mather had spent sixteen years trying to bring the Kings River into the National Park System with protection from water power projects. "With Mather dead," Yard wrote, "the Kings River country, unprotected, is the coming Hetch Hetchy, and Jackson Lake Reservoir will become its new official precedent. Winning the Jackson Lake Reservoir battle now, therefore, will virtually win the Kings River battle also. The fate of the National Park System is in the balance."[20]

In the same release, Yard expressed concern about inevitable cattle drives in the expanded park if Jackson Hole ranchers moved their cattle to summer pasture in the Teton National Forest, as they had long been doing. Also, the proposed park would include the Jackson Hole Elk Refuge, administered by the Biological Survey. He thought it "not the business of the National Parks System to enter into the field of game management for this vast herd, only part of which summers in Yellowstone National Park, the rest being subject to

Arno B. Cammerer, third director of the National Park Service, thought the purism of NPA and its allies was misguided and worked against national park interests.

hunting in adjoining national forests." NPA was worried about precedents. As the association saw it, the national park mission would be dangerously compromised by all this activity. Let the Biological Survey and Forest Service handle these other responsibilities.

Director Cammerer responded to Yard and others who raised these concerns: "Because civilization has moved into the choicest areas faster than they could be established as national parks, some parks must now be carved out of developed areas." Constructing a reservoir in a national park "is a very different thing from the attempt to save a previously violated area from further exploitation."[21] In his view, if there were to be any new parks, some areas historically exploited for commercial purposes might have to be included. Perhaps the area in contention should have been included in Yellowstone National Park in 1872, but because it was not, the task now was to preserve and restore the area. "Should we forego this opportunity for constructive conservation," asked Cammerer, "simply because the natural character of Jackson Lake has been modified?"[21]

The answer from William P. Wharton, president of NPA, was yes. National parks should be natural and wild. Those qualities defined a national park for NPA, as opposed to national monuments, national historical sites, and recreation areas. NPA saw its mission as "nothing less than the saving of the original National Park System as a distinctive and unique American institution."[22] NPA continued to hold to the original definition of national parks as "areas of unmodified natural condition, each the finest of its kind in the country, preserved forever as a system from all industrial use." It had proposed that parks of this type be called "National Primeval Parks" and thought Grand

Teton, given its proximity to Yellowstone, should be one of them. The park should be enlarged, but not to include areas that had been modified by human activity.

In his letter, Director Cammerer asserted that "an ideal is something toward which to work; it should not be something which prohibits us from working."[23] In his view and that of less "purist" conservationists, NPA was taking an extreme position. The Park Service under Albright and Cammerer had embraced a broader mission. Part of that mission was expanding the National Park System to embrace areas judged necessary to "complete" parks. Various schemes to expand and complete Yellowstone had been pursued for decades, and the original Grand Teton National Park and the subsequent expansion were viewed as part of that effort. NPA and the National Park Service often worked in concert; in this case, NPA was viewed by the Park Service and many conservationists as obstructionist. Historian Righter observes that "It was an unfortunate schism between two supportive organizations, for both sides were sincere and informed. In Jackson Hole it was particularly ruinous, for those who opposed park extension for less idealistic reasons were able to exploit this preservation disagreement to their advantage."[24] The expanded park was not to be—at least for a while.

The Battle for Kings Canyon

Yard, arguing against including a reservoir in Grand Teton National Park, was looking ahead to the looming Kings Canyon battle, which he expected to test national park standards once again. The battle came, and NPA again was in the difficult position of opposing, on principle, national park legislation it mostly favored. This must have seemed like deja vu to Yard, who had led the association's opposition to the Barbour bill in the early 1920s. NPA opposition then had been the result of Mather's inclination to trade parts of Sequoia National Park to the Forest Service in exchange for its support for adding the Kings River area to the park. Although Yard was now only editor of NPA publications, his executive attention having been transferred to the Wilderness Society, he remained in the NPA inner circle.

Once again the issue was water power. The 1890 boundary of Sequoia National Park had been extended in 1926 to include the Kern River drainage, and the northern boundary had been readjusted to follow the divide between the Kaweah and Kings watersheds. The park's size increased by 140 percent but it was far from the goal of Mather and other park proponents; it did not contain the Kings River and the Mineral King Valley. The city of Los Angeles and San Joaquin Light and Power Corporation were vying for dam sites in Tehipite Valley and Kings Canyon. Various developments in Mineral King complicated the prospects for preservation there. The Federal Power Commission carefully weighed the power proposals, rejecting them only to have them refiled. The Forest Service, which administered the area, conducted studies and made plans. Preservationists schemed to create a national park. Meanwhile, local opinion favored leaving Kings Canyon and its rivers in the hands of the Forest Service.

Into this stalemate came Franklin Roosevelt's Interior secretary, whom one of his Forest Service enemies, California Regional Forester S. B. Snow, once described as "overambitious, ignorant, egocentric, ruthless, unethical and highly effective."[25] Snow hated yet admired Ickes, because the Interior secretary managed to orchestrate a Kings Canyon National Park, against strong opposition from many quarters including Snow and his Forest Service colleagues. Ickes grated on many people but was a most effective secretary of the Interior and a supporter of national parks. Further, he believed in wilderness parks, as he demonstrated in the struggle for Olympic National Park. He told a meeting of park superintendents in 1934:

> I do not want any Coney Island. I want as much wilderness, as much nature preserved and maintained as possible....I recognize that a great many people, an increasing number every year, take their nature from the automobile. I am more or less in that class now on account of age and obesity. But I think the parks ought to be for people who love to camp and love to hike and who like to ride horseback and wander about and have...a renewed communion with nature....[26]

He acted on these convictions in leading the push for Kings Canyon National Park.

Ickes assigned Park Service Assistant Director Arthur Demaray the task of drafting a bill for Kings Canyon National Park and persuaded Senator Hiram Johnson of California to introduce it in March 1935. "This park will be treated as a primitive wilderness," he told Cammerer in a memorandum. "Foot and horse trails to provide reasonable access will be encouraged, but roads must be kept to an absolute minimum. The state road now being constructed should never be extended beyond the floor of Kings River Canyon.... Accommodations provided must be of simple character and the rates moderate. No elaborate hotels shall be constructed."[27] The bill stalled in the face of strong opposition in California from chambers of commerce, irrigation districts, power companies, farm bureaus, and sportsmen's associations. The Forest Service mounted a campaign against it.

Ickes moved again for the park in 1938 with another bill. Congressman Bertrand W. Gearhart of Fresno, within whose district most of the new park would be located, announced that he would introduce his own bill when Congress convened in January 1939. His bill, drafted by Cammerer and Demaray, made no provision for wilderness management. At Ickes's direction Irving Brant, working as a consultant to Ickes, and Demaray drafted a revision. Their bill stated that the park would be managed as wilderness, with few roads and concessions. The Tehipite Valley and Cedar Grove areas were excluded from the park as potential reservoir sites; they would be managed as wilderness by the Park Service until dams might be authorized by Congress. If the Bureau of Reclamation deemed the sites not suitable for reservoirs, they would become part of the park by proclamation. This bill, with its emphasis on wilderness, seemed one that even purists like NPA could support.

Outspoken Secretary of the Interior Harold Ickes blasted NPA for its opposition to Kings Canyon National Park.

This Gearhart bill was introduced in the House on February 7, 1939, and referred to the Committee on Public Lands, which scheduled hearings in March. The hearings lasted twenty-eight days. Park opponents tried to derail the legislation, but the House passed a revised Gearhart bill on July 18, after which Congress adjourned for the summer. When the NPA executive committee met on February 14, 1939, it approved a resolution to the effect that, because the Forest Service had managed the proposed park area as wilderness for years, NPA should take no action on the matter. It could not support the proposed park.

In mid-March, as the hearings approached, the executive committee drafted a position statement, to be delivered by Executive Secretary Foote. NPA opposed the Gearhart bill because it excluded the Tehipite Valley on the middle fork of the Kings River and Kings Canyon on the south fork. "Mr. Mather might have had his Kings Canyon park some years ago had he been willing to compromise and accept a park that failed to include the justification of its creation—an unspoiled Tehipite Valley and Kings Canyon," said Foote. The bill proposed that the Park Service administer the reservoirs (should they be built) and surrounding land, above the high-water mark, for recreation. The effect of this would be "not only to include reservoirs for irrigation and power within the park, contrary to accepted national park standards, but also…to permit the original condition of these valleys to be radically altered." Foote concluded with a reiteration of NPA's position on the central issue.

> Our attitude is that, if another national park is to be established in that section of California, it should be established free from commercial

uses and artificial features. If that cannot be done, the John Muir-Kings Canyon National Park should not be created for the sake of having another national park. This is doubly important when the effect of such a compromise of ideals on the national park system as a whole is considered. Our Association holds fast to the old standards which have made the system what it is.[28]

NPA's annual meeting in May 1939, which marked its twentieth anniversary, was dominated by the Kings Canyon controversy. The Sierra Club tried to resign from the board, but the matter was tabled. The club had worked hard for a King's Canyon park and was willing to accept a park without the canyons rather than no park at all. It could not accept NPA's uncompromising position. Some NPA members questioned the wisdom of this position, but after discussion, the position taken in the hearings was reaffirmed.

The board also discussed an April 3 news release from Interior Secretary Ickes attacking the association and its president. The Department of the Interior's release was as follows:

Commenting upon activities of certain "name-plate" organizations, Secretary Ickes today said:

"The pseudonymous National Parks Association is once more giving public proof of its true character by opposing the bill before Congress to create the John Muir-Kings Canyon National Park in California.

"I want it clearly understood that the so-called National Parks Association has no connection whatever with the Government, and nobody should be deceived into thinking that it has any relation to the National Park Service of the Department of the Interior. It may be a stooge for lumber interests, but it does not represent, and it cannot speak for, the National Park Service.

"Mr. Wharton, President of this outfit, recently attempted to enroll me as a member. I declined his proffer because I did not want to lend my name to such an organization. My understanding is that Mr. Wharton has 'taken in' some 300 members, who apparently are not aware of the fact that he and his Washington representatives opposed the establishment of the Olympic National Park, which was approved by the Congress and the President, and more recently, the John Muir-Kings Canyon National Park, according to his mouthpiece, the San Francisco Chronicle. In spite of all of the efforts of Mr. Wharton, I am informed that 17 of the leading California newspapers are publicly supporting the establishment of this park."[29]

Wharton's public letter of response explained association positions on Olympic and Kings Canyon park proposals. He replied to the "stooge for lumber interest" accusation.

Kings Canyon National Park was added to the system in 1940.

With their deep canyons, high peaks, and big trees, the Sequoia and Kings Canyon parks comprise a magnificent complex meeting the highest standards.

> You are quoted as saying that the National Parks Association "may be a stooge for lumber interests." This phrase, while not an assertion of definite fact, conveys an impression that is so contrary to the facts that it cannot be overlooked. I call your attention to the fact that twenty-two scientific and educational organizations are represented on our Board of Trustees, and that of the twenty or more members elected at large none, so far as I am aware, are connected with the lumber industry. The conjecture that organizations and men of the caliber and standing of those on our Board could be controlled by the lumber industry, or by any other commercial interest, is so contrary to reason and common sense that I am confident it could not have originated with you.[30]

The accusation did not, in fact, originate with Ickes. Willard Van Name, Rosalie Edge, and Irving Brant all believed that Wharton and NPA had been duped by the timber industry to oppose them in their campaigns for Olympic and Kings Canyon national parks.

This squall with Secretary Ickes passed, and on June 2 the Kings Canyon National Park bill was reported out of committee. James Foote reported that it was "scarcely recognizable as a national park measure," and it had been weakened more than even Ickes liked. Still, the Interior secretary decided to support final passage with the intention of seeking improvement in later legislation; NPA was even more strongly against the measure. Yard undoubtedly represented the views of his NPA colleagues when he wrote to Anne Newman of the Izaak Walton League, "I'm quite sure that the Park Service would be glad to accept the park filled with power dams from the glacial lakes down rather than leave it safe with the Forest Service."[31]

NPA continued to lobby hard against the Gearhart bill, but to no avail. Congress approved the bill in February 1940, and President Roosevelt signed it into law on March 4. NPA's mixed feelings about the outcome are evident in a *National Parks Bulletin* report in July 1940. The "magnificent climax portions of the Middle and South Forks" were omitted and grazing rights for some ranchers would continue. On the other hand, a wilderness park had been created. The secretary of Interior had been authorized to limit the "privileges" he may grant "in order to insure the permanent preservation of the wilderness character of Kings Canyon National Park." NPA was encouraged by this wilderness provision. "Now that we have the matrix of the backcountry," concluded the article, "let us fill this setting with its finest jewels before they have been irreparably defaced! Let us bring Kings Canyon into Kings Canyon National Park!"[32] Twenty-five years would pass before Congress vindicated Ickes's gamble and added the essentially wild Tehipite Valley and Cedar Grove areas to the park. Kings Canyon turned out not to be another Hetch Hetchy, as Yard had feared. Ultimately, the goals of the "purist" NPA would be achieved.

Financial Distress and Organizational Retrenchment

The Depression years were difficult for the National Parks Association. Not only did it take positions unpopular among conservationists, but it struggled financially and organizationally as well. Wallace Atwood stepped down as association president in May 1933, and Cloyd Heck Marvin was elected to replace him. Marvin was president of George Washington University, a post he had held since 1927 and would hold until 1959. No record exists of when Marvin first became active in the association, but when the financial crisis reached the point in 1931 that office rent could not be paid, Marvin offered rent-free space at the university; the offer was gratefully accepted.

Marvin's involvement in association affairs increased during the next few years. He chaired a committee to investigate the placing of the National Park Service in the government reorganization that was discussed in 1932. The committee submitted an extensive report to the 1933 annual meeting that was enthusiastically received. At the same meeting, the Committee on Nominations recommended him for president, and he was duly elected. An enthusiastic outdoorsman, Marvin was interested in national parks and drawn to the association. As its financial crisis deepened during the Depression, he was willing to do what he could to keep it alive. Unable to give much time, he nonetheless could offer space and staff help.[33]

Yard resigned as association executive in 1933 and was not replaced until 1936, when James A. Foote became executive secretary. NPA reached its low-water mark during this period. Although Yard continued as editor, publications virtually ceased for three years. Only one *National Parks Bulletin* was produced, and two releases came from the National Parks News Service. Few meetings were held, and work on association projects, such as the deliberations of committees one and two, nearly stopped. There were no executive committee meetings in 1934 and 1935, only the annual meetings. The record contains little evidence that President Marvin provided much leadership during his tenure. The association had no money and few members; he could tap university resources to keep basic activities going. He told the annual meeting in 1934 that progress had been difficult "with little or no money at hand to meet the financial obligations of the Association, and with some indebtedness carried over from the past."[34] He thought the goals of passing an Everglades bill and stopping the Ouachita project in Arkansas were all the association could hope to do in the coming year.

At the 1935 annual meeting, Marvin was succeeded by William P. Wharton, who had served as a trustee and first vice president since 1925. Wharton was among the association's earliest financial benefactors and had been increasingly active in its affairs. National park standards had been one of his special interests since the mid-1920s. In 1926, he had championed a movement within NPA to remove areas he thought unfit to be national parks—Sullys Hill, Hot Springs, Wind Cave, Lafayette, and Platt—and to examine whether the proposed Great Smoky Mountains, Shenandoah, Grandfather Mountain, and Mammoth Cave were truly fit to be national parks. He had served as co-chairman of the joint committee of NPA and the American Forestry Association that had conducted the outdoor recreation survey of federal lands for the National Conference on Outdoor Recreation in the 1920s. With membership reduced and Yard's attention drawn to the Wilderness Society, Wharton became the power in the association.

Wharton was interested in both forest and park issues. He served for many years as a trustee of the American Forestry Association and advocated close cooperation between it and NPA; and on several occasions, he pushed for a merger of the two associations. An attorney, he lived in Groton, Massachusetts. One of his first acts as president was to hire James A. Foote as executive secretary, with Wharton personally providing his salary. The association could not operate without someone to handle the day-to-day correspondence, membership duties, and legislative monitoring required to keep projects moving.

During his first year, Foote did the routine work well but was not the activist executive Yard had been. He needed to learn the job, and by the summer of 1937, he had done so. He toured parks and proposed parks from Mount Olympus to Mount Katahdin and submitted a long report to the board. He was very thorough in his reviews and made strong recommendations about how the association should stand on the controversies of that year, such as Grand Teton expansion and the Olympic National Park legislation. In contrast to those of the nearly humorless Yard, his reports were salted with humor. Writing of his impressions of Yellowstone National Park, he observes, "If one wants a good time, dancing, and a general night club atmosphere, one should visit Yellowstone." The park is so crowded "You can almost smell cabbage cooking from one end of Yellowstone Lake to the other." In his view, the place

Civilian Conservation Corps workers, here clearing debris from around Jackson Lake, made "improvements" throughout the National Park System.

was overdeveloped. His recommendation—"A load of dynamite and a crew of CCC boys to cart away the debris."[35] Foote offered specific and well-conceived recommendations for solving many problems he encountered. Clearly, by late 1937, he was comfortable in his new job.

Membership and financial problems continued. Foote reported in January 1938 that the association had $598 on hand. At the same meeting, he proposed a publishing venture, an annual book on the national parks. This would be an inducement for members, he thought. Because NPA could not afford to publish its *National Parks Bulletin* regularly, members saw little tangible return for their $3 membership. Organizations such as the American Forestry Association, the American Museum of Natural History, and the National Geographic Society gave members splendid monthly publications. This was neither a new problem nor a new idea. Throughout the association's history, Yard had tried to create a magazine. He had proposed books, but finances had never been adequate. Nor were they now. The *Bulletin* could be published with more regularity—twice a year between 1937 and 1941—but no book was yet possible. The dream of a strong publications program simmered.

Robert Sterling Yard remained active in association affairs. He produced occasional National Parks News Service releases and wrote and edited the *National Parks Bulletin*. He chaired the association's Committee on Legislation, sat on the executive committee, and participated in all association debates and discussions. Yard remained the opinionated, aggressive, and uncompromising conservationist. He generally supported the strongest preservationist position and encouraged strong stands on the issues. In 1938 when Albright, as president of the American Planning and Civic Association, and Director Cammerer publicly abandoned national park standards in the association's view, the board argued about its response. Members did not think Albright should be named, but Yard disagreed. Even though Albright no longer held an official position with the Park Service, his views were influential, and he should be held publicly accountable. Yard lost the argument. Yard was now in his late seventies, yet his energy and dedication to the cause remained impressive. While working with NPA, he was also president and executive for the Wilderness Society, which was emerging as a new, focused, and influential voice in the preservationist debate. His advancing years did not diminish his level of work and commitment to the cause. In 1935, Benton MacKaye wrote to Bob Marshall that "Bob Yard is nothing less than a wonder. You and he make an inspiring team, and Harvey [Broome] and I sleep better nights for having two such dogged watch dogs on the job up at headquarters."[36]

STRUGGLES FOR PRINCIPLE
The Rise of Recreation

Throughout the 1930s, the association remained concerned about the problem of outdoor recreation. NPA held to the conviction that national parks should be managed principally for education and inspiration, not for mass entertainment. An article in the August 1933 *National Parks Bulletin* defined recreation as "outdoor refreshment, physical or mental diversion, entertainment." Education was "orderly understanding imparted by a personal reaction to a natural environment. This makes for intellectual appreciation of nature; it involves the enjoyment of sublimity and beauty; and it leads toward a high conception of the development of man." Inspiration is "liberation of thought or emotion....Its effect is to generate higher thinking and better achievement."[37] NPA repeatedly stated its belief that recreation in the outdoors should be provided to the American people—but not by the National Park Service in national parks. It clashed with Albright and Cammerer over this and worried about the expanding mission of the Park Service.

The automobile pressed change on the national parks, and NPA was not pleased. The Blue Ridge Parkway exemplified the problem. It was designed to provide "recreational" driving to view scenery—and it traversed national parks. President Hoover initiated the Skyline Drive in Shenandoah National Park, and the Roosevelt administration promoted it as part of its public works program. Senator Harry F. Byrd, Sr., of Virginia and others seized the opportunity to promote a parkway linking Skyline Drive to Great Smoky Mountains National Park. This Blue Ridge Parkway became a joint project of the Park Service and the Bureau of Public Roads and was legally assigned to Park Service administration in 1936.[38] The Park Service had long been in the road construction and maintenance business, but a parkway was something new. The road itself was an attraction rather than a transportation route to a natural feature. Compounding its concern about the road's impact on the natural values

NPA did not approve of parkways as national parks, but many visitors came to enjoy the view from Skyline Drive in Shenandoah National Park.

of the parks, NPA thought parkway administration a serious departure from the hallowed mission of the Park Service. Worried as always about precedent, it saw the Blue Ridge Parkway as an open door to future abuse of national parks.

NPA also objected to other Park Service forays into recreation. Congress passed the Park, Parkway, and Recreation Area Study Act of 1936, which directed the Park Service to cooperate with other federal agencies and state and local governments to plan for a coordinated system of public park facilities. The Roosevelt administration bought submarginal farmland and moved farmers from that land to better land. Because some of the land acquired this way was of recreational value, these "recreational demonstration projects" were shifted in 1936 from the Resettlement Administration to the Park Service for administration. They would be planned and developed and turned over to states and municipalities. The Civilian Conservation Corps would do much of the physical work. In 1936, there were forty-six projects in twenty-four states involving 397,056 acres.[39] NPA did not object to the goal of these programs, which was to provide much-needed recreation sites close to urban areas. It was concerned that the Park Service "expansion," as it saw these developments, would at best dilute the agency's ability to administer the true national parks. At worst, it might signal the abandonment of the historical mission of the Park Service.

Yet another problem was the category "national recreation area," not to be confused with the "recreational demonstration area." A national recreation area was an area of national importance that would be developed for recreation and kept under Park Service administration. The most important of these was part of the Boulder Dam project. From the time the idea was proposed in Congress in 1933, NPA opposed Park Service administration of the reservoir behind Boulder Dam, and the concept of a system of "national recreational reservations without standards." A National Park Service-administered area featuring an artificially created reservoir was too far from the original mission of the Park Service. How could it join forces with the dam builders? Despite objections from NPA and other conservation groups, the national recreation area was established at Boulder Dam, and the National Park Service was charged with administering recreation on and around the reservoir.

NPA, enfeebled during the Depression, could not hope to slow the tide of change represented by these recreation initiatives. Under Albright, the Park Service was interested in expansion. Although Cammerer may not have been as ambitious as Albright, the policies of the Roosevelt administration pressured the Park Service to embrace new roles. Interest in outdoor recreation was growing nationally at a time when few people could afford to travel to remote national parks. The need for outdoor recreation close to the people was clear. The American public demanded that government provide new types of recreational services, and agencies such as the Park Service and even the Forest Service could not resist without suffering serious political damage. Against these tides, NPA could only protest.

Routine Defense Continues

While such struggles for principle dominated the association's agenda in the 1930s, defense of existing parks continued. Mining interests invaded Death Valley early in the decade and tried in 1933 and 1934 to open all parks to prospecting. NPA opposed all of these initiatives, and prospectors won no blanket access. Although NPA and other conservationists opposed mining in Glacier Bay National Monument, Congress opened it to mining in 1936.

Perhaps the biggest defensive effort for NPA involved Rocky Mountain National Park. In 1935 a proposal appeared to move water for irrigation from the western to the eastern slope of the Rockies through a tunnel in the national park. After disagreements were resolved between people on both sides of the Rocky Mountain divide, legislation was introduced in Congress in 1937. The Park Service and NPA, this time working in concert, vigorously opposed the project. At a meeting on May 14, 1937, the NPA board outlined its opposition in a resolution. Standards were part of the opposition. "Use of the park for commercial purposes would create a precedent in defiance of standards. . .and lay the National Park System open to economic exploitation." Building a tunnel would "alter natural conditions on the surface in the vicinity" and would "destroy the primitive values of the park." Finally, other routes were available that would not endanger the park.[40] Despite opposition, Congress approved the project in August 1937, and President Roosevelt signed the bill.

Grant Geyser in Yellowstone National Park.

Yellowstone National Park continued to need protection. In opposition to the Rocky Mountain tunnel, NPA had argued that such a project would open the door to similar invasions elsewhere. Yellowstone was a case in point; the association had been fighting irrigation projects there since its inception. At the May 1938 board meeting, NPA President Wharton told the assembly that "would-be despoilers" were already at work, trying to use the Rocky Mountain development as precedent. Bills had been introduced in Congress to authorize the damming of Yellowstone Lake and the construction of a tunnel to divert water to the Snake River for irrigation of Idaho lands. NPA rallied its allies through its National Parks News Service, and in April 1940 reported that this skirmish in the ceaseless defense of Yellowstone could be recorded as a win for conservationists.

The decade of the 1930s was a trying, yet important, period in the history of the United States, the Park Service, and National Parks Association. The nation struggled through the worst depression in its history. Part of the government's response to this crisis affected the national parks and its administrative agency in a peculiar way. Because of public works programs like the Civilian Conservation Corps, money poured into the parks as never before, and the Park Service mission expanded accordingly. These events helped shape the NPA role as a fighter for what it perceived to be the high ground— protection of naturalness and wildness of the big parks that had defined "national park." The hard-working core of NPA—stalwarts such as Yard, William P. Wharton, Henry Baldwin Ward, John C. Merriam, and Albert A. Atwood—found themselves at times at odds with nearly everyone. They were a conscience for the National Park Service, constantly raising troubling questions. They kept the issue of standards sharply in focus for ambitious bureaucrats who might have been dazzled by the opportunities brought their way by the public works programs of the 1930s.

Mather and Yard foresaw a time when an independent group of park enthusiasts was needed to keep the National Park Service on track. NPA saw itself in that role in the debates over standards. Mather's proteges, Albright and Cammerer, were in charge throughout the decade, and they believed they were carrying on his legacy. Mather had sometimes thought NPA too "purist"; the directors who succeeded him shared that opinion. The differences should not be overemphasized, however. NPA and the Park Service worked together on many issues. They shared goals.

In the 1920s, Robert Sterling Yard identified what he saw as three primary threats to the national parks. One was the force of commercialism—companies and entrepreneurs who sought to use the parks for profit. A second was the economic boosterism of communities that fought for "their" park to attract profitable crowds while the federal government paid the bills. The third were enthusiasts for unlimited recreational expansion who sought more parks, regardless of whether they met national park standards. All three threats appeared in the 1930s, and to a level that may have surprised even Yard. He could not have anticipated that the near collapse of the American economy would bring a flood of money and cheap labor to park and conservation projects. With the money came the threats.

NPA's performance was especially remarkable in light of the fact that it operated through most of the decade with low membership, virtually no financial resources, and for three years, no paid executive. The times tested the commitment of the dedicated core members of the association, and they passed the test. They stretched their resources as far as possible, cut back where they could, and survived. Even at its lowest ebb, which seems to have been 1934 and 1935, NPA maintained its stature. Director Cammerer addressed the 1934 annual meeting, as did John C. Finley, editor of the *New York Times*. They may have been few, but active NPA members during this period were influential. Educators, scientists, and businessmen stocked the board of trustees: Wharton, Frederick Law Olmsted, Jr., Victor Shelford, John C. Merriam, Bob Marshall, Huston Thompson, among others. They generously gave their time and money to keep NPA alive, and the association regained its financial health and membership as the decade came to a close.

At the tenth annual meeting, Robert Sterling Yard had reflected at length on the history of the association. His remarks then had a self-congratulatory tone, and deservedly so. William P. Wharton presided over the twentieth annual meeting on May 18, 1939, and Yard was again present. He was seventy-nine years old, still editor of NPA publications. There is no record that he offered historical reflections; in fact, no one looked backward. The meeting's business involved accusations by fellow conservationists of a sell-out by NPA and its "stooges." Wharton and the association had taken their position against Kings Canyon National Park and been attacked by the volatile Secretary Ickes. They continued to be concerned about Cammerer and Albright's perceived abandonment of standards. In their view, a less than ideal Kings Canyon Park was a direct result of this policy change. Two decades of work had brought the association to greater challenges, and at that 1939 meeting, it seemed ready to move forward against them.

Protecting Parks in Wartime

CHANGES IN LEADERSHIP

The 1930s severely tested the association's resolve. It suffered a crisis of leadership when Yard's interests changed and the organization could not afford an executive for a while. Although NPA struggled to survive, a major reorganization resulted in a greatly expanded National Park System and a redefinition of the purpose of that system. The National Park Service was given more responsibilities and, in a period of expansionism, welcomed them. At mid-decade, NPA seemed to be fading from the scene.

The association was rescued by William P. Wharton, whose time and money put the organization back on its feet. Wharton had critics, mostly outside the ranks of NPA, but he came forward when needed. Under Wharton's leadership, NPA remained as uncompromising on the matter of standards as it had been under Yard. Other conservation groups might tactically support less than perfect park proposals and expansion plans; NPA staked out the extreme ground of purism and the ideal park system. Its stubborn adherence to what it perceived as fundamental principles placed it at odds with the most powerful figures in conservation.

As the decade closed, Hitler invaded Poland, and the prospect of global war grew. Defense of national parks seemed a minor concern in the darkening global situation, yet NPA was reviving. Membership began to grow, and some of the dreams of the founders again seemed possible. Several superb "primeval parks" had been added to the system—Great Smoky Mountains, Olympic and Kings Canyon—and the prospect of an Everglades National Park was growing. Threats to the integrity of the system remained, but progress was being made on some fronts.

Mono Rock, Sequoia National Park

The NPA Executive

The turbulent 1930s for the National Parks Association continued through the final meeting of the decade. Robert Sterling Yard had labored as the association's executive for twelve years and was succeeded by Loren Barclay, a young man who barely learned the job before a lack of funds forced his resignation in 1933. For nearly three years, the association could not afford anyone in the role of executive until, in 1936, James A. Foote assumed the post. Foote performed well until 1939, when his work began to fall below the expectations of President Wharton. At the executive committee meeting of October 1939, Wharton brought up "the failure…of the present incumbent to meet some of the requirements of the position."[1] No list of Foote's failings appears in the record, but the problem probably involved soliciting members. Membership had fallen so far in the early to mid-1930s that no mention of numbers appears in association minutes between 1933 and 1943. Membership had dropped to 321 in 1933 and stayed low; it was at 560 in 1943 after Foote's successors claimed several years of modest success in recruiting. Foote seems to have functioned well as a lobbyist, issues analyst, and secretary. He wrote well and maintained the association's network of cooperators. He did not, however, raise funds and recruit members sufficient to satisfy the board. The executive committee unanimously agreed that Foote should be asked to resign, and that ended his tenure with the association.

Wharton announced to the executive committee the following April that he had hired a new executive secretary, Edward B. Ballard. By the annual meeting of May 1941, Ballard could report success in his first membership effort, securing a 45 percent increase in members over the previous year. He had also produced two issues of the *National Parks Bulletin* and had another membership campaign under way. Unfortunately, with war raging in Europe and the American military preparedness growing, he had to inform the board that he had been classified 1A and soon expected to be drafted into the Army. He had already told Wharton of his situation, and the executive committee was actively seeking a replacement. It had, in fact, interviewed a promising candidate, "a young man of 34 years with a wife and child" (and therefore unlikely to be drafted), who was then volunteering and working as an editorial assistant for the American Forestry Association.[2] According to the committee, this young man was fully qualified for the job, and he shared the association's views about national parks. His name was Devereux Butcher.

Ballard was inducted into the Army Air Corps in May 1942 and took his leave at the annual meeting that month. This was also Butcher's first meeting as executive secretary. In his final report, Ballard documented 212 new members and the groundwork for a fund-raising effort looking toward the association's twenty-fifth anniversary in 1944. Although he had not published another issue of the *National Parks Bulletin*, with the help of a board Education Program Committee he had published four releases from the National Parks News Service along with other miscellaneous publications.

Butcher, a native Pennsylvanian, was an artist, photographer, writer, and ardent conservationist. A graduate of the Philadelphia Academy of Fine Arts,

The national parks provided brief respites from "the wearisome tension of war" for many during World War II.

he had studied painting and photography. After a stint as an architectural draftsman, he traveled extensively and did a photographic study of California missions. Family wealth allowed him to pursue his passions and develop his skills, even during the difficult years of the Depression. He married in 1935 and built a stone cabin in the forests of the Delaware Valley, where he experimented with simple living and reflected on the value of wilderness and nature. As his interest and skill in writing and photography grew, he submitted work to the American Forestry Association's *American Forests* magazine, volunteered with the association, and was employed by it as an editor when the NPA job appeared.

Butcher dove enthusiastically into his new job. Robert Sterling Yard, now in his eighties, resigned as editor of publications in June and recommended to the executive committee that Butcher be appointed editor, and he was. Yard would help Butcher as advisory editor. The *National Parks Bulletin* would be changed to *National Parks Magazine* and a new format and editorial policy introduced. The magazine would be published quarterly. Yard's twenty-year dream of an association magazine was about to be realized, although he was too old to contribute much to it. Editorially, the magazine was to be more "popular," to contain not only information about parks, issues, and problems but entertaining articles. Its format would be smaller and glossier, with photographic reproduction of excellent quality. The aim was to produce a publication that would recruit members and reward them for their membership. Yard had always thought a magazine would be an invaluable tool for educating and informing those who were interested in the cause and for building the association. Butcher proved him right.

In the late 1930s, NPA seemed to be at odds with friend and foe alike. It had tangled with Cammerer, Ickes, and even its long-time ally, the Sierra Club. During the fight for a good Kings Canyon National Park bill, the Sierra Club's representative on the NPA board, Duncan McDuffie, resigned in protest over NPA opposition to the Gearhart bill. The board did not accept the resignation and tabled the matter. Early in 1940, McDuffie wrote a letter to Wharton:

> ...as to the Kings Canyon National Park, the Sierra Club from the beginning has been anxious to have the lower end of Tehipite Valley and Kings River Canyon included in the Park but, realizing that this was not possible at this time, supported the Bill for the creation of the Park as the best means of preserving the superb mountain region of the Kings watershed. You can, therefore, count on our willingness to cooperate with you in working, at the proper time, for inclusion of these reclamation withdrawals in the Park.
>
> I feel sure that such difference as that which arose between the Sierra Club and the National Parks Association in reference to the Kings Canyon proposal can be avoided if there is frank consultation between the two organizations before a position is taken.[3]

Fences were mended and McDuffie and the Sierra Club resumed their alliance with NPA. At this stage in their histories the two conservation groups were very different in some ways—the Sierra Club was an outing club and focused on conservation primarily in the Sierra Nevada—yet they shared many goals. Their differences were more tactical than philosophical. Cooperation on park issues, particularly in California, was essential for successful conservation campaigns. The two groups would work together in many fights for the national parks.

The National Park Service

National Park Service leadership changed, and relations with the agency improved. Cammerer resigned because of ill health in August 1940 and was replaced as director by Newton B. Drury. In 1933 Drury had been recommended to succeed Albright but declined because he had other work to complete. When Cammerer asked to be transferred to a position of less responsibility, Secretary Ickes again turned to Drury, and this time, after negotiating with Ickes on several points, he accepted. Drury wanted assurance from Ickes that he would be allowed to concentrate on national park matters and not be drawn off on other Interior Department tasks. (Albright had accepted many nonpark duties, calculating that in becoming an aide to Ickes he would win the secretary's support for the Park Service, which he did.)[4] Drury let Ickes know that he expected him to seek and consider National Park Service recommendations on matters of policy and administration affecting the national parks.[5]

Ickes did not like Cammerer, whose bland, unassertive administrative style annoyed him. Cammerer worked extremely hard, accomplished much, but never won over Ickes. "I am thoroughly persuaded," Ickes noted in his

Newton B. Drury became director of the National Park Service in 1940 and led it through the trials of the next decade.

diary in September 1939, "that I cannot get along much longer with Cammerer."[6] In the view of some observers, the antipathy between Ickes and the director had caused Cammerer's office to be bypassed in some park matters that resulted in morale problems, especially at Washington headquarters. Drury hoped to avoid this problem and restore morale with firm leadership and control, but he would need Ickes's cooperation. Assured that he would have it, Drury accepted the job.

Who was Newton B. Drury? Like Mather and Albright, he was a graduate of the University of California at Berkeley. After a brief career in university teaching and administration and military service in World War I, he became the first executive secretary of the Save-the-Redwoods League. Since 1929 he had also served as executive officer of the California State Park Commission. In these posts, he led a successful effort to preserve thousands of acres of redwoods in California, a task involving political skill, salesmanship, fund raising, and administration of state funds. An admirer described him as "a forceful and eloquent writer and speaker, a man of the highest ideals, combined with sound practical sense, and an executive of solid accomplishments."[7]

Ickes sought an experienced and successful administrator, but he also sought a director who shared his preservationist inclinations and who would disarm the purist critics of the Park Service. Drury had been part of the purist protest against diluting standards. He was a wilderness preservationist and an honorary life member of the Sierra Club. He shared NPA's view that the Park Service, in its rush of expansion since 1933, had neglected the "primeval" parks of the West. Historian Donald Swain summarizes the qualities that recommended Drury to both Ickes and conservationists.

He refused to view his job as an opportunity to aggrandize his position or expand his bureau. A political conservative, the style of retrenchment was congenial to him. To Ickes, Drury held out the hope of new vigor in the Park Service bureaucracy and he symbolized the importance of wilderness preservation.[8]

Drury's appointment was well received by NPA. A *National Parks Bulletin* editorial announced the appointment and noted that "He will take over the helm at a time when the greatly expanded functions of the National Park Service make it imperative to give renewed emphasis to national park ideals so that neither the Service nor the American people will lose sight of the primary importance of the National Primeval Park System."[9] From the start of his administration Drury was open, cooperative, and on good terms with the association. This new era of cooperation is partly attributable to Drury's philosophy, but the times also encouraged a united effort. The war raged in Europe, and the United States was becoming involved in this global conflict. Wartime brought new threats to the national parks.

When the annual meeting of the NPA board convened at the Cosmos Club on May 9, 1940, several agenda items continued the business of the previous decade. Everglades National Park had been authorized in 1934 but was barely progressing toward national park status. A resolution was approved, calling for the state of Florida to expedite land acquisition necessary to establish the park. An oil exploration craze, the Everglades drainage project, and an unsympathetic governor were inhibiting progress. One trustee suggested that the size of the area necessary for formal establishment of the park should be reduced. That suggestion sparked lively debate, and the NPA president was authorized to form a committee to explore the idea.

The association then reaffirmed its support for the Antiquities Act, which was perennially challenged in Congress by congressmen who either opposed a monument or wished to amend the act to allow their pet pork-barrel project to be approved. Congress repealed the president's ability to establish national monuments by proclamation and substituted authorization to create "national recreation areas." NPA opposition to Park Service participation in "multiple land-use administration for economic purposes" was reiterated, and the board approved a resolution opposing any change in the act. In a familiar refrain, President Wharton called for study "of the broad problem of classification of the various types of National Park Service reservations." The diverse units were, he thought, becoming indiscriminately known as "national parks," and critical distinctions and standards were being lost.[10]

WARTIME THREATS
Jurisdictional Transfers and Commercial Opportunism

The executive committee met regularly during the ensuing year, working on a growing number of war-related threats to the parks. Particularly troubling were bills in Congress "To authorize the President temporarily to transfer jurisdiction over certain national-forest and national-park lands to the War Department or the Navy Department."[11] The executive committee instructed Wharton to write Ickes of NPA opposition to this proposal. Wharton described the NPA position, which would be restated many times during the war:

> Because of the small area of the National Parks in comparison with the total area of other publicly-owned lands, we cannot believe that their use for military or naval purposes is now, or is likely to become, necessary to the national defense. Furthermore, on account of the serious and lasting damage to, and perhaps destruction of, unreplaceable natural features and conditions that is sure to result from such use, we believe that only in case of dire emergency should even temporary transfer of national park land to the custody of other agencies be considered.[11]

Proposals for transferring parklands were made more than a year before the bombing of Pearl Harbor and the official entry of the United States into the war. Wharton and his colleagues knew they could not be unpatriotic, yet they also knew that unprincipled opportunists would try to enter the parks to take advantage of the crisis for personal gain, as they had during the First World War.

Newton B. Drury attended NPA's annual meeting in 1941, as he would during most of his tenure as director. At these meetings, he would describe the status of Park Service efforts on various fronts and participate in open discussion. Other Park Service directors had occasionally attended NPA meetings; Drury was more regular than any of his predecessors in his interactions with the association. At the annual meeting, the trustees approved several resolutions, one of which opposed transfer of parklands to the military. NPA's position was that

> only in case of dire necessity should such transfer of a specific park area be made, and then only after every other possible area has been investigated and shown to be unsuitable for the proposed use, and only in accordance with the recommendation of the National Park Service and the Department of the Interior....[12]

Secretary Ickes had responded positively to Wharton's letter in November, and Ickes announced a basic policy much in line with the NPA position. NPA and its conservation allies, the Park Service, and Secretary Ickes would work together to see that demands on the parks would be based on critical necessity rather than on convenience (or greed). Although wartime pressures would require some compromise with national park principles, NPA and the Park Service would work hard to minimize concessions.

The major defense mounted in 1940 and 1941 was against mining interests. NPA's National Parks News Service issued three releases urging strong opposition to efforts by Senator Carl Hayden and Representative John R. Murdock of Arizona to open Organ Pipe Cactus National Monument to mining. NPA presented opening up the park as a dangerous precedent. Mining would be precedent enough. The association really objected to the proposal that, on recommendation of the Interior Department, the designation "national monument" would be changed to "Organ Pipe National Recreation Area," which would permit mining. This less restrictive designation seemed justified, proponents argued, because the area had long been prospected and mined in a small way. NPA saw this as leading to changes in all national monuments to allow mining and other commercial uses.[13]

The "national recreation area" amendment was defeated, but the proposal to allow mining in the monument was approved. In discussing this defeat, Wharton pointed out that it did not actually establish a precedent for mining in the monuments. Proponents of mining in Organ Pipe had cited mining in Glacier Bay as precedent, and mining was also allowed in Death Valley. The door was opening, he thought, and "the position of the national monument system is precarious insofar as mining is concerned." Not only the monuments are threatened, said Wharton, but so are the "national primeval parks." Some of these were already open to limited commercial uses, among them Mount McKinley (mining), Rocky Mountain (irrigation), Yosemite (grazing), and Olympic (logging). The association should be ready, for "under war conditions, pressure for economic exploitation is constantly increasing."[14]

Director Drury spoke to the 1942 annual meeting on May 14 and talked at length about war-induced threats to national parks. Pressure from mining, grazing, and logging interests was increasing. The price of military victory would be high and, in his view, "will involve sacrifices of certain national park values. It is the responsibility of the National Park Service and conservation leaders to prepare themselves to measure the degree of such sacrifice justified by the needs of the Nation." The park most threatened at that moment, he thought, was Olympic. With the agreement of conservationists, the Park Service was fighting to keep loggers out of the park. Drury would hold to his policy "unless it is indicated by proper authority that these resources within the park are absolutely essential to the defense of the Nation, that no alternative exists and that their use is…a last resort."[14]

Drury reported that park visitation was "tapering off fast." April visits were down by 48 percent from the previous year. The Office of Defense Transportation had ordered the elimination of sight-seeing by automobile or bus. (Visitation in 1941 had reached an all-time high of more than 21 million and dropped by 55 percent in 1942).[15] Despite reduced visitation, Drury assured the group there would be no reduction in the effort to keep the system intact, to protect it from fire and sabotage, and to provide a good experience to visitors who could reach the parks. The Park Service and Interior Department were staunchly opposing a proposal to quarter enemy aliens in national park areas, he reported. Drury expressed his view that in wartime, the spiritual value of national parks was greater than ever. The trustees were reassured by all of this, and much of the meeting's business produced resolutions involving war-related threats to the parks.

William P. Wharton published an editorial titled, "The National Parks in Wartime," in the summer 1942 issue of *National Parks Magazine*. What, he asked, might be the highest contribution of national parks in wartime? Pointing out that less than 1 percent of the total area of the United States, Alaska, and Hawaii was in the "so-called 'Federal Park System,'" and that this 1 percent had been designated "for the sake of preserving greater values," he answered his rhetorical question:

> ...the highest contribution of the national parks in wartime is the same as it is in peacetime, but of even greater importance. In the wearisome tension of war and the let-down of the following peace, people turn to the fundamental and immutable things to renew their courage and their hope. In the primeval parks and monuments these fundamental and immutable things can be found singularly unchanged—an earnest [affirmation] of the continuity of life and hope. Surely these havens of refuge, tied to the eternal past, must and will be kept inviolate, to the end that not only this generation, but those which follow it, may feel the healing power and inspiration they possess.[16]

The burden of proof must be on those who would threaten such fundamental values "to show that wartime needs can be met only by use of lands inside the parks." Even as he wrote, the threat was growing in Olympic National Park.

Lumber Interests in Olympic National Park

The lumber industry had fought long and hard against an Olympic National Park and leaped at the opportunity the war presented. It would attempt to recoup some of its losses of the previous decade, claiming that national security required spruce from the park. William B. Greeley, former chief of the Forest Service, long-time park foe, and head of the West Coast Lumbermen's Association, spoke for the industry in 1943 when he wrote in the June 9, 1943, *Seattle Post-Intelligencer* that "The Olympic Peninsula National Park [sic] should do its part towards victory by giving up certain of its fine grade, old growth timber to the war effort. The principle of the draft should extend from our boys to our resources. Nothing is too sacred to do its share."[17] Carsten Lien notes that Secretary Ickes led the successful drive to create the park, then committed a fundamental error. "He left in place to administer the Olympic forests the same local Park Service crew that had been persistently working to keep the bitterly won rain forests out of the park."[18] According to Lien, this crew sympathized with the timber industry on the Olympic Peninsula, which had overcut timberlands and overbuilt lumber mills. The mills needed the big trees in the park, which the local Park Service officials were willing to provide.

Led by Greeley, the lumbermen first went after forestland in the Queets Corridor that was scheduled to be purchased and added to the park. The Park Service made stipulation agreements with landowners that land acquired for park purposes was subject to logging. Next, with Queets trees headed to the mills, the lumbermen turned to the park itself. Greeley wrote Donald Nelson, director of the War Production Board, requesting his assistance "in obtaining

a moderate supply of spruce logs from the Hoh, Queets and Quinault water-sheds...recently placed in the Olympic National Park."[19] The spruce was needed for airplane production, Greeley said.

NPA resolved in May 1942 to look at each threat to a park and determine whether national interest required the encroachment. The Olympic spruce logging did not pass the test. Spruce was available in Alaska, British Columbia, and even outside the park in the Pacific Northwest, although loggers had not spared much.[20] NPA and others suspected that landowners were holding their timber, including spruce, hoping to get trees off public land. Their privately owned trees would return more later when the tax climate of peacetime allowed more profit.[21] Finally, the strategic importance of spruce was in question. Aluminum was replacing wood in most American aircraft production by 1940.

Drury attended the NPA annual meeting in 1943. He called the threat to Olympic the most serious confronting the Park Service, and he admitted that some parklands might have to be sacrificed. Drury had visited the area and met with representatives of lumber interests, who claimed they needed the park timber to sustain their mills. Drury said he told them that sustaining mill operations was not sufficient reason to sacrifice park timber—the only reason was the winning of the war.

By this time, Drury had decided to approve logging in the Hoh extension of the park but did not share this with NPA. His recommendation had been rejected by Ickes. By June 1943, he seemed willing to sacrifice 77,045 acres of park timberland, but Ickes again rejected the idea.[22] The drive to log the park was finally stalled late in 1943. NPA did its part with its *National Parks* magazine editorial and a National Parks News Service release. Ickes knew he had the conservation community solidly behind him. Lien's analysis of the affair concludes that the Olympic raid was defeated despite Director Drury and the Park Service. Drury had reluctantly conceded that part of the park would have to be sacrificed, but Ickes disagreed. Ickes once again rescued Olympic National Park, but this time NPA was with him.

Grazing in the Parks

Other commercial interests looked for opportunities to invade the parks, and NPA invariably voiced its opposition. Ise notes that stockmen, "always patriotic in war time and anxious to do their full part to feed the nation," did their best to gain access to parklands.[23] Pressure was particularly strong in California, but Ickes and Drury, backed by conservationists, kept grazing in major parks to pre-war levels. Drury took to magazines and radio, arguing that forage in parks was too little to make much difference, that grazing damage from the First World War was so severe that some areas were still recovering, and that American men were fighting and dying for just such values as those embodied in the national parks.

National Park Service Director Drury had preservationist credentials earned in his long effort to protect the California redwoods. He is pictured here (fourth from the right) at Bull Creek Flat in 1925 with Mather and Secretary Work (to Drury's right).

Park superintendents like Colonel John White at Sequoia joined in this campaign, rejecting any grazing permits that came their way and mounting their own publicity campaigns. As Sequoia National Park historians Dilsaver and Tweed note, the struggle there was won by a coalition of the Park Service, preservation organizations, biologists, and even some tourism groups.[24] NPA reiterated its position in resolutions, *National Parks Magazine* articles, and news releases on the "burden of proof question." It reaffirmed its "unalterable opposition to all private economic exploitation of the National Primeval Park and Monument systems, whether that exploitation affect timber, forage, minerals, water or any other materials."[25]

The Park Service Wanes

NPA was also concerned about appropriations to the National Park Service, which declined dramatically as the war ground on. It could do little but publicly express support for the Park Service and call for budget levels sufficient for the agency to maintain basic levels of resource protection and other services. Appropriations dropped from $21 million for fiscal year 1940 to $5 million in 1943 and stayed there until 1947.[26] Visitation also dropped dramatically, from 21 million in 1941 to 6 million in 1942. The 1941 level was not reached again until 1946. In a resolution to the Senate subcommittee on Interior Department appropriations, the association expressed concern that proposed reductions in the 1944 appropriation "will seriously handicap the Service in carrying out its responsibilities to preserve for the future the national parks and monuments and other reservations committed to its care." It was also fearful that the cuts would damage the Park Service's ability to provide services to the armed forces and other war workers who visited the parks "in this time of strain and stress."[25] Wharton, Butcher, and others carried their concerns to Capitol Hill. Director Drury expressed his appreciation for their effort on behalf of the Park Service.

JACKSON HOLE NATIONAL MONUMENT

One war-unrelated issue at the top of the NPA agenda involved Jackson Hole and Grand Teton National Park. In the late 1930s, NPA had taken its unpopular stance against expanding Grand Teton National Park to the east. Reservoirs and sage flats were not of park quality, they argued. Along with other factors, this schism among conservationists had frustrated the Park Service's Grand Teton expansion plans at that time. Much of the land to be added to the park had been purchased by John D. Rockefeller's Snake River Land Company. Since 1930, Rockefeller had offered 32,000 acres of land to the federal government, and Congress had refused the gift.

By 1940, the prospects for an expanded park had grown dim. The Park Service began to consider presidential proclamation of a Jackson Hole National Monument under provisions of the Antiquities Act. This idea had been around since the Albright administration but had never been formalized.[27] As hope for an expanded park receded, this monument approach took on new life. On November 27, 1942 Rockefeller wrote Secretary Ickes that if the

Some of those who opposed the Grand Teton expansion contended that the sage flats in the foreground should not be included in the park.

government would not accept his land gift, he would consider other ways to dispose of it. Rockefeller knew Ickes and the Park Service reluctantly supported the national monument idea (preferring an expanded park) and probably wrote the letter to give them the leverage they needed to plead their case with President Roosevelt.[28] No one suffered illusions about the political uproar a presidential proclamation would generate. Ickes warned Roosevelt that some in Congress would see the move as a circumvention of their will and would likely move to strip the president of the power to establish national monuments.

President Roosevelt was persuaded by Ickes, and on March 15, 1943, he signed Executive Order 2578, a proclamation establishing Jackson Hole National Monument. The new monument brought 221,610 acres to the east of and adjacent to Grand Teton National Park under Park Service control. Of this, 99,345 acres were transferred from the Teton National Forest. Jackson Lake and the smaller Two Ocean and Emma Matilda Lakes comprised 31,640 acres of water surface in the monument. Rockefeller's 32,117 acres were included.

The reaction Ickes had predicted burst on the president, the National Park Service, and the monument. Wyoming's Congressman Frank Barrett introduced a bill "To Abolish the Jackson Hole National Monument." In the Senate, Pat McCarran of Nevada and Joseph O'Mahoney of Wyoming co-sponsored a bill to repeal the Antiquities Act of 1906.

The National Parks Association was faced with a dilemma. Its position against including reservoirs in national parks was unchanged, yet much of the area in the new monument should, in its view, be part of Grand Teton National Park. The Antiquities Act must be defended. The association was supporting the Park Service and its director and did not relish the prospect of opposing them openly on this issue at a critically important time for national parks. The issue generated much discussion and some disagreement among NPA leaders.

Executive committee meetings on April 14 and 24, 1943, were devoted largely to this dilemma. Concern was expressed about Jackson Lake and about the elk refuge. Would elk hunting be reduced, allowing the elk herd to grow to proportions that would damage habitat in the surrounding area, including Yellowstone National Park? Some voices urged strong public opposition to the monument, even support of the Barrett bill; others counseled caution. NPA President Wharton was dispatched to meet with Director Drury about the matter. The executive committee drafted a resolution to present to the annual meeting:

> That the Trustees of the National Parks Association regret that the Jackson Hole National Monument, including Jackson Lake Reservoir and other commercial developments, was recently created by Executive Order without public consideration. We believe that such consideration should have revealed that the area in question is not of suitable character to constitute a national monument, and that its creation may open the door for the whole area, contrary to national park standards, to be included in the Grand Teton National Park—a proposal to which this Association has been in the past, and is now, strongly opposed.[29]

Three days after the drafting of this resolution, the trustees gathered at the Cosmos Club for the 1943 annual meeting. Director Drury discussed the Jackson Hole situation. Some might disagree with him, he admitted, but he thought the area in the monument "one of the great natural areas of the country." Directly addressing the association's greatest concern, he said that although he would not have included an artificial lake in the monument for itself, he was willing to accept it to preserve more park-quality area. He urged opposition to the Barrett bill.[25]

When the business meeting convened, Wharton presented the draft resolution. Several trustees spoke against it. Trustee Theodore Palmer of the American Ornithologists Union argued that the association should take an aggressive role in resolving the long-standing disagreement between the government and the people of the area. He was not concerned about the new monument and thought the association should address the larger issues. Wharton disagreed with Palmer. The issue was maintaining national park standards, he said, and that should be the association's main concern. Trustee Harold Anthony of the National Research Council bluntly stated that the proposed resolution was anti-conservation. He did not think the association "should promote the throwing out of an area simply because man had left his imprint on it." The distinguished geologist and trustee Francois Matthes thought Jackson Hole "unquestionably qualifies as an area of great scientific interest" and was appropriately a national monument. A motion to table the resolution was "almost unanimously" approved.[25]

A National Parks News Service release in September summarized the association's position on Jackson Hole National Monument. The Barrett and O'Mahoney bills should be opposed; the Antiquities Act should be protected. The Barrett bill would only encourage "selfish interests" and do nothing to solve such Jackson Hole problems as a shrinking tax base and managing the southern Yellowstone elk herd. The monument's boundaries should be revised in the future to exclude irrigation reservoirs and other "commercial features," but in the meantime the anti-monument legislation should be opposed.[30] Another release went out when the Barrett bill was on the floor for action in June 1944. NPA had not abandoned its concerns about threats to standards involved in the Jackson Hole case, but the voices within the association urging a moderate and constructive approach to the situation had been heard.

The fall issue of *National Parks Magazine* carried an article on Jackson Hole National Monument by the eminent biologist and conservationist Olaus Murie. "Jackson Hole," wrote Murie, "is not merely a sky-piercing range of mountains for tourists to point their cameras at. It is a country with a spirit of its own."[31] This spirit may be found not only in the mountains but in the elk, the Snake River bottoms with their deer, in "a wildly calling loon swimming across the reflection of Mount Moran" in Jackson Lake. Murie eloquently made a case for the monument, even for the parts of it judged by the purists to be of insufficient value for a national park.

> Here are the sage flats that some have objected to; the "featureless, monotonous" country to some people; but others, who are sensitive to the "Spirit of Place," recognize it as an intimate part of Jackson Hole, without which this area would not have the esthetic or historic meaning it now possesses.[32]

Murie departed from the purist line NPA had followed on the Jackson Hole issue. Was this article published to present a viewpoint different from the association's? Or did it signal a shift of the association's position? Under Devereux Butcher, *National Parks Magazine* was broadening its scope. There were editorial limits to what the magazine would print (no articles were run supporting the Barrett legislation, for instance), but the pages were open to a respected and articulate conservationist like Murie, whose writing would appear in the magazine often in years to come.

After Murie's article, NPA resolutely opposed legislation to abolish Jackson Hole National Monument that Congressman Barrett repeatedly introduced. There were no more statements that standards were being compromised by the monument. When a bill was introduced in 1950 to fold Jackson Hole National Monument into an enlarged Grand Teton National Park, NPA

Olaus J. Murie argued for Grand Teton expansion in National Parks *magazine.*

lined up in support, although the park would include Jackson Lake. Murie must have persuaded the purists to mute their objections. One exception was the venerable Robert Sterling Yard who, as president and permanent secretary of the Wilderness Society, was expending his fading energy primarily on behalf of that organization. He attacked the monument and the Jackson Reservoir in the October 1943 issue of *The Living Wilderness*.[33] To the end of his life, Yard remained uncompromising in his defense of national park ideals. Struthers Burt, a dude rancher and writer who had been fighting for the park for more than two decades, responded to Yard's attack with regret that one of his most intimate and admired friends "must be becoming senile."[34] The decades-old dream of a Grand Teton National Park that contained the lakes, viewpoints, and elk bottoms of Jackson Hole, a dream held by John D. Rockefeller, Jr., Horace Albright, and Struthers Burt, among many others, was finally realized. President Harry Truman signed the act creating the expanded park on September 14, 1950.

STANDARDS AND CLASSIFICATION

Through the early 1940s, the problem of standards simmered as more pressing issues came and went. The internal NPA debate over Jackson Hole National Monument brought it to the surface. Perhaps Yard's reduced involvement resulted in a dampened fervor about the matter. Still, William P. Wharton and Devereux Butcher had not lost their conviction that clear standards for national parks were fundamentally important. The concept of "national primeval parks" as a distinct category had been floated in 1936, and Wharton and other NPA members regularly used the term in reference to Yosemite,

Yellowstone, Glacier, and others. As the Park Service expanded and its administration covered new categories of units in the system, NPA again saw need to clarify definition and classification of these new areas. It worried about diluting national park ideals as the Park Service and the public referred to diverse National Park Service-managed units as "national parks." At its 1941 annual meeting, the association called for "clear cut distinctions and definitions for different classes of National Parks, National Monuments and other reservations to bring about a better public understanding of their primary values and functions."[35]

The Park Service was then working on this problem, and a reclassification bill had been introduced in Congress. The war, however, focused the Park Service's attention elsewhere. As years passed, arguments over commercial entry in parks and monuments, a perceived increasing vulnerability of monuments to commercial invasion (Organ Pipe Cactus in particular), and the Jackson Hole situation all reminded NPA leaders of the need to address standards and classification.

At the 1944 annual meeting, Wharton pointed out that "the question of classification is closely connected with standards; we must have standards if we are to classify intelligently."[36] A clear definition was needed not only for "primeval national park" but for "national monument." Monuments, said Wharton, were a "heterogenous assortment" including such diverse areas as the Statue of Liberty, Appomattox Court House, Muir Woods, Jackson Hole, Death Valley, and Glacier Bay. A resolution was passed calling for a study of the standard and classification problems, and a committee was created to do the work.

The Committee to Study National Park Standards and Classification began meeting in October 1944. Its meeting of February 17, 1945, was attended by Director Drury, who brought with him several high-ranking Park Service officials. After a lively discussion of park and monument objectives and functions, Drury talked at length about his agency's definitions of the various categories, and of the goals of each. He addressed national parks, national monuments, historic sites, national parkways, and national capital parks. He assured the committee that "development for recreational use as usually referred to [that is, outdoor sports and activities] must be subordinated to the preservation and interpretation of significant natural and historical values." He explained that the Park Service saw its planning mission as including examination of primitive, natural, scientific, and historic areas to assess whether they are of "such outstanding national values as to justify being brought within the system."[37] The Park Service group reiterated the historic mission and standards of the Park Service and seemed to tell the committee that it need not worry; the historic standards were working. Drury concluded with hope that the committee would write a statement "along the lines" of his talk.

The committee's report was presented to the board at the 1945 annual meeting. It was based on the Camp Fire Club of America's statement that NPA and other park advocate groups had earlier embraced. The committee believed that the Camp Fire Club statement had been directed toward what

NPA could not define standards for national monuments such as Saguaro—the variety of areas with monument status was too diverse.

NPA in the 1940s was calling "primeval" parks. Its members decided they could not address the national monument problem, stating only that "These standards should apply also to National Monuments that are of similar character and purpose as the National Primeval Parks."[37]

The 1945 rewrite was not a major departure from earlier statements on standards. The principal difference was the addition of "primeval" to the description. The statement says, in effect, that there is a specific category to which the standards, developed and defended over time as "national park" standards, apply. These are distinct standards that do not apply to historic sites, parkways, and national recreation areas. They *may* apply to some national monuments.

The committee's inability to address the special problem of monuments may have reflected the divergent thinking, stimulated by the Jackson Hole situation, in NPA at the time. All could agree that some monuments would meet primeval park standards—Glacier Bay National Monument was the classic example—but all could also argue that some monuments were *not* of primeval park caliber, such as Appomattox Court House. And there were areas in between, such as Jackson Hole. In the end, the committee could say only that the standards should apply to some monuments. More work was needed to resolve the problem of monument standards.

Although most changes from the earlier Camp Fire Club version were rearrangements and rewording, several were significant. The early statement said that "wilderness features within a park should be kept absolutely unmodified."[38] The revision stated that wilderness "shall be kept unmodified except insofar as the public shall be given reasonable access to outstanding spectacles." This suggests a slight retreat from preservation, allowing exceptions. Yard never would have agreed to these exceptions, but his final illness prevented him from participating in the discussions. NPA had not retreated from defending the "primitive" in national parks, but its policy had softened to recognize that some developments, such as trails, might be appropriate.

A change in wording about wildlife reveals another shift in thinking. The original statement said that each park would be a sanctuary for "all wild plant and animal life within its limits, to the end that no species shall become extinct." The new wording read that a park would be sanctuary for plants and animals "originally within its limits, to the end that all native species shall be preserved as nearly as possible in their aboriginal state." This change reflected growing awareness of the threat exotic species brought to indigenous communities in parks. In 1941, NPA had publicly opposed proposals to introduce exotic deer species into the vicinity of the proposed Everglades National Park.[35] The association was also concerned about the Lamar bison herd in Yellowstone and elk in the Grand Teton and Yellowstone region. Preventing extinction was no longer sufficient. Concern now must extend to preserving natural communities.

The push for defining standards and reclassifying areas subsided after the committee made its report. Drury told the association he saw nothing in the report that was not in accord with the views of the Park Service.[39] The clarification sought was hardly achieved; no dramatic new directions were taken. The exercise amounted to a reiteration of the historic position of the association, supporting original ideals of preservation with minor modifications to reflect changing ideas and challenges. The report was published in the fall 1945 *National Parks Magazine* as "National Primeval Park Standards: A Declaration of Policy." Thousands of reprints of this "Declaration" were distributed.

TURNING POINTS FOR NPA
Growing Membership

The year 1945 marked a turning point for the National Parks Association. The Depression had damaged membership severely, but by 1945 membership reached an all-time high of 2,300. The association's executive always had the responsibility of raising money and recruiting members. The years during the 1930s with no executive allowed membership to drop so low that it was slow to recover. James Foote was not able to do much with this problem, but Edward Ballard and especially Devereux Butcher brought NPA back to solvency and membership strength. Between September 1943 and May 1945, Butcher increased membership from 560 to 2,300. Mailings were sent to lists of likely prospects, and the returns were excellent, even surprising. *National Parks Magazine* was published quarterly beginning in 1942, and this was an attraction. Butcher successfully recruited schools, colleges, and libraries, using the magazine as an incentive.

National Parks Magazine

National Parks Magazine was a great success. Butcher was an excellent editor, as Yard had been, but he produced a magazine that was more pleasant visually and more interesting to read than the *National Parks Bulletin*. Yard had made the *Bulletin* the mouthpiece of the association. Nearly every article in it related to a current issue or association initiative, which was necessary, and he did most of the writing. Butcher's magazine attracted more contributors, some of whom were excellent writers. The scope was broader, and readers not only could catch up on national park issues but read entertaining, well-illustrated, and informative treatments of various park-related topics. Butcher, committed to good reproduction of photographs, insisted on excellent-quality black-and-white printing of the magazine. He also recruited contributors more easily than Yard because he could assure writers that the magazine would come out on schedule and that their work would be published in a timely manner. Readers were informed of park-related legislative activity in "The Parks and Congress" section of each issue. This service began with the summer issue in 1942 and was an economical, effective way to help readers follow the complex legislative maneuvering affecting national parks. Butcher tried to schedule timely treatment of key legislative developments, but the level of park-related activity in Congress at times outstretched the magazine's resources. Nonetheless, the readers of *National Parks Magazine*, most of whom were association members, had far more information on a broad range of issues than ever before.

This growth was not without problems. At the 1945 annual meeting, Butcher told the trustees that he was overwhelmed—he and his assistant could not cope with the work. He would no longer work seven days a week; his health was suffering. An echo of Robert Sterling Yard could be heard here. Yard had many times made the same plea but had never gotten relief. Butcher simply told the board that if it wanted the association to continue to grow, provide services to members, and do the necessary political work, he had to have help. He also reported that although the association was in the black and healthier than it had been for a long time, its financial condition was fragile. There was no reserve. Publishing each magazine issue drew the association's cash to virtually nothing. A drop in membership or some other crisis might bring things to a halt. The trustees responded by forming a finance committee and assuring Butcher that they would find help for him.

Devereux Butcher was proving a worthy successor to Robert Sterling Yard. During this wartime period, he emerged as a strong executive, well organized and aggressive. His reports indicate that he cared deeply for the parks. He loved to visit parks and traveled to them to examine issues first-hand. The war made traveling difficult, and when he did travel, he often financed the trip from his own resources. Butcher worked well with Wharton, who praised his work highly at each annual meeting. Like Yard, Butcher was conscientious and did a prodigious amount of work. Yard left no record of his thoughts on his young successor. One cannot help but speculate, however, that he was pleased to see his dream of a magazine come to fruition and to see NPA under the executive leadership of a capable and committed young man who passionately loved the national parks. Butcher's son Russ recalls how his father would visit Yard at his home and the two men would discuss editorial and park policy. They would talk for hours about standards and how they should be defined and defended. Yard was Butcher's mentor, and Butcher resolved to carry on the defense of standards to which Yard had devoted himself so uncompromisingly.[40]

Robert Sterling Yard died in 1945 at age 84 after thirty years of service to NPA.

Passing Generation of Leaders

The year 1945 saw the passing of a generation of National Parks Association leaders. The first to go was Robert Sterling Yard, who died on May 17. Yard was active until the last year of his life, though he suffered his final illness for nearly a year before he died. For twenty-six years he was a stalwart in the association. At a time of life when many men think of retirement, Yard responded to the summons of his friend Stephen Mather and launched a new career. In national park work, he said later, he had found his mission, his passion.

As the story of NPA so far illustrates, Yard was a man of high energy, deep commitment, strong opinions, and active mind. He was the quintessential purist, damned by some and praised by others for his untiring adherence to the ideals of national parks as the finest examples of nature and scenery in the American landscape. He pushed NPA as far toward these ideals as he could, then helped create another organization to promote the goal of preserving wilderness outside park boundaries. Without abandoning NPA or the cause of national parks, he reached beyond them, following the lead of Bob Marshall, Aldo Leopold, and others to seek protection of wild nature outside the parks.

Perhaps because he came to park and wilderness work late in life, or perhaps because of his sometimes cantankerous personality, Yard never achieved the prominence in the conservation movement of some of his contemporaries. He was not a charismatic leader like Bob Marshall or an original thinker like Aldo Leopold. He was a worker in the trenches, an executive, a detail man. He would do the day-to-day work necessary to launch an NPA or Wilderness Society. By the time he died, he could be satisfied that he had set two important organizations on a solid path toward realizing the preservationist goals he had so doggedly espoused.

In his eulogy to Yard in *The Living Wilderness*, Horace Albright wrote of a trip that he, Mather, Yard, and others had taken to California in 1915 to confer with park superintendents. As Albright recalled, Yard said little through most of the meetings, then rose to say a few words:

> Mr. Secretary and friends, I have no business to be here, as I think you all know, for I am a tenderfoot. I have no right to be standing up here talking to mountaineers....I have been hammering the streets of New York for a good many years. It is a dozen years, at least, since I have cast a fly in any water inhabited by anything of a finny character.
>
> Nevertheless, I have got the stuff inside of me. When I am in the woods I feel closer to God than anywhere else....I have not qualified for the Rocky Mountains. But I know I shall qualify, because the qualification for the mountains, as I well know, lies inside of one, lies in the soul, and not in one's accomplishments. So it is that I, the treader of dusty streets, boldly claim common kinship with you of the plains, the mountains, and the glaciers.[41]

Yard worked for the next thirty years to substantiate this claim, and Albright testified that he achieved his goal. Summing up his view of Yard, with whom he had so often differed over the years, Albright wrote that "If ever there is a National Park Roll of Fame, or a list of National Park Immortals, Robert Sterling Yard's name will be inscribed high on the tablets.[41]

The second NPA leader to follow Yard from the scene that year was John C. Merriam, who died on October 20 at the age of seventy-six. Merriam had been an NPA trustee for twenty years and one of Yard's strongest allies. He, too, was a purist, deeply committed to the preservation of wilderness, but he took an even less pragmatic approach to that goal than did Yard. Merriam reached his peak of involvement in NPA activities in the late 1920s and early 1930s with his work for the National Conference on Outdoor Recreation and his service on the educational committees set up by NPA and the National Park Service. An eminent scientist, he played the role of philosopher during his involvement with NPA. He attempted to develop a rationale for preservation and founded that rationale on the educational and inspirational values he found in the parks. More than anyone else in NPA ranks during the 1930s and 1940s, he tried to reveal how science and poetry could come together in landscapes to lead people to a deeper understanding of nature and the human place in it. For Merriam, as for Yard, the national parks were "temples" and "museums," not playgrounds, and he worked hard to explain how this was so. Like so many other NPA leaders, Merriam was active in other related organizations. He was a founder and served as president of the Save-the-Redwoods League for twenty-four years. When his career took him to Washington as president of the Carnegie Institution, he sought fellow park advocates and found them in the ranks of the National Parks Association.

Merriam brought great intellect and prestige to the group. He and Yard shared a deep interest in what they called the "primitive." Merriam, the thinker and scholar, complemented Yard, the writer and activist, and together they greatly advanced the causes of national parks and wilderness.

The final NPA leader to die in 1945 was Henry Baldwin Ward, who passed away suddenly soon after writing a testimonial to his friend and colleague Yard for *The Living Wilderness*. President of the Izaak Walton League from 1928 to 1930, Ward had long served on the NPA board as a representative of the league. He had served as vice president for the association since 1933 and emerged as one of the most regular and active leaders during the difficult years of the Depression and Second World War. Ward had enjoyed a long, distinguished career as scientist and university professor and administrator. He seems to have retired to conservation work in 1933 and devoted himself tirelessly to the league and NPA. Although no single achievement in his history of NPA activity leaps from the record, he seems to have stood with Yard, Merriam, Wharton, and several others as the "anchors" of the association during difficult times. When finances and membership were low and the association seemed alone in its defense of standards, men like Ward worked to keep the organization alive to emerge into better times.

The first half of the decade of the 1940s saw the National Parks Association revive. A vigorous new leader, Devereux Butcher, appeared. William P. Wharton continued to provide steady leadership and carry on the tradition of purism in his approach to national park development issues. NPA allied itself more with the Park Service than it had for many years. The common cause of defending the national parks against wartime threats, combined with the more compatible leadership of Newton B. Drury, brought a period of cooperation between the association and National Park Service. New challenges loomed in the post-war period, and NPA met them with renewed strength and purpose.

154   *Guardians of the Parks*

Battles in the Post-War Period

THE NEW LANDSCAPE

The war in some ways provided relief to the national parks and to NPA. Trends in the 1930s suggested to the association that the National Park System was, in its terms, in trouble. Development was damaging park resources. Recreation was rising as a purpose for national parks. The ambitions of the National Park Service were moving the agency into new activities that distracted it from what the association thought should be its central mission—protection of the "primeval" national parks. The war stopped all of this, at least temporarily.

War brought its own threats, but with a National Park Service director who generally agreed with association views, NPA was able to work with the Park Service in a largely successful effort to minimize war-related impacts. Visitation was very low, and in some cases park resources could even recover from impacts of intensive pre-war activity. Park development virtually stopped as American attention and resources were diverted to the war effort.

When Germany and Japan were defeated, the situation changed rapidly. People returned to their parks, and a burst of domestic economic energy, which had been contained by the Depression and the war, was released. The result was a return of the previous assaults on the parks and initiatives with even greater potential for damage. The conservation movement, which had lapsed into relative dormancy during the war, responded with new energy, organization, and leadership. NPA was a part of this revival.

Tehipite Dome rises more than two thousand feet above the Tehipite Valley in Kings Canyon National Park.

Changes at the Park Service

The national parks came through the Second World War relatively unscathed, thanks in part to the vigilance of the National Parks Association and its allies. Not so the National Park Service. Its ranks were severely depleted by military service and budget cuts. Permanent, full-time positions dropped from 4,510 on June 30, 1942, to 1,974 on June 30, 1943. Headquarters was moved from Washington to Chicago in 1942 to make room for military functions.[1] Reduced visitation took some of the pressure off this smaller work force during the war, but by the time hostilities ended in August 1945, tourists were returning to national parks in large numbers. They returned far more quickly than did staff or budget dollars, stretching park staff and allowing facilities to deteriorate. Drury and his headquarters staff stayed in Chicago until 1947 and from there struggled to cope with an avalanche of post-war problems.

Harold Ickes did not last long as secretary of the Interior Department when Harry Truman became president. He clashed with the new president and left office on February 15, 1946. His remaining years were spent in typical combative fashion as a columnist for the *New York Post* and *The New Republic*. Truman replaced him at the Department of the Interior with Julius Krug of Wisconsin. Krug had served a term as chief of Wisconsin's public utilities commission, had been a staff member at the Tennessee Valley Authority, and served as chairman of the War Production Board. The Truman administration intended to give priority to water and power development, and Krug's credentials in this arena aptly suited him for this new post.[2] National park advocates were concerned that Krug knew little about that part of his new domain, and NPA's Butcher and Vice President Charles Woodbury called on him on March 26. They explained the origin and purpose of the association and presented their special focus that Everglades National Park be established as soon as possible. Krug told them that his chief concern was making the parks available to all the people, especially those with lower incomes. He thought the parks were closed to such people and favored creating a more extensive system of campgrounds throughout the national parks.[3] More promotion directed to this group was necessary, he thought. Krug did not share Ickes's special interest in the national parks; Butcher and Woodbury left the meeting anxious about their agenda.

"Land Grab"

One of the first challenges to the new Interior secretary and to NPA and other conservationists was what writer Bernard DeVoto called the "land grab." In a post-war climate of unbridled enthusiasm for economic growth and development, the stockmen, lumbermen, and miners of the West launched a new campaign against government management of natural resources. They thought they were being denied their "vested rights and interests" by a growing natural resource bureaucracy in the Forest Service and the Bureau of Land Management, created when Truman brought the Grazing Service and General Land Office together in a new agency. In Congress, led by that friend of Grand Teton National Park, Frank Barrett of Wyoming, the "land grabbers" joined under the banner of economy to cut the budgets of the bureaus that administered land and resource use.[4] For the National Park Service, this meant opposition to necessary budget increases.

Although the Bureau of Land Management and Forest Service took the brunt of this attack, the national parks were not exempt. The congressional election of 1946 brought the Republicans to power and lifted Western congressmen to influential roles, which they used to pursue reduced regulation and freer access to public resources. Bills were introduced to open the parks to mineral exploration and lumbering. In 1947, the Eightieth Congress saw five bills introduced to gain timber interests access to Olympic National Park, a bill to abolish Jackson Hole National Monument, and a proposal to reduce the size of Joshua Tree National Monument. Growing concern about communism was used to support these and other "land grab" bills. According to the argument, if free enterprise was to be built as a defense against totalitarianism, national parks were a luxury the nation could hardly afford.[5]

The push for Everglades National Park was threatened by real estate and oil operators. Across the American landscape, but particularly in the West, "federal landlordism" was under attack as an obstacle to prosperity and even democracy. DeVoto described the movement:

> A few groups of Western interests, so small numerically as to constitute a minute fraction of the West, are hellbent on destroying the West. They are stronger than they would otherwise be because they are skillfully manipulating in their support sentiments that have always been powerful in the West—the home rule which means basically that we want federal help without federal regulation, the "individualism" that has always made the small Western operator a handy tool of the big one, and the wild myth that stockgrowers constitute an aristocracy in which all Westerners somehow share.... To a historian it has the beauty of any historical continuity....[6]

DeVoto opened his campaign with this in January 1947 and continued to write about this problem until his death nearly nine years later. His writing proved an invaluable resource for conservation groups trying to protect parks and other public land.

NPA joined the Izaak Walton League, Sierra Club, and Wilderness Society in the defense against this movement. The Washington staff watched legislation carefully and described it in "The Parks and Congress" section of *National Parks Magazine*. The assault on Olympic National Park was of particular concern. At the 1946 annual meeting, Butcher reported that the timber industry was claiming a shortage of building materials for veterans housing because Olympic National Park timber was inaccessible. The Civilian Production Authority favored opening the park to logging, and Butcher met with the authority's director to explain NPA opposition. Butcher told the trustees that the U.S. Chamber of Commerce advocated opening the park and quoted its government affairs manager of the Chamber as claiming that selective logging had not harmed the national forests. "I am sure," said the Chamber of Commerce executive, "the same practice could be carried on in the park to the advantage of the commercial users of forest products and still retain all of its primitive appearance for tourists to view."[7] The association, of course, rejected this view and went on record as "emphatically in opposition" to logging in Olympic or any other part of the National Park System.

When five bills proposing the modification of Olympic Park boundaries were introduced into the Eightieth Congress, NPA vehemently opposed them all in congressional hearings, letters to key congressmen, and magazine articles and press releases. The association's new field secretary, Fred Packard, testified at a House Public Lands Committee hearing in Port Angeles, Washington, in September 1947:

> The National Parks Association is not opposed in principle to alteration of the boundaries of national parks, and has, in fact, advocated such action where it seemed desirable. The national parks are national property, and any action taken in regard to a national park is of concern to all of the citizens of the nation. The boundaries of the national park should be altered, in our opinion, only where the national interest will be served, and where the park itself will benefit from the change, and they should not be altered in response to pressure from local groups against the broader interest of the nation as a whole. In the present case, the Association believes that the proposals to withdraw the lands concerned in these bills are a result of local pressures, and that their realization would be seriously detrimental to the national good.[8]

One bill proposed to create a commission to study the forests of Olympic National Park and recommend what areas should be opened to logging. The commission would consist of five representatives from local towns, and one each from the Washington State Conservation Commission, the National Parks Association, the National Park Service, and the U.S. Forest Service. NPA opposed the proposed commission because it was clearly weighted toward the timber industry. In his testimony, Packard argued that instead of a survey to exclude forest land from the park, the restoration of cut-over land around the park should begin, and boundaries should be "rounded out" by additions.

National Parks Magazine ran articles and editorials on the Olympic park problems. Logging in Olympic, said one editorial, "could be an entering wedge leading to the eventual disintegration of the national park system in response to local demands. Accordingly, the National Parks Association is opposing all efforts to invade the park."[9] When Packard testified against the commission idea, he argued that the Park Service could do the survey, assuming it would come out defending the wilderness park. He was wrong. For various reasons, the Park Service supported the removal of 56,000 forested acres from the park.

National Park Service Director Drury was accused by his preservationist colleagues of selling out Olympic National Park but defended by them when he lost his job in the battle over Dinosaur National Monument.

This proposal, supported and even promoted by Director Drury and introduced by Washington congressmen Fred Norman and Henry M. Jackson, brought a firestorm of protest from park advocates.[10] Writing in *National Parks Magazine*, Olaus Murie recognized the dilemma of the National Park Service but opposed its position. "Is it not logical," he asked, "that the administrators would find it difficult to determine what is the public will? How much primeval land do we want? It must be a perplexing problem to know just how much boundary revision can be permitted in the public interest, without truly interfering with the true purpose of a national park."[11] For Murie, the proposed revisions constituted an attack on Olympic National Park. There was no question that "we who are defenders of the public interest and the interest of future generations in America's precious primeval parks, must earnestly resist this attack."

Olympic Park defenders, led by Rosalie Edge, Irving Brant, and Harold Ickes, fought off the attack. At the Port Angeles hearing, Assistant Interior Secretary G. Girard Davidson stated that "it is the position of the Department of the Interior that the forests of the Olympic National Park should be preserved in their natural state for the enjoyment of future generations."[12] The Park Service withdrew its report proposing transfers to the national forest. Secretary Krug consulted with his friend, Governor Wallgren of Washington, who as a congressman had sponsored the successful Olympic National Park bill; Wallgren advised him to oppose the timber interests. He proposed that if any park tracts had to be transferred to national forests, equal areas from that reserve should be added to the park.[13]

In the end, the proposals to reduce the park died in Congress. Krug promised that as long as he was Interior secretary, there would be no boundary changes to reduce any park or monument. Little harm was done except to the reputations of Drury and the Park Service.

The principal topic of discussion at the 1947 NPA annual meeting was the threat posed by land grabbers. Wharton explained why he thought the association should rally to defend the entire public land system:

It is hard enough to protect the National Parks against threats of commercial exploitation as it is. What then would happen if the surrounding federal lands were surrendered to the tender mercies of private ownership, free to use them as they choose? I believe it is imperative that the whole great system of federal reservations be maintained intact, and their resources conserved for the public benefit. National Parks are a part of that system. They will suffer if other parts are impaired.[14]

The association resolved to continue and, working with other groups, to increase its efforts to beat back this threat. Pressure on land management agencies eased somewhat, but as historian Elmo Richardson describes the situation, "Like other aspects of the 'land grab,' threats to the national parks did not die at this point in time; they merely lay waiting for a more favorable political climate."[15]

Many thought the climate would change in favor of "land grabbers" in the presidential election of 1948. Republicans would win the Oval Office, control Congress, and the "land grab" package would be delivered. Truman's upset of Thomas E. Dewey averted this threat. Western political forces would have to regroup and press their cause another day. Truman retained Krug at Interior, and park advocates breathed a sigh of relief.

Expansion at NPA

The National Parks Association was growing and operating smoothly. William P. Wharton continued as an active president and Butcher as an effective executive secretary. Membership grew, passing the 3,000 mark late in 1946. The work load increased, and the board decided in April 1946 to add another staff member to reduce the pressure on Butcher. Fred Mallory Packard was hired and began work in the fall. Because the association did not have adequate funding to ensure the position's permanence, Packard agreed to work for six months, a year, or as long as funds were available. He would be with the association until 1959.

After four years of service, Packard had come out of the Navy a lieutenant commander. He had nurtured a keen interest in conservation for many years. After he earned a degree in biology at Harvard, he landed a job with a motion picture photographer for the Pan American Union. The union was doing a film on the Pan American Highway, and Packard traveled to Guatemala and Mexico. The filming was not demanding, and he spent much time watching birds. During his bird explorations, he encountered Rosalie Edge, who was on a bird-watching vacation in Guatemala. They sailed home on the same ship, and Packard queried Edge about her work and about conservation in general.

In 1938, Packard was hired by the National Park Service as a wildlife technician. Although his specialty was birds, he was assigned to work with Adolph Murie on a study of bighorn sheep in Rocky Mountain National Park. When Murie was sent to Alaska to study wolves, Packard found himself alone on the sheep project. He worked on sheep for the next two years, earning his masters degree in biology at the University of Colorado along the way.

When his sheep work was done, he volunteered for Rosalie Edge and was rewarded with paid work in California. Edge's Emergency Conservation Committee was trying to save the South Calaveras sequoia from loggers. Packard's immediate supervisor was Aubrey Drury, brother of the soon-to-be Park Service director, and his funding was provided by Willard Van Name. Here was an ironic crossing of paths—Edge and Van Name were ardent critics of NPA, and Edge was fresh from lambasting Yard and NPA for their stand on the Kings Canyon National Park bill. When Packard later joined the NPA staff, Edge's relations with NPA improved greatly.[16]

National Parks Magazine continued to expand. Butcher devoted the July and October issues in 1946 to brief articles describing the national parks and monuments. As he told the board in May 1946, his intention was to combine these two issues and the statement on national park standards in a reprint. He believed this publication would interweave useful information on the parks and monuments—such as how to get there and where to stay—with information "to help people realize the dangers threatening the parks and monuments." This "book" would, he hoped, become a useful tool in advancing the association's work. The trustees embraced the idea. By the following spring, Butcher's book took shape. Butcher expected the book to do well and told the trustees in May 1947 that he was already planning issues of the magazine devoted to archaeological monuments of the Southwest and the national parks of Canada. He hoped to turn both of these into books as well.

Exploring Our National Parks and Monuments was published by Oxford University Press in 1947 and, as expected, sold very well. This first of eight editions featured black-and-white photographs, many by the author. The book described each park and monument, mentioning its prime attractions, principal access routes, and the superintendent's address. A bit of history and brief description of persistent problems and issues rounded out the entry for each unit. A frontispiece to the second edition, published in 1949, quoted Robert Sterling Yard and summarized the association's purpose: "The national parks and other areas of primitive wilderness with their virgin forests and their original plant life and wildlife have been bequeathed to us by the generation we have succeeded. We, too, must fulfill our trust during our time, and deliver these superb areas unimpaired to the generation following ours."[17]

Yard had dreamed of this book. He had hoped to produce it for the association in the early 1920s. His *National Parks Portfolio* had given a boost to the National Park Service and national park movement when he worked for the agency, but he never found the means to produce a similar book under the auspices of the association. Just as he realized Yard's dream of a magazine, Butcher now brought Yard's book idea to fruition, and respectfully quoted Yard at its beginning.

Fred Packard proved a capable addition to the NPA staff and was given the title of field secretary. Gradually he assumed many of the executive secretary's duties. Butcher remained the chief executive but was increasingly occupied with the magazine, his book projects, and travels to parks. On those trips he built a strong network of contacts, studied the problems facing park managers, and photographed the sights. He oversaw the budget and membership campaigns; Packard was responsible for the NPA office, legislation, and contact with Congress and federal bureaus. A document in the NPA files dated April 14, 1949, and titled, "Functional Reorganization," leaves little doubt that Butcher was relinquishing duties while trying to retain executive control. In October he told the executive committee that he was thinking of resigning as executive secretary. He did so in May 1950. Packard became executive secretary, and Butcher was named field representative and continued as editor.

Butcher's change of position required agreements regarding his book-publishing projects. He had prepared *Exploring Our National Parks and Monuments* as editor of NPA publications. This project, which the association supported, first appeared in NPA's magazine. On the other hand, Butcher had contributed much of his own time and many photographs to the project. The board agreed that NPA would retain copyright to any future editions of the

Devereux Butcher was executive secretary of NPA from 1942 to 1950 and edited National Parks *magazine from 1942 to 1957.*

book, and Butcher would be given full credit for his contributions. A separate agreement specified that two books Butcher was preparing would be marketed by the association but that Butcher would own the copyrights. Royalties would be split evenly between Butcher and the association.

Devereux Butcher gave much of himself to NPA. He took his vacations in the parks, meeting Park Service people, examining problems, and gathering photographs for his articles and books. He usually traveled at his own expense. NPA work allowed Butcher to develop as a writer and editor. Gradually he moved more toward writing, finally deciding that writing should come first. Although he was reluctant to relinquish his administrative responsibilities, he realized that he could not be both executive and writer. A choice was necessary, and he chose the latter path. Butcher continued as editor of *National Parks Magazine* until 1957 and remained a trustee of the association for several years beyond his resignation from that post.

The Butcher "administration" proved to be difficult but good years for the National Parks Association. World War II had impact on voluntary associations like NPA, on the National Park Service, and on the national parks themselves. They were lean and trying years for everyone, but Butcher teamed with William P. Wharton to provide strong, focused leadership. Butcher's membership work brought the organization back from the brink of dissolution. His extensive traveling, writing, and speaking brought the association to the attention of many, both inside and outside government. Occasionally during his years as executive, trustees suggested that the association expand its scope to embrace broader conservation concerns, but Butcher's passion was the national parks, and that is where he kept the organization's focus. This proved wise. People identified the association with national park issues more strongly than before.

The Butcher years were also marked by relatively peaceful relations with the National Park Service. The association did not always agree with Park Service policy and action, but there was mutual respect between NPA leaders Wharton and Butcher and Director Drury and his staff. Butcher knew Park Service people throughout that organization and understood their problems and dilemmas. A true devotee of park ideals in the mold of Robert Sterling Yard, he was an ardent defender of standards, though not as strident as Yard had sometimes been. As editor of *National Parks Magazine,* he enlisted some of the most eloquent conservation voices of the time such as Olaus J. Murie and Sigurd F. Olson to argue the issues. He cultivated relations with leaders of other conservation groups, and their views often appeared in the magazine.

The Butcher years were a period of cooperation during which the association was ready and willing to help on national park battle fronts, even though it seldom took the lead. In the struggle against the invasion of Olympic National Park by timber interests, for instance, Rosalie Edge and her Emergency Conservation Committee could lead while the association lobbied in Washington, published articles on the issue in the magazine, and testified at hearings. There is no way to gauge NPA's level of contribution in these fights, but Butcher and Wharton could throw the association's reputation in Washington behind an issue and help move it along.

Butcher expanded Yard's educational approach to national park advocacy and protection. Yard had believed that if people knew of the parks and monuments and understood their natural and human history, they would support them. If they knew the parks' problems, they would defend them. A major part of Yard's approach was to disseminate information to potentially interested and politically active groups. Butcher carried on this tradition with his magazine and books. He sought media exposure for the parks and the association. When the magazine achieved a regular publication schedule (it was published quarterly beginning in 1942), he solicited subscriptions from public libraries and schools and universities. He reported in 1946 that 468 public libraries and 486 schools and universities were "members" of the association. Articles such as Murie's piece on Jackson Hole were reprinted in the thousands and widely distributed. On the road, Butcher gave illustrated lectures to audiences large and small. Although little was said in NPA publications and board discussions of the association's educational mission, that mission had not been abandoned when Robert Sterling Yard moved on. Butcher carried on the educational work.

Fred Packard joined Butcher at NPA none too soon, for the work load was steadily increasing. Although the "land grab" dominated conservation politics immediately after the war, other issues and problems required NPA attention. One was the Park Service budget. The association worked for years on this, lobbying in Congress to restore staff and for funds to repair the deteriorating facilities of many parks. This campaign had little success. The park system appropriation leaped momentarily from its annual wartime level of $5 million to $26 million in 1947, only to drop back to $10 million in 1948. Director Drury attended the association's annual meetings, offered thanks for its efforts on this front, and urged it to keep the pressure on. At the 1949 annual meeting, NPA President Wharton told the trustees that lack of funding was the most serious threat to the parks. The problem would continue for years. Although appropriations for parks increased in the early 1950s, visitation increased even more, and the Park Service could not seem to catch up with its backlog of needs.

The Inholding Problem

One budget-related item of particular concern to the association was state and private inholdings in park areas. These lands within parks and monuments might be used for purposes "at variance with national park standards."[7] In 1946 there were some 609,000 acres of this land, and the association took the position that all of it should be traded for or purchased and added to the parks and monuments at an estimated cost of $20 million. The first mention of this issue in association records appears in a resolution in 1946 supporting a $350,000 appropriation to Interior in 1947 to begin the acquisition program. Toward the end of the war, the Park Service had created a Real Estate Branch within its Lands Division to survey the inholdings. By 1946 it had completed its analysis, set priorities for acquisition, and requested funds to begin the process.

Each time the Park Service made a request for land acquisition funds, strong opposition appeared in Congress in the guise of economizing. Western congressmen opposed addition of any land to federal reservations. When appropriation for inholding acquisition came up, NPA lobbied and testified in favor. Success was modest—$200,000 was appropriated in 1947, another $200,000 in 1948, and $300,000 in 1949. Drury wryly commented that this was a beginning, enough "to buy all the private lands in the parks in 100 or 150 years if prices did not rise."[18] Senator Hugh Butler of Nebraska introduced legislation to authorize $1.25 million yearly until a total of $20 million was reached, the tax impact to be offset by payment of 25 percent of revenues collected in the affected state. The association was enthusiastic in its support of this measure, which was debated in the Eightieth and Eighty-first Congresses, but it went nowhere. Despite the inadequate appropriations, the Park Service was able to reduce the private inholdings from 609,000 to 410,000 acres between 1946 and 1951.[18]

For many years the association supported an effort to create a Quetico-Superior International Wilderness in northern Minnesota and adjoining parts of Ontario. The wilderness would be in the Superior National Forest. The project was jeopardized by the development of private inholdings within the proposed wilderness such as fly-in fishing and hunting resorts. Land exchanges and acquisition were needed to halt this development. NPA joined the Quetico-Superior Council, the Izaak Walton League, and other conservationists in a bid for funds to acquire the inholdings. The Izaak Walton League raised an emergency revolving fund of $100,000 and between 1945 and 1948 purchased fourteen properties that it presented to the Forest Service. In 1946 bills were introduced in Congress for funds to complete the acquisition program, and after two years of debate in which NPA participated, the Thye-Blatnik bill (named for Minnesota's Senator Edward Thye and Representative John Blatnik) was passed authorizing an appropriation of $500,000 for land acquisition. In June 1949 the first payment of $100,000 was struck from the Department of Agriculture budget but restored to a compromise $75,000, thanks to the leadership of Senator Thye.

The association also joined a large coalition supporting a bid to control air traffic over the Superior Roadless Area. Air traffic to remote resorts was destroying the wilderness quality of this exceptional canoe country. The Department of Agriculture requested that President Truman establish an air space reservation over the area. The Department of Commerce ruled against the request on the ground that the proposed reservation was not consistent with the public policy of promoting aviation and would set a dangerous precedent. Agriculture Secretary Charles F. Brannan, joined by Interior Secretary Krug (and then Oscar Chapman, who replaced Krug late in 1949), appealed directly to the president. Sigurd Olson, a writer and wilderness advocate from Minnesota, described the situation in *National Parks Magazine* in the fall of 1949. The issue awaited Truman's decision even as Olson wrote.[19]

"December of 1949 was a fateful month in the long effort to preserve the wilderness character of the famous canoe country of the Quetico-Superior region," wrote Olson in the spring 1950 issue of the magazine.[20] President Truman had signed an executive order establishing the air space reservation—"a hard won milestone and a turning point in the national attitude toward all remaining preserves of primitive terrain," as Olson wrote. Some of those preserves were in national parks. Although NPA had not led the effort to protect the Quetico-Superior wilderness, it saw the relevance of joining that campaign even though no national park was involved. The precedent set there would be important to all wild places, no matter who managed them. NPA gained from its involvement in another way; Sigurd Olson joined the board in 1950 and remained one of its leaders for a decade.

ENCROACHING DEVELOPMENT

NPA was drawn into these issues because, in the post-war period, protecting national parks in general and wilderness or "primeval" parks in particular could not be confined to activity within park boundaries. The powerful drive to log, mine, graze, and develop the American landscape, stimulated by rapid economic expansion, threatened to invade the parks and all wild areas outside the parks. The more perceptive conservationists knew that protecting the "primeval" in nature required new strategy. Some of the NPA leadership perceived this need, Sigurd Olson in particular, and participated in this drive. The Sierra Club's David Brower and the Wilderness Society's Howard Zahniser led the effort, but NPA played a strong supportive role in the struggle for statutory wilderness that was beginning in the late 1940s.

Rapidly growing crowds in national parks after the war were another threat. Visitors flooded the parks after wartime restrictions on travel were lifted. Butcher, visiting the parks, immediately saw this problem and began documenting it. In the association's view, the difficulty was not merely one of numbers but of where people went, what they did when they arrived, and what the Park Service was doing (or not doing) to disperse the crowds and soften their impact on the park environment. Yosemite was a special concern; crowding, overdevelopment, and "nonconforming" uses were dramatic problems there. In 1946 President Wharton wrote Director Drury that "We suggest and urge that the long-time objective for correction of the present evils of serious over-crowding in Yosemite should be to remove from the valley floor all structures for housing and entertainment of visitors, and to provide the necessary accommodations on other available sites." Problems in the valley arose from "unnecessary entertainment facilities and the general prevailing atmosphere of jazz" in the valley, he stated.[21] *National Parks Magazine* ran an article on the Yosemite problem by William Colby, who had long watched the valley as a Sierra Club leader. Colby echoed Wharton's concerns, saying there was much inappropriate "jazz" and "ballyhoo," including dancing, going on in the valley. "Dancing comes closer to the borderline. Personally I do not object to it in moderation."[22]

Wharton met with Drury about the matter, but the Park Service took no action. The crowding problem grew, occupying the Park Service in coming decades and continuing as a major concern to NPA. This issue, as much as any, would lead the association into a broad search for ways to relieve pressure on the parks. It would explore wilderness preservation, new parks, recreation in national forests, even regional planning in its search for ways to reduce pressure on the parks.

Water Projects in the Parks

Park issues of long history continued to boil during this period. The Olympics, Grand Tetons, and Everglades demanded constant attention. Wyoming's Congressman Barrett repeatedly introduced bills to abolish Jackson Hole National Monument. The association's backing of the monument was solid as the debate raged, and when legislation was introduced to add the monument to Grand Teton National Park, the association was in favor. In this case, its concerns about standards had been laid to rest. The trustees thought a larger issue was finding a way to compensate the Jackson Hole community for the loss of tax revenue from the acquisition of private land for the park. Meetings with Park Service Associate Director Arthur Demaray resolved their concerns, and the association wholeheartedly supported the enlarged Grand Teton National Park. NPA celebrated when this long struggle ended with Truman's signing of the park bill in September 1950.

The Everglades situation remained complex. The act authorizing the national park had been approved by Congress in 1934, but progress toward its establishment was painfully slow. Raising funds and acquiring land were arduous processes. By 1941, only a few small isolated tracts in the proposed area of the park had been turned over to the federal government by the state of Florida. Then, in 1944, oil was discovered near the park, and land acquisition became even more difficult and expensive. Pro-park Florida Governor Spessard Holland, about to leave office, negotiated an agreement by which the state issued conditional deeds covering 847,175 acres to the federal government in return for protecting wildlife until the lands could be acquired. Congress approved a measure confirming this agreement.[23]

The Everglades park was a constant NPA meeting topic during this period. Board members feared a move was afoot to accept a park smaller than originally conceived. Drury, in fact, told one NPA meeting that the Park Service would accept a smaller park, which did not please members. The Florida legislature appropriated $2 million in 1947, and on June 20 Interior Secretary Krug signed an order establishing 710 square miles in south Florida as the Everglades National Park. This was good news, said the association in a press release, but "Association members will recognize the present area as far too small."[24] Drury told the association in January 1948 that it was a beginning, but there was much to do to make Everglades a real national park. Another news release was issued in April, urging members to lobby the House Subcommittee on Public Lands in support of a bill authorizing the acquisition of private lands within the new park. The effort for the park was intense, and Drury could report to the 1949 annual meeting that land acquisition was finally going well. The problem of oil reservation within the park was settled in 1949, and in 1950 Interior Secretary Oscar Chapman, who replaced Krug in 1950, enlarged the park from 464,000 to 1,228,500 acres.

Everglades National Park was approaching the size envisioned by proponents of the 1934 legislation. Still, 128,000 acres of private lands remained in the park, and the Army Corps of Engineers annually brought forward plans to control water flows outside the park at great peril to its natural systems. Devereux Butcher toured the new park early in 1948 and stopped in Jacksonville to discuss with the Corps of Engineers its plans for south Florida. "The fantastic array of dikes and canals proposed is reminiscent of the 'canals of Mars,'" he reported.[25] A new "primeval park" of great significance had been established, but the struggle to maintain its natural integrity was just beginning.

A Flood of Dam Proposals

In the late 1940s and early 1950s, the issue of greatest concern to the association and to the conservation movement generally involved other plans hatched, at least partly, by engineers. From its beginning, the association had been fighting plans to develop water resources in parks and monuments. After the war, this development pressure intensified. A decisive struggle was fought over an obscure and little-visited canyon-gashed national monument in the northwest corner of Colorado and northeast Utah, Dinosaur National Monument.

Dam building was an American obsession, born of beliefs that nature could and should be controlled, that water was a "resource" used to advance economic development, and that with ingenuity and engineering, water could be moved anywhere it was needed or diverted from places it was not wanted. Dams were built to control floods, produce electricity and, especially in the West, to make deserts bloom. Starting early in the century, a society was built

The effort to keep dams out of Dinosaur National Monument dominated the conservation agenda of the early 1950s.

in the American West on a kind of Christian ideal of greening the desert.[26] No one expressed this ideal more bluntly, as Wallace Stegner has said, than the Mormon hierarch John Widtsoe: "The destiny of man is to possess the whole earth; the destiny of the earth is to be subject to man. There can be no full conquest of the earth, and no real satisfaction to humanity, if large portions of the earth remain beyond his highest control."[27] Parks were just such "portions," especially by mid-century, as resource exploitation and development spread across the American landscape.

Pursuing this ideal required organization. In 1902, Congress passed a Reclamation Act and created a Reclamation Service, which metamorphosed into the Bureau of Reclamation in 1923. The aim was simple—to "reclaim" deserts of the West and transform them into gardens by catching and transporting water. This bureau joined the U.S. Army Corps of Engineers in its work of straightening, damming, and otherwise controlling American rivers. The Army Corps had been doing this for more than a century, its efforts concentrated largely in the East and Midwest. Although these two dam-building bureaucracies seldom worked in concert, they were a formidable force in the West after the Second World War. Writer Marc Reisner has called the West "a dam builder's nirvana, full of deep, narrow canyons and gunsight gaps opening into expansive basins."[28] Some of the best of these canyons, gaps, and basins, by the post-war period, happened to be in national parks and monuments.

The first post-war dam skirmish, a preview to the Dinosaur battle, involved Glacier National Park. The Columbia River flooded in the spring of 1948, causing considerable damage on the lower reaches of the river. A Bureau of Reclamation survey of dam sites in the West, especially on the Columbia River system, was in progress and the subject of a large political fight over river basin planning and management. The flood offered a stimulus to step up plans for dams on the system. First among them was a proposal for Glacier View Dam on the north fork of the Flathead River, which forms the western boundary of Glacier National Park. Local citizens enthusiastically supported the idea. Montana politicians united in support, and Congressman Mike Mansfield quickly drafted legislation to authorize construction. NPA and its allies rallied in opposition. This was another classic park invasion. The dam threatened to inundate 19,460 acres of the park, destroying some of the best wildlife foraging areas within park boundaries.[29]

The crisis was reported in *National Parks Magazine* and opposed editorially. At a hearing in Kalispell, Olaus Murie, representing both NPA and the Wilderness Society, testified against the project. Drury stated the Park Service's strong opposition. The hearing revealed divided opinion in Montana and opposition from reclamationists in other states, who were pressing for their own schemes.[30] Drury reported to conservationists in April 1949 that the dam was to be eliminated from Columbia River Basin planning. "The wide interest of all conservationists in the preservation of Glacier National Park has been most gratifying," wrote Drury. "Without it the importance of keeping Glacier National Park inviolate would not have received such recognition."[31]

NPA claimed leadership in organizing conservationists to oppose Glacier View Dam, but warned members that this was only a skirmish in a growing war. "Many headaches and the waste of great sums of money could be avoided if, once and for all, the officials of the Bureau of Reclamation and the Army Engineers would realize that the national parks and monuments have been set aside by law to be preserved as nature made them." Readers were told to watch for articles on the Bridge Canyon and Kanab Tunnel projects proposed for the Grand Canyon and big dam plans for Dinosaur National Monument. "Our struggle with the two federal bureaus is far from ended, and it is imperative to remain alert and on the job until the slate is wiped clean of park-destroying projects."[31]

NPA had been against water projects in parks since its beginning. The Falls River Basin project to dam Yellowstone Lake, originally proposed in 1920, had come up repeatedly until its final defeat in 1938. The Bechler Basin project to build a reservoir in Yellowstone's southwestern corner had been proposed many times since the 1920s. The Mining City Dam on the Green River in Kentucky, authorized in 1934, had drawn NPA attention because of its threat to Mammoth Cave National Park. Dams proposed for the Kings River had created problems for proponents of a Kings Canyon National Park. A struggle over the Colorado-Big Thompson project in Rocky Mountain

National Park had been fought and lost. In the mid-1920s, Robert Sterling Yard, flushed with victory at a defeat of Senator Walsh's plans to dam Yellowstone Lake, had looked ahead to a time when the association could rest, educate, and be free of the burdens of protecting the parks. Now, in the late 1940s, the challenge was greater than Yard might have imagined. The Grand Canyon was the target of plans for a Bridge Canyon Dam below the boundary of what was then a national monument. This dam would create a reservoir the length of the monument and upstream into the national park. A proposal to divert the Colorado River above the park through the Kanab Tunnel was seriously proposed. This idea was so outrageous that it had been squashed, at least temporarily, by order of the secretary of the Interior. Five dams that would affect Kings Canyon National Park were considered.

Elsewhere in the country, the proposed Southern and Central Florida Flood Control Project would affect flow of fresh water into the newly established Everglades National Park, and surveys were authorized for dam sites within Big Bend National Park. Finally, and of great concern for the precedent they would set, the Bureau of Reclamation proposed dams in Dinosaur National Monument—Echo Park and Split Mountain dams—as part of a massive water development scheme called the Colorado River Storage Project. This engineer's and Western irrigationist's dream would harness rivers stretching from Wyoming into New Mexico and drain parts of six states. The critical strategic battle between developers and conservationists would be fought over the Dinosaur projects.

Showdown at Dinosaur

The story of this momentous conservation struggle has been told many times and need not be retold in detail.[32] The struggle for Dinosaur is especially significant in conservation history for two reasons. First, as Stephen Fox has pointed out, the conservation movement "came of age during Echo Park."[35] Diverse parts of the movement came together as never before and "beat down powerful federal bureaus and private commercial interests."[33] Rosalie Edge said that "I am amazed at the aroused spirit of conservationists and their greatly increased numbers. Something should be done to unify the thousands who constitute a seething mass opposed to the Echo Park dam."[34] Conservationists emerged from the fight over Dinosaur a much better organized and politically stronger movement. Fox correctly attributes this strength largely to the leadership of David Brower and the emergence of the Sierra Club and Wilderness Society on the national conservation scene.

This struggle was also significant because it prompted the movement to provide *permanent* protection to such places as Dinosaur. Protection had long been an NPA goal—Yard had recognized early in the 1920s that merely reacting to attacks on the parks would not be enough to protect them. But NPA had never come close to achieving this goal. The emergence of the powerful conservation movement in the 1950s made such protection possible with a tool that had not occurred to the association's leaders, except perhaps to Yard— statutory protection of wilderness. "Instead of merely reacting defensively to attacks on wilderness," writes Fox, "after Echo Park the movement took the offensive. The shift from protective response to aggressive initiation was symptomatic of a movement suddenly stronger and more popular than ever before."[35]

Dinosaur National Monument had been established to protect an eighty-acre dinosaur graveyard. Enlarged to 209,744 acres in 1938, it included some of the West's wildest and most beautiful canyons. The 1938 proclamation contained a provision that reclamationists claimed allowed construction of dams in the monument. In the late 1940s, when the Bureau of Reclamation drafted proposals for dams there as part of the Colorado River Storage Project, the stage was set for a showdown. Drury objected to the dams. Conservationists objected at hearings in 1950. Bernard DeVoto assailed the idea in the *Saturday Evening Post*, but Interior Secretary Chapman gave his approval to Echo Park Dam. Money for the project, however, would have to come from Congress.

Members of the conservation community, including NPA, banded together in an emergency committee. They allied themselves with private power interests opposed to federal power projects, and vowed to stall congressional authorization of the Colorado River plan as long as it included Echo Park Dam. David Brower of the Sierra Club and Howard Zahniser of the Wilderness Society led the fight. Opponents testified at hearings, wrote articles, and organized float trips through the monument. *This Is Dinosaur*, a striking book edited by Wallace Stegner, was published. After five years of maneuvering, in a compromise worked out by Zahniser, a bill approving the Colorado River project passed Congress without Echo Park and Split Mountain dams and with a stipulation that the project would not violate any national park or monument.

The Role of NPA

What role did the National Parks Association play in this struggle? The association continued its long-time work, joining the coalition to fight the dams, educating with *National Parks Magazine*, testifying at hearings, lobbying on Capitol Hill. Fred Packard presented the association's position at a hearing held by Interior Secretary Chapman on April 3, 1950. NPA was opposed to dams in Dinosaur or any other unit of the national park and monument system. Any new projects on the nation's rivers should be deferred, argued Packard, until a report was submitted by the president's Water Resources Policy Commission "and there has been established a national policy regarding the uses to which our rivers should be devoted."[36] Chapman ruled in favor of the project despite considerable testimony against it.

Packard prepared a long letter to the Water Resources Commission in which the association's position was stated unequivocally:

> The National Parks Association urges that the Water Resources Policy Commission reaffirm in its final report the importance of safeguarding the national parks and monuments against any intrusion, direct or indirect, by water development construction projects. This can best be done, in our opinion, by including in the Commission's report the rec-

ommendation that it is to be the national policy that no projects shall be surveyed, authorized or constructed within the boundaries of any existing or new national park or monument, or that will affect the natural features of such reservations, except in the most dire national emergency.[37]

The association added that "similar protection for Wilderness Areas and Primitive Areas within the national forests be recommended as a national policy."[37] Packard's letter, early in the struggle, was a comprehensive statement of the coalition's general position on the matter.

In February 1951, the association convened a meeting of conservation leaders to hear the arguments of civil engineer Ulysses S. Grant III. Grant countered the arguments from Reclamation engineers that the Echo Park site was desirable because of its minimal evaporation rate; the advantage was only an insignificant 0.5 percent over alternative sites. A meeting was arranged with Chapman, who listened to Grant's arguments and responded that he would continue his investigation of the matter. Grant was later invited to be part of a task force to search for alternatives to the Echo Park site (over objections from the Bureau of Reclamation).

National Parks Magazine was an important tool in this struggle. Issue after issue carried editorials and articles explaining the problems.[38] Excellent photographs, most by Devereux Butcher, illustrated the beauty of Dinosaur's landscape. Butcher worked furiously throughout the Dinosaur battle. He and his family visited the monument in the summer of 1950. They camped on the beach in Echo Park, looked down from Harper's Corner, and toured the monument with the superintendent. Butcher was gathering photographs and ideas for the battle. He took the threat to Dinosaur to heart and, despite having several writing projects under way, he focused his field and editorial efforts for the association on the Dinosaur problem.

Grant described "Alternative Sites for Dinosaur Dams" in the fall issue of 1951, and venerable conservationist Arthur H. Carhart contributed a long article in the winter of 1952. During the struggle over Dinosaur, numerous bills relating to the issue were introduced in Congress and were tracked in the magazine's "The Parks and Congress" section. The analysis of this legislation in the magazine was extensive.

Association leaders testified in all of the major hearings on congressional bills authorizing the Upper Colorado project, three of which were introduced in the House in April 1953 and one in the Senate. In December, Interior Secretary Douglas McKay, who had replaced Chapman earlier that year, recommended to President Eisenhower the building of dams in the national monument. Packard testified for the association in March 1954. He offered NPA's position in Senate hearings again in July, and Sigurd Olson made a powerful presentation to Senate hearings in March 1955:

The founders of the national park system would be shocked to realize what is proposed in 1955, the most serious and threatening attack yet launched against these great reservations. Should Echo Park Dam be built, it will serve as a precedent that may well make it possible to construct other dams in the Grand Canyon, in Kings Canyon, Yosemite, Glacier and Mammoth Cave national parks, and others. Let no one think this danger is not real, for many of the projects have progressed beyond the blueprint stage and need only a precedent to set the new pattern for them. If Echo Park dam is built, or any other, the sanctity of the entire national park system will be endangered.[39]

Others offered technical and historical objections to the Dinosaur dams; Olson emphasized the spiritual values involved in the decision before the Senate: "A whole philosophy is endangered by this one act, an emerging concept of regard for the beauties of a primitive scene, a realization that there are certain benefits that are beyond price or practical considerations."[39] This theme would echo through Olson's writings and through a decade of discussion of wilderness and national park protection.

NPA rejoiced when the Dinosaur fight was finally won, after five years of struggle, in 1955. There would be no dam at Echo Park. It had cause to celebrate a string of victories. The Mining City Dam that had threatened Mammoth Cave was prohibited by the Omnibus Rivers and Harbors flood control bill of 1950. The Bridge Canyon and Marble Canyon dams threatening the Grand Canyon were, for the moment, beaten in Congress. The Kanab Tunnel project seemed dead, and projects proposed by Los Angeles on the Kings River were rejected. The prospect of adding Kings Canyon to the park bearing its name was alive. All of this success was surprising to some. A conservative Republican administration was in power, but the conservation movement was growing stronger by the year. "The utilitarian notion of nature for man's use gave way to a preservationist's forbearance," writes historian Fox in summing up this period in conservation history.[40]

The long struggle over Dinosaur changed the conservation movement. It brought new players, new ways of doing business, and framed new issues. Conservationists rallied as never before around Dinosaur and the sanctity of the National Park System. NPA, which had often worked quietly with little help on park issues, became one among many groups engaged in this cause. It used familiar methods of educational articles, letters, and testimony at congressional hearings, but it was part of an effort orchestrated by up-and-coming conservation leaders like Brower and Zahniser. Brower emerged as the charismatic field general, leader of a powerful coalition of conservationists. An experienced editor, he used film, books, advertisements, and public relations in new and very effective ways.

Dinosaur also brought wilderness to the top of the conservation agenda. As Fox argues, the Sierra Club and Wilderness Society became leading organizations in the struggle because they had good leaders, but also because the issue of wilderness, which was central at Dinosaur, was "the central concern of both groups."[41] The post-war development boom, the land grab, and rapidly growing technological ability to alter nature threatened to overwhelm

America's last remaining wild and natural areas. The strongest policy for protecting wild and natural areas so far established by Congress was the protection provided national parks and monuments. The time had come, thought the Sierra Club and other conservation groups, to draw the line at the parks and monuments. As Sierra Club President Richard Leonard observed, a dam at Hetch Hetchy in 1913 had been countered by the Organic Act establishing the Park Service in 1916. A dam in a national monument would be a terrible modern precedent. The battle was not only for the National Park System. "This lesson might well carry over to the wilderness areas of the national forests."[42]

The Loss of Drury

One casualty of the Dinosaur struggle was Newton B. Drury. When Interior Secretary Chapman asked him to step down as director of the National Park Service in 1951, NPA voiced strong objection. Drury had worked more cooperatively with the association than any previous Park Service director. Although NPA and Park Service had not always agreed on issues, Drury expressed his concerns openly to the association, usually in his regular visits to the annual meetings. NPA viewed Drury as a cooperator, as a bureaucrat who had many interests to balance and had done so fairly and effectively. Drury was a strong national park advocate for whom they had nothing but respect. In a long article about Drury in *National Parks Magazine*, NPA trustee Waldo Leland asserted that park defenders "have looked upon him as a stalwart defender, within the government, of the integrity of the national parks. They have recognized his honesty, his singleness of purpose, his reasonableness, and his devotion to the ideals which they themselves hold."[43] Leland thought that Drury's dismissal raised many questions about the Truman administration's priorities and about the future of the Park Service.

Chapman's public reason for asking Drury to leave was that he wanted to reward Associate Director Arthur Demaray for his long service by appointing him director before his imminent retirement. This nice sentiment fooled no one. NPA and others thought Drury had been fired for publicly opposing his boss on the Dinosaur dams. Drury and the Park Service adamantly opposed proposals for dams in the monument and said so in public hearings.[44] The Park Service was told to tone down its opposition, while the Bureau of Reclamation went strongly ahead with its promotion of the dams. Drury made his objections known to Chapman. The secretary's decision on Dinosaur "was in keeping with the utilitarian expectations of most Americans and was certainly a fulfillment of the development emphasis of the Truman administration," writes historian Elmo Richardson. "Yet Drury and his associates were acting as if near treason was involved."[45] Chapman decided Drury had to go.

The director did not go willingly. The association and others tried to save him, even taking the case to President Truman. For political reasons, Truman may have been behind Chapman's decision to approve the dams. Replying to

National Park Service directors Connie Wirth, Newton B. Drury, and Horace Albright (left to right) led the agency from the 1920s through the 1960s.

Irving Brant's warnings about trouble ahead over the Dinosaur matter, Truman said, "It has always been my opinion that food for coming generations is much more important than bones of the Mesozoic period."[46] He would offer no help. Leland, working through the National Parks Advisory Board, of which he was a member, proposed that conservationists would let the matter die if the secretary would pledge a new watchfulness over the parks and monuments. Chapman ordered that no further studies of water projects involving parks and monuments would be done without his approval, and every area under the Park Service's jurisdiction would be protected from the adverse effects of water projects. With this, the conservationists backed off, and Drury went back to California to direct the state park system.[47]

NPA promptly appointed Drury to its board and passed a resolution commending him for "his courageous and vigorous defense of the nation's heritage of unique works of nature."[48] It also assured Demaray and his Park Service associates of its desire "to cooperate fully and heartily with them in maintaining the standards and ideals which have been upheld by Director Drury and his predecessors."[48] Demaray, agreeing to postpone his retirement briefly until a new director could be named, was appointed director on Drury's departure; he served only a few months. Conrad L. Wirth, a twenty-year veteran of the Park Service, succeeded Demaray in December 1951.

GROWTH AND TRANSITION AT NPA

The National Parks Association continued to evolve and keep its eye on numerous issues while the long Dinosaur struggle unfolded. It stepped up efforts to attract new members, and Packard reported a membership of 5,400 to the 1952 annual meeting. Sales of Butcher's books were brisk. Strong new voices were heard in board discussions. Sigurd F. Olson was elected vice president in 1951. A new trustee named Anthony Wayne Smith, a labor attorney who was also chairman of the Congress of Industrial Organizations (CIO) Conservation Committee, had much to say. A larger office was necessary, and the association moved to rooms adjacent to the Wilderness Society. Olaus Murie, serving as an NPA trustee and as president of the Wilderness Society, urged a closer tie between the two groups. He told the NPA executive committee in April 1952 that "there is increasing growth of similar philosophies by the two organizations, and closer community of interest."[49]

The scope of the association's interest was wide. The topics discussed at the 1952 annual meeting illustrate this: the Echo Park Dam; invasion of national parks for economic purposes (uranium in Capitol Reef National Monument); pumacite mining in Katmai; the Buttle Lake Dam in Strathcona Provincial Park, British Columbia (opposed); Kings Canyon Dam and Grand Canyon Dam threats; the reviving "land grab"; Park Service appropriations; vandalism in the Everglades and in national parks in general; threatened reductions to the Gila Wilderness in New Mexico; too many substandard areas in the National Park System; the need for a stronger naturalist division in the Park Service; architecture in parks and monuments (faddish, mediocre, and too modernistic); and conversion of Petrified Forest National Monument to national park status (opposed). After a wide-ranging business meeting and lunch, a "discussion meeting" convened. Among those present were new Park Service Director Wirth, his assistant director, and three division chiefs. Forest Service Chief Lyle Watts and Clarence Cottam, assistant director of the Fish and Wildlife Service, represented their agencies. Howard Zahniser represented the Wilderness Society and the Sierra Club. Discussion touched many topics over four hours, including lively exchanges of questions and answers.

William P. Wharton stepped down as NPA president at the thirty-fourth annual meeting in 1953. He was seventy-three years old and had served as association president since 1935. A trustee since 1925, Wharton had been an association vice president before assuming his eighteen-year presidency. In the early 1920s when the association was struggling to survive its first years, generous financial support from Wharton, by Yard's account, had been critical to solvency and survival. Wharton had assiduously attended NPA board meetings for nearly thirty years. He had chaired national committees, most notably the Joint Committee on Recreational Survey of Federal Lands of the American Forestry Association and the National Parks Association, to the National Conference on Outdoor Recreation. Throughout much of his career with NPA, he had also served as a trustee of the American Forestry Association, providing a liaison between these two influential conservation groups. This dual allegiance caused trouble with some conservationists like Willard Van Name and

William O. Douglas, avid hiker, conservationist and Supreme Court justice, leads a hike along the Olympic Peninsula coast north of LaPush.

Rosalie Edge, who regarded the American Forestry Association as merely a conservation front for the timber industry.

Wharton had admired Yard and followed him in strong advocacy of national park standards. As Yard's voice began to fade in association discussions, Wharton carried on the tradition of "purism." Yard was the professional, Wharton the amateur and volunteer. He pursued the agenda with less intensity than Yard, partly because of temperament and partly because he lived in Groton, Massachusetts, where he maintained a law practice and thus could not devote as much time to the cause. Wharton seems to have been a low-key leader, able to work with diverse colleagues, watchful of executive management of the association, yet able to give Foote, Butcher, and Packard the freedom they needed to do their jobs. He deserves enormous credit for seeing the association through a most difficult period in its history.

Wharton could relinquish the reins with satisfaction and confidence in NPA's future. Membership stood at more than 5,000. *National Parks Magazine*, now forty-eight pages, had a circulation of nearly 8,000. Reserve funds were established and growing. Executive Secretary Packard had the Washington office running smoothly, and Field Representative Butcher traveled the country, keeping the association apprised of park problems and opportunities from first-hand observation. Relations with conservation allies were good, as they were with the National Park Service and even the Forest Service. The time seemed right for Wharton to step down, to pass the leadership burden to a younger man, Sigurd F. Olson.

Olson had joined the NPA board in 1950 and become vice president in 1951. Author, teacher, guide, ecologist, and conservation activist, Olson made his home in Ely, Minnesota. For thirty years he had traveled the Quetico-Superior country of Minnesota, mostly by canoe, and become a leading advocate of wilderness protection in that area and elsewhere. Olson published his first article in the *Milwaukee Journal* in 1925; ever since, he had been developing his skills as an outdoor and conservation writer. NPA members first noticed him in 1946 with the publication of his article, "We Need Wilderness," in *National Parks Magazine*. By 1953, this was considered a classic statement on wilderness values. His first book, *The Singing Wilderness*, had appeared in 1951.

Olson ventured from his beloved Minnesota after the war, crusading for wilderness in the Quetico-Superior and in general. He carried the wilderness banner for the Izaak Walton League and as a consultant to the Wilderness Society. He served as an advisor and consultant to President Truman's Quetico-Superior Committee and was a leader in the effort to achieve an international management agreement for the wilderness canoe country along the international boundary. A series of articles in *National Parks Magazine* informed readers of the struggles for wilderness in the Quetico-Superior. As an NPA board member, he pressed the association into the Quetico-Superior effort even though it involved no national park. In his view, the challenge of protecting nature transcended public land categories. Wilderness preservation, whether in national forests, national parks, or elsewhere, offered the best hope for the protection of nature. The board of the association seemed to agree with that view when it elected him president.

The Sigurd Olson Years

T he National Parks Association found itself adjusting to a changed world in the 1950s. The American economy was expanding, and development was threatening the nation's wilderness and parks as never before. As post-war prosperity allowed more people to buy automobiles, they were visiting national parks in ever-growing numbers. Parks were crowded, and people's expectations of their park experience were changing. They wanted more and better roads and facilities. The association continued to argue that the first responsibility of national park management should be to preserve and protect park resources.

The politics of conservation was also changing and forcing the association yet again to examine its mission. Other conservation groups were broadening their missions and reaching into NPA's traditional domain. How should it respond? A strong and coordinated conservation movement was working to create a national wilderness preservation system that embraced lands outside the National Park System. NPA had been puzzling over wilderness since the days of Merriam and Yard. Wilderness outside parks should be protected, but what might be the implications of this new system for parks? New leaders were emerging on the conservation scene. They were taking their organizations in new directions and using new tactics. NPA had to consider how its business should change in this new environment. Would the methods and missions that had served the organization well continue to do so?

Campers crowded into the national parks in the 1950s as illustrated in this scene from Yosemite National Park.

A NEW LEADER TAKES CHARGE

Sigurd Olson assumed the NPA president's chair at the thirty-fourth annual meeting, held May 21, 1953, at the Textile Museum in Washington. The tenor of the meeting was positive. William P. Wharton stepped down of his own accord after long service, and was delighted to pass the gavel to someone of Olson's stature. He was certain the future of the association would be prosperous. The budget was healthy, staff was working well, and membership had doubled during the past two years. In his remarks, Wharton praised Olson: "His great vigor and dedication to the cause assures a militant but discreet advance to much greater accomplishments than the National Parks Association has ever achieved before."[1] Wharton had given a significant portion of his life to "the cause" and could look back over virtually the entire history of the association.

Olson invited comparison with Robert Sterling Yard. Yard had not been an outdoorsman of Olson's stature, nor had he attached himself to a place in the way Olson had to his beloved canoe country of Minnesota. Yet the two men were soul mates. Yard had believed the spiritual and educational values of national parks were above all other park values. So did Olson, although he usually expressed his belief in wilderness rather than national parks. To Yard and Olson, wild nature was essential to civilization. Olson wrote in *National Parks Magazine* in 1946 that "Wilderness to the people of America is a spiritual necessity, an antidote to the high pressure of modern life, a means of regaining serenity and equilibrium."[2] Yard had expressed similar views in his park writings. Olson wrote of wilderness, a concept that transcended parks. He was a philosophical descendent of Thoreau and Muir and, more recently, Yard, Leopold, Marshall, and John Merriam. So enamored had Yard become of the wilderness idea that he moved his primary allegiance as an active conservationist to the Wilderness Society during his final years. National parks were important to him, and he continued to work for them, but the larger concept deserved more attention. Now Olson, a leading wilderness advocate, was assuming the leadership of Yard's National Parks Association.

There is another parallel between the two men. Yard had been a writer; so was Olson. Neither man was primarily a writer, but both had turned to the pen as a powerful tool to pursue their causes. And both found their voices rather late in their careers. Olson had recently achieved enough success as a writer to leave his career as an educator and devote himself full time to writing and conservation work. More than anything, he wanted to write, but he knew the wilderness and park protection cause demanded more of him. So he answered the call of the association and all of the traveling, testifying, politicking, and meeting the role required. Olson's son has written that, in his view, his father sacrificed himself in the demanding and hectic role of conservation leader when he would have preferred to live the quiet life of a writer. "He liked simple things and simple living," wrote Robert Keith Olson. "And he wanted to write about it. That's all."[3] Sacrifice it may have been, but Olson, like Yard, believed so strongly in the cause of park and wilderness preservation that he was willing to do what he could to advance that cause, even at his own expense.

Sigurd F. Olson established his conservation credentials in the effort to preserve wilderness in the canoe country of Minnesota and served as NPA president from 1953 to 1959.

"Militant but Discreet"

One difference between Yard and Olson was style. Wharton characterized Olson's style as "militant but discreet." No one ever called Yard discreet; he could be combative to a self-defeating extent. Olson was more akin to Wharton—quiet, a rather gentle man, given to thoughtfulness and occasional bursts of eloquence, who preferred diplomacy to confrontation. Wharton knew that the conservation arena in which the NPA president must work contained large egos and strong personalities. Congress was full of such people. DeVoto and Brower were of that ilk, as was Devereux Butcher. Although the association was strong and its prospects good, its leadership would require a special sort of person, and Wharton seems to have thought Olson the right man for the job.

Olson convened the business meeting, and Executive Secretary Packard reported on his activities. Like his predecessors, Packard complained of his work load (personally typing some 200 letters a week), but unlike them, he could report that relief was imminent. With the help of an anonymous donor who had offered a matching fund challenge and a successful campaign to acquire the match, more staff could be added to free him for his primary duties of researching and tracking issues and legislation, testifying, lobbying, and generally managing the association's affairs. He reported that association finances were improving and that reserves were accumulating. For the first time in its history, the association was not hanging on the edge of insolvency.

Packard had attended the third biennial Wilderness Conference convened by the Sierra Club. There he reviewed discussions of the proposed Arctic Wilderness Reserve in Alaska's Brooks Range and told the group that the quality of the area was equal to the quality of the finest national parks. It should be preserved in primeval condition. He also sat in Devereux Butcher's seat on the Advisory Committee on Conservation to the secretary of the Interior, Butcher being in the field. The legislative agenda was full, and Packard briefly reviewed the status of key issues. The Echo Park issue ground on. Bills had been reintroduced to build Bridge Canyon Dam, and a new attack was planned on Olympic National Park. Of most concern to him was the revival of the land grab, which had been defeated in 1946 and 1947. At the urging of Western grazing interests, bills had been introduced to emasculate the ability of land management agencies to regulate land use in areas under their jurisdiction. He warned the trustees that a new battle on this old front was looming. Packard was instructed to inform congressional leaders of the association's position on the matter. It "is definitely opposed to pending and suggested proposals to turn over reserved public lands to state or private ownership or control."[1]

An Expanded Annual Meeting

When Olson assumed association leadership, he decided to take more advantage of the annual meeting. Not only might business be done, but trustees could interact more with each other and with other players in the conservation struggles. After dinner on May 21, association business having been completed, Editor and Field Representative Butcher presented a slide show to trustees and guests. Butcher continued to travel throughout the National Park System, and he illustrated both the glories and problems of that system in his presentation.

The next morning, the trustees convened with representatives of eleven national organizations and leaders of key agencies for a meeting. Olson presided, and the agenda moved from national park to wildlife refuge and national forest issues. Park Service Director Wirth, Assistant Director Ron Lee, and Herbert Evison, chief of the Office of Information, led discussion of national park issues. They reviewed problems at Everglades and Olympic national parks and discussed how to help visitors gain the most from their park experiences. As usual, issues of park values and standards boiled beneath the surface of this cordial discussion. One topic pointed toward the future: What should be done in "buffer zones" around the parks? Could they be used to locate, or relocate, developments that should not be inside the parks? Could there be cooperative agreements between contiguous agencies that would decrease pressure for inappropriate development within parks?

Albert Day, director of the Fish and Wildlife Service, led a discussion of wildlife refuges, and Forest Service Chief Richard McArdle spoke about national forest issues. The chief took a pro-wilderness position, arguing that the 14 million acres in the Forest Service's wilderness system were worth more in their natural state than they would be if they were exploited for their resources. He also supported the concept of managing buffers around national parks to avoid affecting the experience of park visitors with logging and other intrusive activity.

Discussion continued all day. The meeting adjourned in late afternoon and was followed by a buffet dinner and motion pictures. One picture had been produced by the newly appointed NPA field representative in photography (whose title was soon to be changed to "director of motion pictures"), Charles Eggert. "This is Dinosaur" was a documentary on the much-contested national monument. The meeting was an unqualified success. The Sigurd Olson administration of NPA was off to a cordial start.

PROBLEMS WITH THE PARKS
"Financial Anemia"

Perhaps the greatest among many challenges facing the national parks and their friends was the lack of financial resources to meet increasing demands on park resources. Visitation rose steadily, from 37 million in 1951 to nearly 55 million in 1956, while appropriations remained about the same.[4] Park facilities were antiquated, insufficient, and in some cases, badly deteriorating. The ranger force was too small, seriously underpaid, and overworked. Butcher documented these problems in his travels and described them repeatedly to NPA trustees.

Bernard DeVoto once again raised his voice in defense of the parks. He devoted a *Harper's* column in October 1953 to the Park Service's "financial anemia." The park ranger in the field cannot, he said, tell the public that the Park Service is an "impoverished stepchild of Congress" and that congressional neglect has brought the National Park System to the "verge of crisis," but it was true. Rolling out anecdotes, facts, and figures to make his case, DeVoto argued that if Congress continued to be stingy with its purse for national parks, "the system must be temporarily reduced to a size for which Congress is willing to pay." Close Yellowstone, Yosemite, Rocky Mountain, and Grand Canyon, among others. Seal them, assign the Army to protect them, and hold them secure until they can be reopened with adequate staff and resource protection. If people cannot visit their treasures—Old Faithful, Half Dome, the Great White Throne, and the Bright Angel Trail—they will be moved to force Congress' attention to the "nationally disgraceful situation" of the national parks.[5]

NPA and others who shared DeVoto's concerns debated remedies to this park overcrowding and deterioration: increase the number of parks (but that would take money and stretch the Park Service even further); disperse the use to other areas, such as national forests, state parks, and Army Corps and Bureau of Reclamation recreation areas (all of which would take money that would not go to national parks); reduce National Park Service responsibilities, allowing it to place priority on the natural and scenic parks, turning management of lesser sites, historical areas, recreation sites, and parkways over to state, local, and even private agencies. NPA remained a haven to "purists" to whom this last idea still appealed, Devereux Butcher foremost among them. All of these measures would help, but most observers of the park situation, including NPA, believed that a significant infusion of money was necessary to rescue the system's quality.

Mission 66, controversial though it may have been in NPA circles, was popular with the ever-growing crowd of visitors and the park managers struggling to serve them.

Many factors had led the park system to this situation, National Park Service leadership included. Newton B. Drury had many strengths, but good relations with Congress was not one of them. His administration had not been able to cultivate the congressional support necessary to address the chronic funding crisis.[6] His successor, after the brief and honorary tenure of Demaray, was Conrad Wirth, a landscape architect and planner who had been a Park Service official since 1931. Wirth had long overseen the agency's Branch of Lands and had been in charge of the extensive Civilian Conservation Corps activity in national and state parks. He had developed a reputation as a good administrator who could garner the support of the wealthy and influential for the national parks. He had been a central figure in acquiring support from Paul Mellon for a major survey of the nation's seacoasts and the purchase of lands necessary to create Cape Hatteras National Seashore.

NPA and its allies were distressed when Drury left the National Park Service and uncertain about Wirth. He did not come with Drury's preservationist credentials. They feared that his civil service careerism might hamper his independence, and that he might emphasize the recreation rather than the preservation mission of the National Park Service.[6] As he settled into the director's role, however, Wirth seemed attentive to NPA and other park supporters and appeared to possess the political skills to tackle the funding problem. NPA watched him closely to see that, in doing so, he did not compromise the standards for which it stood so firmly.

Mission 66

Wirth did not succeed immediately in devising an effective strategy to address the funding crisis, but early in 1955 a scheme came to him. "Think big!" he told himself, and "Mission 66" was born. What would be necessary, he asked, to meet the current needs for park protection, staffing, interpretation, use, development, and financing? Beyond that, what was required to develop the parks so they could accommodate the anticipated numbers of visitors, and how could this all be done by the golden anniversary of the National Park Service in 1966?[7] He proposed a thorough study of the situation.

Wirth presented "Mission 66" to the NPA annual meeting in May. He explained that his approach was based on the belief that Congress would be more willing to fund parks if it understood the overall scope and objectives of the national park program, rather than be approached on a park-by-park basis. All park master plans were on the table, and studies already revealed the need for change in many of them. The trustees listened carefully and pressed Wirth on the issue of winter ski development in the parks, about which they had been concerned for several years. Wirth had authorized installing T-bar ski lifts at Rocky Mountain and Mount Rainier, a decision NPA opposed. The association promoted cross-country skiing as a more suitable winter alternative, but Wirth told them he did not think the T-bars posed much threat. Cross-country skiing was fine for the fit and experienced but would not meet broad winter recreation needs. Wirth did not think the ski development matter significant, but Butcher thought otherwise. In his field report to the meeting, Butcher stated:

This resort development trend is a threat to the integrity of the entire park and monument system. Unless we can hold the line now, Mr. Yard's words will come true: that it will be but a matter of time before we have reduced the system to the level of commercialized playgrounds, and that we will then have a national park system in name only.[8]

Butcher did not think Wirth had the correct priorities, and he and other NPA leaders resolved to watch this Mission 66 program carefully and critically.

Congress liked Wirth's idea. To the delight of NPA and other park supporters, funding for national parks increased dramatically from $33 million in 1955 to $49 million in 1956, $68 million in 1957, and $76 million in 1958. By 1965, funding would exceed $128 million.[9] The goal was to improve public support for the park system by meeting as wide a range of park user needs as possible. This could be achieved by reconstructing substandard park roads and constructing new ones, and by constructing and improving trails, campgrounds, picnic areas, parking lots, campfire circles and amphitheaters, comfort stations, visitor centers, and interpretive exhibits. Utilities and administrative and service buildings could be improved, and substandard housing for park personnel upgraded. Training centers for national park professionals could be built and training programs developed.

This ambitious agenda was guided by a statement of principles with which NPA largely agreed. The principles placed the preservation of park resources foremost, stated the goal of development to be prevention of overuse and damaging impacts, advocated information and interpretation services as a means of enhancing visitor appreciation and understanding of park values, and said concessions should be provided only in limited and appropriate ways. Wilderness areas should remain undeveloped, and inappropriate activities, such as spectator events, should be excluded.[10] Nevertheless, the plans provided for extensive construction; NPA would have to be watchful.

A year later Packard could report that he had been studying Mission 66 closely, including plans for individual areas, and they seemed good except in one regard. Some of the proposed road building seemed unnecessary and damaging. Even Butcher reported his view that "it is gratifying to comment [on] a program the Park Service has developed that is so excellently conceived and so far-reaching in its implications as to encourage the belief that the national parks and their administration are going to improve for a long time to come."[11] Reviews remained generally positive a year later, but concerns grew. One NPA resolution read, in part, that "Mission 66 provides opportunity to correct conditions that have arisen during the years when adequate funds were not available, and, if the basic philosophies expressed in Mission 66 are adhered to, it will enable the National Park Service to achieve its goal."[12]

TURMOIL WITHIN NPA
Ignoring the Standards

Public resolutions from NPA suggested that, although it was critical of some Mission 66 elements, the association was generally pleased with the way things were going. Behind the scenes, however, Mission 66 and other issues were generating one of the greatest confrontations in NPA's relatively placid organizational history. For some time, Devereux Butcher had been disenchanted with what he perceived to be NPA's inadequate response to critical national park issues. In late 1954, he had editorialized in *National Parks Magazine* that the "termites" were at work undermining national park policy. He feared that these termites, usually commercial interests, might be finding sympathetic listeners in the Eisenhower administration. Mission 66 had not yet been conceived, but Butcher wrote:

> It is enough that we build roads adequate, but no more than adequate, to enable the public to see the areas. Some parks already have too many roads. But this is no reason for causing further disfigurement with tramways, chair lifts, swimming pools, golf courses and honky-tonk, which already are available all across our country for those who want them.[13]

Butcher's reports from the field expressed growing distress over activities such as the plan to build the eye-catching Shrine of the Ages on the south rim of the Grand Canyon, inappropriate architecture in many park buildings throughout the system, the T-bar in Rocky Mountain, and road building in many parks. He was frustrated at what he believed to be an inadequate response from the association leadership to these problems.

Butcher Departs the Staff

Butcher's frustration reached a boiling point in the spring of 1957. He decided he would resign because of what he saw as an association attitude of "keep the peace at all cost" and a "growing tendency on the part of our leadership to ignore the standards." In a letter to the NPA board, he attacked Packard, Olson, and Wirth. He suggested that NPA was not militant enough, that it was being hoodwinked by Wirth into silence on critical issues, and that he, who wished to use more aggressive tactics, was being muzzled by the timidity of Packard and Olson. In his view, their desire not to alienate anyone was achieved at the expense of the park protection work they should be doing. He cited a conversation he had with Yard in 1942, in which Yard had warned, "If Wirth ever reaches the top, watch out." Now, with Mission 66, T-bars, and other developments emanating from Wirth's administration, "We are seeing today how well Mr. Yard understood." In Butcher's view "The handwriting has been on the wall in bold letters since 1953. Without seeking the opinions of others, the Service is making a headlong rush to allow commercialism into the parks....It is the duty of our Association to explain fearlessly to our membership, to our allies and to Congress what we consider to be a misuse of the parks, and that we do this even though it means criticizing the Park Service."[14]

All of this prompted a sharp response from Olson at the executive committee meeting in May 1957. He defended the NPA approach. The issue was not one of personalities, he said, but of policy. The NPA policy was to use persuasion, to try to improve understanding, and to achieve its goals without confrontation. He did not think the Park Service was "a bunch of rascals trying to bring standards down." Although NPA did not agree with the agency on everything, channels of communication must remain open. By talking with Park Service officials, "we can argue with them and plead with them, and threaten them with the only kind of action we know how to threaten with, that is public opinion." If NPA were to resort to name calling, unfair criticism, and attacks on the integrity of individuals, doors would be slammed and important opportunities lost.[15] Olson reaffirmed his approach, which was diplomacy and discretion. He wanted to build cooperative relationships with government agencies on a foundation of trust. This had been his approach since he assumed the NPA presidency.

This showdown between the militant Butcher, disciple of Robert Sterling Yard, and the more moderate Olson and Packard had been brewing for years. All three shared the same goals; their disagreement was over the most effective means to attain their ends. As a result, although the executive committee

Connie Wirth conceived and promoted Mission 66 and successfully increased appropriations to the Park Service during his tenure as National Park Service director.

refused to accept Butcher's resignation, it launched a review of staff responsibilities, created an editorial advisory committee for the magazine, ruled that reports from the field should be strictly internal documents, and asserted that the executive committee and trustees would set policy. Butcher was effectively rebuked and soon resigned as field representative and editor.

What was happening? Butcher had served the association loyally for fifteen years. Much of the time he had worked for little or no salary. His record of achievements was impressive. He had built the membership significantly during his tenure as executive secretary. Under his editorial leadership, *National Parks Magazine* had become an excellent periodical. He had represented the association well on many issues, and his books had extended the association's reach and reputation. Believing that the real story of the parks could be learned only in the field, he had traveled the parks extensively, seen scores of problems first-hand, and developed wide contacts in the Park Service and with other conservation organizations. His commitment to the cause was exceptional.

At the foundation of the dispute was the old issue of standards. Like Yard, Butcher believed there could be no compromise of standards. As Yard had believed that national parks should be protected and were not primarily for recreation, so did Butcher. As Yard had worked to articulate and defend standards, so had Butcher. And most important, as Yard had little patience for compromise, neither did Butcher. The association interacted with two National Park Service directors during Butcher's tenure as an NPA staff member. He had no problems with Drury, because Drury essentially agreed with the NPA position on standards. Wirth, however, did not. Like Albright before him, Wirth thought the parks were for people and that recreation was a major mission of the Park Service. This attitude was reflected in one of Mission 66's guidelines, which stated that "Substantial and appropriate use of the National Park System is the best means by which its basic purpose is realized and is the best guarantee of perpetuating the System."[16] In 1964 Wirth told writer Michael Frome that "My basic philosophy has been that parks are for people—for people to use and enjoy, but not with the right to destroy."[17] Butcher was a preservationist, cut from the classic "purist" mold, who believed the Park Service was going too far with its development work under Wirth's leadership. He desired a stronger organizational stance on this than his NPA colleagues were willing to support.

NPA had rewritten and reasserted its defense of standards in 1956. Butcher published them as "A National Policy for the Establishment and Protection of National Parks and Monuments" in the January 1957 issue of *National Parks Magazine*.

> National parks are spacious land and water areas of nation-wide interest established as inviolable sanctuaries for the permanent preservation of scenery, wilderness, and native fauna and flora in their natural condition. National parks are composed of wilderness essentially in a primeval condition, of areas of scenic magnificence, and of a wide variety of features. Their unexcelled quality and unique inspirational beauty distinguish them from all other areas....[18]

National monuments should "preserve specific natural phenomena of such significance that their protection is in the national interest; they are the finest examples of their kind, and are given the same inviolate federal protection as the national parks." The policy statement went on to emphasize national significance, need for adequate area, intent of Congress (preservation), and purpose of the system (protection and education). Commercial activity was judged inappropriate, as were "amusement attractions" and "mechanical intrusions." Roads should be minimal, and any buildings should blend with the environment. The policy concluded with the statement that "Any infraction of these principles in any national park or monument constitutes a threat to all national parks and monuments."

This statement was strong and specific. It reached back to the Camp Fire Club statement of the 1920s, to the work of NPA committees one and two in the early 1930s, and was consistent with positions taken on standards throughout NPA's history. Butcher believed that asserting and defending these "principles" should be the core of the NPA mission. Mission 66 was proceeding contrary to these principles and should be vigorously opposed. Butcher thought Olson and Packard were not being aggressive enough in opposing ill-conceived Mission 66 projects.

Olson, in his discussion of the matter with the NPA board, indicated that Butcher had the right to disagree with NPA policy, but that disagreement should not be public or take the form of letters like that which he had sent to the board. (Butcher even threatened to publish his letter in the magazine, but withdrew that threat.) Devereux's son Russ, who has followed in his father's footsteps as a staff member in the association, says that his father thought of Mission 66 as his downfall with NPA, believing that Wirth had friends on the board who conspired to control him and ultimately to force him out of the association's leadership. The elder Butcher had liked Olson's ideas on wilderness and was instrumental in persuading him to take on the NPA presidency. He had also encouraged Fred Packard to join the staff. Ironically, both men turned out, in Butcher's view, to be "compromisers" and overly diplomatic types. Russ Butcher believes that his father became too uncompromising, too rigid, and thus could not abide those around him who did not rise to defend standards as strongly as he did.[19]

In Russ Butcher's words, Olson and Packard were "people oriented." They believed in sitting down with their adversaries and trying to resolve their differences. One cause of Devereux Butcher's frustration in the spring of 1957 was a meeting Olson and Packard arranged with Wirth and other Park Service officials in February. The outcome, according to Devereux Butcher, was that the association agreed to stop criticizing the Park Service. Olson's response to this interpretation of the meeting describes how he preferred to do association business:

> Important things done are the result of men sitting down together, trusting each other and trusting each other's loyalty. We are working toward the same goal, and a man is not a scoundrel because he draws a government paycheck. We should be able to charge him with being wrong if he is wrong and go after him militantly...sit down and talk it over in calm spirit; get reactions, get reasons, and find out the facts. If it is impossible to get any further, one recourse is to depend on public opinion which we have done again and again. Otherwise we shall fail to achieve positive results.[15]

Insulting and attacking one's adversary was not, in Olson's view, a productive approach.

At the board meeting the day after Olson responded to Butcher's letter, Horace Albright, again a trustee, took the floor. He expressed his concern about "certain destructive criticism of Mission 66, made in public statements of the Association." Although opinions might differ, for instance, on the architectural style of Jackson Lake Lodge at Grand Teton National Park, too much criticism of it would be to "disregard the work the Rockefeller family has done to rescue Jackson Hole from exploitation." Further, he thought the fuss about ski-tows and problems in Yosemite to be overblown. The Badger Pass area was cut-over land when Mather brought it into the national park and was better off than before, even with a ski-tow. Yosemite Valley was certainly improved over "the situation prevailing in the 1890s, when cattle and sheep grazed there, dust covered the valley floor, a slaughter house was in operation, and many other such facilities, since removed under the Park Service program, existed."

Conditions were so much better that he thought objections to current development "picayune."[20] Another NPA generation was hearing from the outspoken and influential Albright, who undoubtedly saw the influence of his old adversary, Yard, on Butcher's activities. Albright was a Wirth supporter and may have been the trustee whom Butcher later believed was responsible for his ouster.

Contention Endures

Butcher's departure from staff did not end the association's work on issues involving Mission 66. The executive committee discussed the matter at length in its September meeting. William Wharton expressed concern about whether the association was "standing firm against certain recreational developments appearing in the Mission 66 program." Packard read portions of a report from the relatively new Western Field Office criticizing developments in that region, and the group discussed a letter to Wirth about Mission 66 from trustee Olaus Murie. The association had no comprehensive approach to dealing with the threats from Mission 66 and could not agree on how to mount one. Finally, the president was authorized to appoint a special committee to determine how to counter "the present trend toward artificial and incompatible amusements, toward wholly unsuitable architecture, and other developments in the national parks and monuments inconsistent with accepted national park standards."[12]

Murie wrote another long letter to Wirth about Mission 66 in January 1958. He expressed concern about the "thoughtless developing" in the country, including the parks. He was writing, he said, "to be helpful in avoiding a low mass average, and in building our park ideals on a high level."[21] NPA trustees were scheduled to meet later in the month, and in a letter to Wirth, trustee Horace Albright said he had examined Murie's letter and was ready to "'take on' Murie if you should want me to do so." Albright expressed uncertainty about whether he should remain on the NPA board, and asked Wirth to advise him on that question.[22] Albright objected to the critical tone of NPA responses to Mission 66.

Albright remained on the board and responded to the concerns of Murie and others at the board meeting on January 23. He warned trustees that they might lose sight of the larger objectives of Mission 66 in their obsession with specific problems. After all, NPA had long sought the financial support that Mission 66 had provided. He thought the Park Service was under "expert direction" and doing the right thing. Other trustees argued with Albright. B. Floyd Flickinger, a retired Park Service veteran, seemed to sum up the views of many trustees when he observed that "there are things one cannot do in church, and in the same way there are things which should not be done in national parks."[23] Association concern about Mission 66 continued to grow and would soon focus on one project—the Tioga Pass Road in Yosemite National Park.

Devereux Butcher also continued to serve on the board for several years but was bitter about his perceived mistreatment by its members. He was replaced as editor of *National Parks Magazine* by a young Californian, Bruce M. Kilgore. With degrees in wildlife conservation and journalism, Kilgore had spent summers as a seasonal ranger. Before joining the association, he had worked with The Nature Conservancy in Washington as an editorial and publicity assistant. Because editing the magazine was not a full-time job, he helped Packard with the growing executive duties of the association, particularly its promotion of the Wilderness Bill then before Congress. Some of Butcher's field duties were assumed by the association's Western office and its "staff."

C. Edward Graves of Carmel, California, had contacted the association in 1953 and volunteered his services to open a Western office. A retired librarian, Graves was an avid outdoorsman, conservationist, and photographer who, like Butcher, had traveled the national parks of western America. He had compiled picture sets accompanied by interpretive scripts that he marketed to schools and colleges under the trade name "Colorful America." Graves agreed to operate without salary and with minimal support. He would travel his region, representing the association at meetings, providing liaison with other conservation groups, and monitoring park problems.

THE PRESERVATIONIST AGENDA
The C&O Canal

While the association endured its internal stresses, it stayed focused on many park issues. The Echo Park fight continued until the dam was defeated, then became a campaign to "upgrade" Dinosaur from a national monument to a national park. Other challenges loomed closer to home. Dam builders were eyeing the Potomac River, and highway advocates had long been pushing a proposal to build a parkway on the bed of the old Chesapeake and Ohio Canal from Great Falls to Cumberland, Maryland. Commercial use of the canal stopped in the 1920s, and at the behest of Franklin Roosevelt, it was acquired by the federal government. By the early 1950s it had obtained a "patina of naturalness," a ribbon of relatively wild land 200 feet wide and 189 miles long.[24] The National Park Service thought it a good site for a scenic parkway. The association analyzed the road proposal and recommended against it. The canal offered natural, wildlife, and recreational values close to the heavily urban Washington area and should be preserved for those values. New NPA trustee Anthony Wayne Smith took a personal interest in this matter and, with Howard Zahniser of the Wilderness Society, pushed the Park Service to prepare a plan to protect the canal right-of-way.

Supreme Court Justice William O. Douglas, a strong conservationist, took up the issue and organized a hike the length of the right-of-way. Sig Olson, Anthony Wayne Smith, and Olaus Murie joined a dozen other walkers for this epic hike, which gained extensive media attention. A C&O Canal Committee was formed, with Smith representing NPA. Boosted by the power and prestige of Justice Douglas and national media attention, the campaign to protect the canal was successful. The Park Service abandoned its plans for a parkway, and

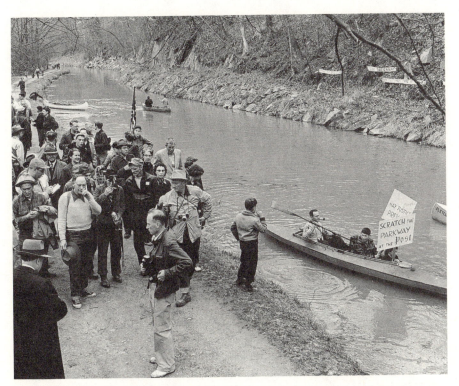

Justice William O. Douglas, here leading a hike along the C&O Canal, lent his prestige to the successful effort to preserve this historic waterway from development.

in 1961 Congress designated the Chesapeake and Ohio National Monument (changed to national historical park status in 1971). This C&O Canal episode was the first of a series of major NPA involvements in issues in the Washington, D.C., area.

Securing the Wilderness

Wilderness preservation was the major conservationist initiative at this time. NPA had fought to protect the "primeval" in national parks since it began its work in 1919. Robert Sterling Yard and John Merriam were especially interested in protecting wilderness values, and Sigurd Olson's personal crusade was wilderness preservation. Although NPA focused on parks, the center of activity for wilderness protection had shifted to the national forests. Aldo Leopold, Bob Marshall, and others had advocated protecting wild lands beyond the borders of parks; at times, wilderness preservationists had thought wilderness protection more likely outside than inside national park boundaries. The Park Service had never espoused interest in the idea of special wilderness protection, believing that its mandate of 1916 adequately defined its mission. It argued that special wilderness legislation would be redundant and unnecessary.

The Wilderness Society and the Sierra Club led the campaign for a statutory national wilderness preservation system, with NPA playing a supporting role. In this effort to achieve wilderness preservation, Sig Olson and Olaus Murie were important players, Murie primarily representing the Wilderness Society. The society and the association shared offices during much of the 1950s, and the two staffs interacted regularly. Howard Zahniser attended association board meetings, and he was the principal architect of the push for a wilderness bill. He called on Packard, Butcher, and other association leaders to help, and together they drafted and redrafted the legislation and lobbied for it. As the struggle for the wilderness bill stretched on year after year, the NPA board repeatedly and publicly supported the legislation.

NPA did not come to its supporting position on the wilderness bill without some soul searching. As late as 1956, Butcher argued that the national parks already had legislative protection directed toward preserving natural conditions and wilderness. If the provision in proposed legislation to include national parks were retained, "it might well have the effect of saying that the big tracts of wild lands are to remain intact, but let the bars down as regards developed places to the point where these eventually would become villages with all kinds of urban developments."[25] Ultimately, the association decided the gains would offset any losses and spoke strongly for the proposed wilderness system, which would include parks and other areas. The prevailing argument was that park wilderness areas would protect the majority of national parklands from development more securely: park wilderness would be protected by law.

Among opponents to the wilderness legislation were the Forest Service and the Park Service. At hearings on an early draft of the bill, Wirth worried that "the inclusion of the National Parks in a general system of wilderness areas...will have the effect of placing National Parks on a less firm foundation of protection than has already been provided by federal law."[26] The real objection may have been to the effect of wilderness designation on agency flexibility. According to Michael Frome, the agencies had broad power "not simply for administration but for determining the character and use of the land."[27] Neither agency, especially the Forest Service, wanted to lose any of these powers. In the struggle for the wilderness bill, NPA attempted to convince the Park Service that it should not oppose the legislation, that a wilderness act would be another tool it might use in addressing its mandate to "preserve and protect" park resources for future generations. The association was only marginally successful. The campaign for the wilderness proposal was protracted, involving many hearings between June 1957 and May 1964. President Lyndon Johnson signed the Wilderness Act creating a National Wilderness Preservation System on September 3, 1964.

The Glen Canyon Compromise

On November 2, 1955, the Upper Colorado River Commission formally rejected the proposal to build dams at Echo Park and Split Mountain in Dinosaur National Monument. The long battle was won at last. A new Hetch

Hetchy had been avoided. This victory was not won without cost, as NPA was aware. Back in 1949, the Bureau of Reclamation planning report that had first alerted the association to the threat to the monument had suggested several possible sites for storage reservoirs: one at Bridge Canyon near the western end of the Grand Canyon; another in Glen Canyon in the upper Colorado basin; and three sites far up-basin on the Green River in Flaming Gorge, Split Mountain, and Echo Park. The association had opposed the dams in the national monument on its long-standing principle of the inviolability of national parks and monuments. Its opposition had not extended to Glen Canyon, partly because no one knew anything about the canyon and partly because it was not in the National Park System and therefore not of concern. In fact, an argument against the Echo Park dam had sometimes been that Glen Canyon was a preferable site. Build a higher Glen Canyon dam, the argument went, and there would be no need for the Echo Park dam.

As the struggle progressed, some preservationists, including NPA leaders, realized that Glen Canyon itself was of value. At the invitation of the Sierra Club, Fred Packard had taken a trip through Glen Canyon in 1955. The group's members came out of the Canyon impressed and depressed by what they saw. In his report to the trustees, Packard wrote:

> There is no question in our mind that Glen Canyon is fully qualified for national park status. It is one of the most incredibly beautiful areas that we have ever seen, unique in character, and possessing almost infinite interest. It is completely wild, there being no settlements, houses or other developments between Hite (near Capitol Reef National Monument) where we entered, and Lee's Ferry, below the Arizona line where we came out. It is the choicest part of the vast Escalante Wilderness.[28]

Packard admitted that conservation organizations had not opposed Glen Canyon Dam, though they had worried about its impact on Rainbow Bridge National Monument. He hoped there might be some way to block the building of the dam.

The project was approved by Congress only when Echo Park was removed from the Colorado River Storage Project, new provisions stipulated no dams in the National Park System under the act, and protection was given Rainbow Bridge National Monument. Glen Canyon Dam was the centerpiece. Some conservationists, including David Brower and several NPA leaders, now wanted to look for a way to block Glen Canyon Dam but could not garner support from their organizations. Sad though the prospect of losing magnificent Glen Canyon might be, a deal had been made—no dam in the National Park System, no further opposition to the Colorado River Storage Project. NPA trustees talked the matter over. Because Glen Canyon was not a national park, NPA should not oppose the project. They did not want NPA to be accused of "being excessive in its demands beyond the proper scope."[28]

NPA fought to protect Rainbow Bridge from the rising waters of Lake Powell.

The Glen Canyon situation bothered many who had fought to defend Echo Park. They knew that, in Glen Canyon, they were about to lose a place that should have been a national park. Glen Canyon easily met all of the standards, but no one had pursued a park there. Others believed they had served the greater good in defending the National Park System. One Sierra Club leader said that if there had not been a Grand Canyon, then Glen Canyon should have been preserved. Many leaders of NPA and other organizations believed a necessary compromise had been struck and that the credibility of their organizations depended on honoring it. Richard Leonard of the Sierra Club observed that "Congress would have been convinced that the preservationists were unreasonable and were urging that the entire Colorado River be unused and just allowed to flood away into the Gulf of California."[29] Credibility was important, for there were many more issues to address. Glen Canyon Dam would be built, its floodgates closed in 1963 to begin the inundation of one of the West's most magnificent canyons, and NPA would demand that Congress honor its pledge to protect Rainbow Bridge. The Dinosaur decision would, in the end, damage the National Park System after all.

A LEADERSHIP CHANGE

In the summer of 1958, a major shake-up in organizational leadership occurred when Fred Packard left his position as executive director and was replaced by trustee Anthony Wayne Smith. Tony Smith had been discovered by Fred Packard at a meeting of the Congress of Industrial Organizations Regional Development and Conservation Committee in Chicago in 1951. Smith was the committee's executive secretary and the meeting's chairman. He expressed a desire to establish committee connections with national conservation organizations. Packard reported this to the NPA executive committee, and Smith was elected a trustee in June 1952. Born in Pittsburgh in 1906, Smith was a graduate of the University of Pittsburgh and Yale School of Law. He had served a brief stint as secretary to Governor Gifford Pinchot of Pennsylvania, had practiced law with a firm in New York City, and from 1937 had served the CIO as assistant general counsel. He was a smart, tough, and aggressive attorney who had worked for several years to interest labor in conservation matters, with some success. Smith injected himself into board discussions and emerged as a leader in issues involving the greater Washington, D.C., area such as the C&O Canal campaign and opposition to dams on the Potomac River.

A Volunteer Initiative

One of Smith's first tasks was to guide the association's sponsorship of an educational program for young people in the national parks. His handling of the matter got his administration off to a rocky start. Elizabeth Cushman, a student at Vassar College, approached the National Park Service in 1955 with the idea that "Volunteer students can help our understaffed national parks—and at the same time derive great benefit from the experience."[30] She had come up with the idea in her senior thesis and aspired to create a vehicle to help both parks and students. She proposed to set up a program that would bring high school and college students to national parks during the summer, where they could work with park personnel on projects ranging from construction and maintenance to research. The Park Service, intrigued with the idea, recommended that Cushman seek the sponsorship of the National Parks Association, which she did. Fred Packard told NPA trustees in May 1956 that he had encouraged Cushman, that she had explored the idea with park superintendents and university faculty, and that everyone was greatly impressed with Cushman's well-developed proposal. The Park Service had agreed to administer the program in the parks, and the association agreed to select participants, help with fund raising, and provide whatever assistance it could. The intent was to work toward a trial for eight weeks in the summer of 1957. The project would be called the Student Conservation Program.

Packard reported the following December that plans were moving forward for two pilot programs in Olympic and Grand Teton National Parks. Cushman and another young woman, Martha Hayne, did most of the project's work as volunteers. Funds had been received for the Olympic program but not yet for Grand Teton. (Shortly after Packard's report, Laurance Rockefeller made a grant of $7,400 to NPA to cover the Grand Teton program.) Packard was enthusiastic, telling the executive committee that here was a way NPA could help directly with the national park manpower shortage, while building "a reservoir of potential Park Service personnel and young conservationists."[31]

With financial support from many sources, including Rockefeller, The Conservation Foundation, The Old Dominion Foundation, the Wilderness Society, and other groups and individuals, Cushman and Hayne had raised $14,000 for the project. Fifty-three high school boys, college and graduate men and women from sixteen states, representing twenty-two colleges, worked at Olympic and Grand Teton national parks.[30] Cushman and Hayne, reporting on the summer experiment, summarized the park staff's response that the program "permitted the accomplishment of work that could not otherwise have been done with present personnel."[32] At Olympic the students had constructed nature trails, studied the park, worked on wildlife research, and helped organize a museum. The Teton group worked as assistants in all divisions. Packard reported to the NPA board that the experiment was an unqualified success. Although the intent had been to operate the program for one year and then, if it was successful, transfer responsibility to the Park Service, he recommended that NPA continue its sponsorship. The Park Service was not ready to assume complete responsibility for it, and Packard was so impressed by Cushman and Hayne that he believed they could successfully do nearly all of the work for a second summer. They would do so as volunteers.

The two young women again pulled the program together, raising funds from more than twenty sources, screening participants, and ensuring arrangements in Olympic and Grand Teton national parks. Interest in the experiment was widespread, with states especially interested in adopting the approach in their state parks. Newton B. Drury, now directing California state parks, expressed his intention to develop a program there. The second season was as successful as the first.

Second Thoughts at NPA

After reviewing the 1958 season, Cushman, Hayne, new staffer and editor Bruce Kilgore, and new Executive Secretary Smith met in early October to agree on how they should approach the Student Conservation Program in the future. Smith believed the program's administration in the NPA organization had been too loose and insisted it be more explicit and contractual. NPA would continue to sponsor the program, but its involvement depended on funding. Cushman and Hayne would report to Smith. The SCP would be a "department" of NPA, and the association would take 20 percent overhead from funds granted to the program. Smith would "direct" the program, and clear contractual agreements would be written. NPA would attempt to raise salaries for Cushman and Hayne, but they would continue as volunteers if salary funds were not forthcoming.

This meeting suggests that Smith had his doubts about the association's involvement with SCP. He was bent on improving NPA's administration and thought the arrangement with SCP needed reform. The problem of liability bothered him, and he was looking ahead to a legal arrangement that would

Sigurd Olson is seated third from the left at the 40th annual meeting of NPA in 1959. Devereux Butcher is directly behind Olson and Anthony Wayne Smith is seated second from the right.

protect the association and its trustees from risk. He did not think the association should take any financial risk in the matter. Still, the program was a winner. Later in the month, a letter came in from Director Wirth commending SCP and pledging continued Park Service support. Relations with Wirth and the Park Service were rocky on other fronts at the time, and this program provided a source of good-will.

Smith reported to the annual meeting in May 1959 that everything was set for another summer of SCP programs. Cushman still carried the administrative burden as a volunteer. Hayne had already moved on, and Cushman could not continue at her work level. In Smith's view this could not continue. A paid administrative staff was necessary. He recommended that the program be sponsored for one more year—through the summer of 1959—and if better administrative arrangements could not be found, that it be discontinued.

Smith had reason to continue the association's involvement with SCP for a while. NPA and other conservation organizations were having problems with the Internal Revenue Code that threatened their tax-deductible status, and "We relied heavily on the SCP in showing that our major purposes were educational in respect to our tax-exempt status, and unless and until we have an alternative educational operation in actual existence, we cannot afford to drop the SCP."[33]

At the annual meeting in May 1960, Smith reported that funds had not been raised to cover the cost of operating the Grand Teton part of SCP. He recommended that it be discontinued. Olson objected. A vote resulted in a tie, which was broken by the presiding officer (Olson) in favor of operating in the Tetons in the summer of 1960, even at cost to the association. Funding came in late and covered the cost, but the vote revealed a sharp split on the board about continuing. Smith presented a detailed report on SCP and recommended that the board decide then whether it would sponsor the program in 1961. Although he clearly thought the association should separate from the SCP, he laid out the benefits and costs to the association and insisted the board should decide. "Risks of the kind we have taken in the past in connection with the SCP have been regarded as acceptable temporarily in view of the publicity value of the Program, its importance for tax purposes, and the contributions made to the parks."[34] The tax problem had been somewhat resolved, and Smith thought the financial risks of SCP in its current form would increase.

The board decided to continue SCP, contingent again upon more solid funding. The association could not wait until the eleventh hour, as it had for several years, for financial support. If there was funding for multiple years, the association should continue its sponsorship. When Smith reported to the executive committee in November that the funding applications for SCP had been rejected, its members decided to discontinue NPA affiliation with the program.

The association never seemed sure about its involvement with the Student Conservation Program. Initially, it committed to one year, but that first year was so successful, and Packard was so enthusiastic, that it committed for another year. Smith did not share Packard's enthusiasm, but he saw practical value in continuing because of tax problems. He thought SCP drained resources that could be better used for other association work. No one denied that the program was valuable; some just did not think it should be an NPA project.

Devereux Butcher, for instance, in his angry letter to the board in 1957, wrote that "From time to time, NPA has become involved in certain 'respectable' but time-consuming projects, such as photographic exhibitions and the Student Conservation Program....With the parks needing all our attention and financial resources to defend them, we can ill-afford such diversions."[14] Smith may have shared this view. Once again, NPA struggled with its priorities, believing that education was important and part of its mission, but not comfortable committing resources to it.

The argument over the Student Conservation Program prompted the association to examine its educational mission. B. Floyd Flickinger, a retired college professor, Park Service historian, and long-time NPA trustee, proposed an ambitious educational agenda to the executive committee in March 1959. NPA, he thought, should be *the* leader in park education: should work with higher education on teacher and adult education programs about parks; should publish popular and scientific books on park matters; should sponsor a national conference on national park problems, standards, and education; and should have an education director to oversee all of this. The Student Conservation Program should be part of this educational effort and should be the responsibility of the proposed director, who would report to the executive secretary.[35]

In his report to the board on March 5, Smith raised questions about the association's educational priorities. Should it pursue the approach of the National Audubon Society, which ran summer camps that reached far more students and teachers than the SCP model? Should the NPA focus be espe-

Elizabeth Titus, founder of the Student Conservation Association under NPA auspices, accepts an award from Interior Undersecretary Nathaniel Reed.

cially on teachers because of their "multiplier" effect on many students? Should the focus be to provide educational materials to schools and colleges—as other conservation groups were beginning to do—and on slides and motion pictures about national parks? Charles Eggert had been director of motion pictures for several years, but had received little financial support from the association. "If we are to spread the gospel of the national parks, of nature preservation generally, and of the relationship between man and nature in industrial society," said Smith, "we obviously need to enlarge our operations along all these lines."[36] Flickinger agreed to work on educational program development as an unpaid consultant.

When the executive committee met a few months later, all was not well on the education front. Smith presented an education program proposal, to which Flickinger objected, accusing the executive secretary of ignoring the proposal he had offered just months earlier. Smith's plan had been drafted without his consultation, said Flickinger, and he was supposed to be the educational "consultant." Smith had analyzed the budget and concluded that Flickinger's ideas were simply beyond the means of the association, at least until it could raise substantial money for educational programming. To resolve disagreement about this, Flickinger proposed that a committee be formed to look into the question of educational priorities. Despite Smith's objection, the executive committee approved the idea.

At this point the Student Conservation Program remained an NPA enterprise, with board support, but the argument about education indicated that the program was in jeopardy. It was a pawn in a power struggle within NPA. If Smith prevailed in this struggle, SCP's association with NPA would be tenuous. Only a strong, reliable funding base would guarantee NPA support, and that was not to be during the association's phase of the Student Conservation Program's history. Cushman had presented her idea at the beginning of a rather tumultuous period in association history, and this turmoil was a factor in the demise of the SCP-NPA relationship. NPA was still a small organization (about 12,000 members in 1959) with a broad mission, and the heated argument over how to use its limited resources led to its retreat from this educational initiative.

Despite the association's misgivings, and ultimate withdrawal from SCP, it made an important contribution to education and to the parks by helping Elizabeth Cushman launch her project. NPA essentially allowed Cushman to conduct a demonstration project in two parks, and she demonstrated that her approach was very helpful to parks and students. When the association withdrew its sponsorship, SCP did not die. Under the leadership of Elizabeth Cushman Titus (she married in 1960), it incorporated and continued to place students in parks and other public land units. The organization became the Student Conservation Association in 1964, and the program has prospered for thirty-five years, placing thousands of volunteers in national parks and forests and other public lands. While advancing their education, they have rendered excellent service to natural resource managers and to the public lands. The Student Conservation Association served as a model for a federal program, the Youth Conservation Corps, which extended the approach to thousands of high school students, and for "YCC" spin-off programs of various sorts in many states.[37] Although NPA stumbled into this opportunity and was reluctant to commit to it, the assistance it provided the Student Conservation Program is one of its most significant and lasting educational contributions.

Tony Smith prevailed in this argument over how NPA could best pursue its education mission. He established a Conservation Education Center for the Greater Washington Region in 1961. Initially, its programs were mostly lectures, but as the center evolved, it included field trips and motion pictures. Focused primarily on national parks, distinguished speakers addressed audiences sometimes numbering several hundred. Rachel Carson, who had recently published *Silent Spring*, spoke in October 1962. Smith credited NPA with being the first to bring this important writer to Washington. The center continued for more than a decade, its programs reflecting the evolving interests and activities of the association. Should he need to display the educational nature of NPA's work, Smith could cite the work of the Conservation Education Center.

GROWTH AND CHANGE IN CONSERVATION

The 1950s were a period of growth for all conservation associations, NPA included. When Olson became president in 1953, there were slightly more than 5,000 members; when he stepped down in 1959, membership exceeded 12,000. Other groups, especially the Sierra Club and Wilderness Society, also grew dramatically. The issues of the decade, particularly the proposed invasion of Dinosaur National Monument and protection of wilderness, had captured the attention of the American people. Diverse groups had earlier divided up the conservation challenge into separate territories: National Audubon Society, birds; Wilderness Society, wilderness; Sierra Club, California issues. Now they came together for the Echo Park challenge. Seventeen groups, including NPA, fought the dam. As Stephen Fox notes, "In a *realpolitik* sense the movement came of age during Echo Park.... The conservationists, perhaps to their own surprise, beat down powerful federal bureaus and private commercial interest. They even influenced the cabinet politics of an unfriendly administration."[38] The conservation movement was growing stronger.

Although the association struggled internally late in the decade, the world of national park and conservation affairs was full of politics and action. Mission 66 was moving ahead, and conservationists were concerned. Ronald Foresta has described how the program was viewed by those interested in park protection: "Mission 66 was the essence of those things they detested; it was the sacrifice of preservation to mass use; it was catering to the lowest common denominator of park taste; it was the submergence of the unique qualities of the individual parks under the weight of seeming interchangeable, undistinguished development plans."[39]

On another front, Zahniser and others were working to draft and pass a wilderness bill. *National Parks Magazine* ran an article by the Sierra Club's David Brower in 1958 sharply criticizing the National Park Service stance on wilderness. It generated internal dissension; Albright, as usual, did not think the association should support open criticism of the Park Service. It also drew angry response from the agency leadership. The association reaffirmed its policy of publishing articles as "one contribution to the discussion of the wilderness policy of the Service."[40] Wirth was invited to respond in the magazine but declined.

Prompted by concerns about the increasing demand for outdoor recreation resources, the association had long supported the idea of a commission to review the nation's outdoor recreation needs. Even with Mission 66, the national parks were overrun. State and local parks were similarly under pressure, and urban areas expressed growing concern about the lack of open space and adequate planning.[41] Congress approved an Outdoor Recreation Resources Review Commission in 1958, and the group was occupied by a project that would have significant implications for the American outdoor recreation system and the National Park Service.

Away from the world of congressional and agency politics, a writer worked on a project that would change the very nature of the "conservation movement." Rachel Carson had launched her inquiry into the effects of pesticides on the environment. The fruit of her effort, the best-seller *Silent Spring*, would catapult environmental concerns into the public eye. Carson's concerns transcended national parks, and she had no involvement with the National Parks Association, but her work and other changes in the business of protecting nature would have a profound effect on its future.

Anthony Wayne Smith Takes Charge

O ne of the pervasive themes of NPA history is its constantly shifting relationship with the National Park Service. Pressures of this relationship built up and threatened the unity and focus of the association in the 1950s. Olson and Packard's approach to Conrad Wirth and his Mission 66 was attacked by Butcher, which in turn led to organizational self-examination and an internal power struggle. For many years, the association had been a strong advocate of increased funding for the National Park System and the Park Service. Ironically, Wirth's achievement of this funding goal created problems for the association by forcing it once again to consider its response to basic questions. What should be the principal purpose of national parks? Whom should parks serve, and how?

Other external forces were compelling changes in the way the association approached its mission and its work. New leaders and new tactics were changing the conservation movement. A post-war development boom increased pressure on parks and wilderness and forced a more aggressive conservationist response. Increasing affluence and mobility led to growing demand for services in the national parks. Congress responded by appropriating increased funds to the Park Service. The outdoor recreation scene was changing, sparking a national debate reminiscent of that in the 1920s. In this new scene, what role should be played by the national parks? How would this new situation affect the traditional concept of national parks, in which NPA had invested so much of its energy?

In the late 1950s, the association thus faced a daunting array of internal and external challenges. Once it had resolved leadership and organizational

The Grand Canyon became the focus of park protection struggles in the 1960s.

Anthony Wayne Smith was chief executive of NPA from 1959 to 1980.

issues, it could turn to assertion of its mission in the changing political environment in which it found itself.

NEW MODEL OF LEADERSHIP
A Strengthened Executive

The National Parks Association changed when Anthony Wayne Smith became its executive secretary. Throughout its history, the NPA president, executive committee, and board had been integrally involved in the organization's affairs. A strong executive like Yard might have ideas that would set direction, but the board always set policy and guided association affairs. Board committees and individual board members assisted the staff in addressing the NPA agenda. The president, elected by the board, was especially active, often writing important letters to officials and meeting with them, usually with the assistance of the executive secretary. All of this changed with the Smith administration.

First came the unusual elevation of a trustee to the lead staff position. This had never happened before and raises questions about why it happened now. There seems to have been some jockeying for power within the board. After the confrontation between Butcher and Olson in the summer of 1957, the board authorized Olson to appoint a Special Committee on Staff Functions. Spencer Smith was named chairman, and he appointed trustees B. Floyd Flickinger and Donald A. McCormack to join him. When they reported to the executive committee in January 1959, they proposed a strengthened executive secretary. Several positions were abolished, the executive's salary was raised to $8,400, and the appointment was specified as annual. The committee recommended that Packard continue but be reviewed carefully to ascertain if he was suited to the newly defined role. The executive committee also recommended that its size be reduced from fifteen to seven members, a change approved at the full board meeting in May.

The special committee continued its scrutiny of Packard and, at a special meeting of the board on June 25, recommended that he be relieved. A search began immediately for a replacement. A special committee composed of trustees McCormack and Michael Hudoba was created to review applicants; one month later, they reported to the executive committee that they had "put out feelers" for applicants and had received one application they had not yet carefully reviewed. In a letter to the executive committee, McCormack said, "I believe the man you are considering, Mr. Anthony W. Smith, meets the requirements we have thought are needed for this job....why look for something you have already found?"[1] The newly streamlined executive committee offered the position to Smith.

Smith was not only a member of the board but of the executive committee throughout. He excused himself from deliberations about his suitability for the post, but Smith was certainly well positioned to affect the outcome. He accepted the appointment, but with stipulations that raised concerns with Olson, who recommended (in absentia) to the executive committee in September that it not draft a contract with Smith at that time. The committee did so anyway. The contract specified that Smith's salary would be $12,500, rather than the $8,400 agreed to nine months earlier, and that he would be appointed for five years with an automatic five-year renewal.

Why was Packard relieved of his executive duties? His days were numbered when the Special Committee on Staff Functions was appointed in 1957. Sigurd Olson was a strong Packard supporter, but some trustees were concerned about Packard's management of the association. In 1957 William Wharton wrote Charles Woodbury that matters had reached a point at which he felt compelled to share his concerns about Packard with the president. Packard's planning had become unsystematic, and in Wharton's opinion, he and the NPA operation were disorganized. The budget was not being well managed, and he feared this might lead to trouble.[2]

Woodbury replied that he, too, was concerned, and something must be done. Wharton shared his concerns with Olson, who went immediately to Packard. Later in January 1957, Wharton wrote Butcher that Packard "must realize that he is, to some extent, on trial, and let us hope he will concentrate on matters of primary importance."[3] The situation did not improve. In August, Wharton wrote Woodbury that he was hearing from several quarters "that the NPA fails to stand up and take the lead in opposing continued introductions of artificial amusements which help to build up the number of visitors who wouldn't otherwise be attracted to them. In this and other respects, these people feel that the NPA is becoming a stooge of the NPS."[4] Butcher's concerns that NPA was "soft" on the Mission 66 issues were shared outside the organization. Some trustees thought Packard primarily responsible for this, although Olson was given some of the blame.

Discontent with Packard continued to grow. Olson told the 1958 annual meeting that he would serve one more year as president, and Wharton, Woodbury, and other trustees worried about a successor. They did not think Spencer Smith, the vice president, should move into the presidency because, as a professional lobbyist in Washington, he would have a conflict of interest. No one else among board members seemed suited for the role. Newton B. Drury, retiring as director of California state parks, was recruited, but he declined. Ultimately, the board convinced Olson to stay another year, but even that process brought Packard trouble. He sent a memo to the nominating committee under Olson's name suggesting a slate of officers with Olson as president, giving the appearance of Olson renominating himself. He retracted this, but the damage was done—trustees did not think he was doing the job. They asked him to resign.

The reasons for Packard's fall from grace are not clear in the record. Butcher's public revolt seems to have shaken some trustees' confidence in the NPA leadership team. The situation provided an opportunity to those on the board who wanted a leadership change, Anthony Wayne Smith and Woodbury among them. A new standard for conservation leadership was being set by the Sierra Club's David Brower and others, and some may have agreed with Butcher that more aggressive NPA leadership was necessary. There is no confirmation in association records of the financial difficulties claimed by some Packard critics.

New Executive Secretary Smith wasted no time in clearing his path of obstacles. He charged that Packard had misrepresented himself to a potential donor to the Student Conservation Program as "director" of that program and recommended to the executive committee that Packard, who was to be assigned other association duties when he stepped down, be separated entirely from NPA. Olson protested that he had authorized Packard to do this, but the committee authorized Tony Smith "to inform Mr. Packard that he will be given no further duties as a staff member and is not entitled to act as a representative of the NPA."[5] Olson's attention was drawn increasingly to his writing, and although he remained NPA president until the following May, Smith had become the power in the organization.

Restructuring

The new executive secretary moved immediately to make changes. He reworked the relationship with the Student Conservation Program. He decided to change the size and format of the magazine and increase it from six to twelve issues annually. He would change the dues structure and mount an aggressive membership drive. The Western Field Office would be closed, and he would create "action groups" around the country to harness the grassroots resources of the association. NPA's approach to its mission would be adjusted to comply with new tax laws, and it would undertake more education to solidify its status as a nonprofit educational organization.

Sigurd Olson officially retired from the NPA presidency at the board meeting of May 21, 1959, and Victor Cahalane was elected to succeed him.

Trustee Victor H. Cahalane succeeded Sigurd Olson as NPA president in 1959 but resigned because of a power struggle with Anthony Wayne Smith.

Cahalane was a highly respected wildlife biologist who had been chief of the Biology Branch of the National Park Service for many years. When he retired from that post in 1955, he became assistant director of the New York State Museum. An active conservationist, he had served many organizations in addition to NPA. Spencer Smith was reelected vice president, B. Floyd Flickinger secretary, and Donald A. McCormack treasurer. Executive Secretary Smith was soon at odds with Flickinger over educational programming. He went his way without consulting Flickinger, whom he had pledged to involve in educational matters.

Open conflict between Tony Smith and trustees erupted at the board meeting in November. A motion was offered to authorize Cahalane to contact other conservation organizations about all of their shared tax problems. Smith objected, arguing that NPA should not be seen as a leader in any effort to change IRS rules. Further, this was his responsibility anyway, as stipulated in his contract, and only he should be making such contacts. The motion was defeated. At this point, some of the trustees expressed their view that the executive committee was too close to the executive secretary and that the by-laws should be amended to ensure greater separation and clarify that policy was to be made by the board. Smith again objected, arguing "that there cannot be two executive authorities in this Association if it is to function."[6] Smith won again, and the question was referred to the executive committee over the objections of William P. Wharton and several others of the "old guard."

Another exchange of letters between Wharton and Woodbury in 1959 reveals the internal struggle that occurred for control of the association. Wharton complains that he and other trustees do not receive minutes of the executive committee meetings. Tony Smith's excuse is that he fears "leaks" of confidential information. Smith also proposes by-law changes that Wharton thinks will give him too much power:

> There are several changes in the proposed By-laws which seem to me undesirable, especially the wording which would give the Executive Secretary (or is it Director?) complete authority to do as he wants— the contract which was made with him went altogether too far in that direction.[7]

Wharton thought the by-laws should stay as they were, although he knew that Smith was using language in his contract to pursue his agenda beyond the bounds stipulated in the by-laws.

Woodbury offered a defense of Tony Smith:

> One or two things should be remembered here...the feeling was general that we should, for heaven's sake, get some one in as executive secretary who could and would be a competent executive, and who would get the organization past the "playing everything by ear" period and bring us back before disaster to financial and administrative soundness.
>
> Another thing that should be remembered, Will, is that Tony didn't seek the job. The office sought the man.[8]

Woodbury told Wharton that he hoped the infighting would end soon; Wharton responded that he hoped so too but "It is hard to see how the 'fighting' within the Association can be wholly ended while Tony is in the saddle, and is so obstinate in his approach."[9]

As an executive committee meeting approached in April, Wharton wrote Woodbury that he thought Tony Smith had taken complete command of association affairs. Spencer Smith was his ally, and Tony Smith had decided that the president had no authority between board meetings. The chairman of the executive committee, Spencer Smith, was "the man to deal with," in Tony's words. The two Smiths had "just brushed Vic [Cahalane] aside....Spence and Tony have had a more or less hand-picked Executive Committee, which is not really representative of the Board. What is now needed, in my opinion, is an Executive Committee which respects the Board and works with it, and a President who can be chairman of the Executive Committee and 'exercise general supervision over the affairs of the Association,' as provided in Article V, Section 2 of the present By-laws."[10]

All of this came to a head early in 1960. At the executive committee meeting in February, Cahalane protested that the executive secretary had gone beyond his instructions and his authority in contacting prospective board members. Cahalane openly accused Smith of misusing his position to create a board sympathetic to his position. Smith was also, said Cahalane, promoting by-law revisions that would solidify his authority, and fifteen trustees were opposed to his proposals. Cahalane could not, however, defeat Smith, who

Trustee Charles G. Woodbury was an active NPA trustee for more than two decades.

had rallied support among the other trustees. Finally, Cahalane had enough. At the April 19 executive committee meeting, he resigned as president. Spencer Smith became acting president, and on a technicality, a new nominating committee was formed to replace one appointed earlier by Cahalane. Smith's committee nominated a slate of association officers, headed by Clarence Cottam as president, and they were elected. Cahalane, supported by Devereux Butcher and other trustees, made a last-ditch attempt to elect members of their faction to the executive committee, but failed. Tony Smith had won and was now the power of the association. Cahalane dropped off the board, and Butcher soon followed. Olson and Wharton remained trustees for four more years but wielded little power or influence.

The End of Amateurism

This episode marks the change in NPA's approach to business with the rise of Anthony Wayne Smith. From this point on, Smith was its spokesman. In his reports, he would tell the board that it should make a particular decision, but invariably he set policy, and the board went along. Throughout the history of NPA, the board had been "activist" and carried out some of the association's business. No longer. Tony Smith and his staff would do all the business. Staff was accountable to him, and he could control everything.

This transition may have been necessary in the association's history, although perhaps it could have been more peaceful. NPA was much larger and required a bigger staff and better organization than ever before. The conservation scene was being dominated by executives, by professionals like David Brower of the Sierra Club, Howard Zahniser of the Wilderness Society, and

Carl Bucheister of the National Audubon Society. Historian Stephen Fox has argued that the amateur tradition in conservation has been the strength of the movement, "the driving force in conservation history."[11] This tradition remained strong in conservation groups that had a grassroots organization across America, such as National Audubon, but the professionals were taking over in Washington, D.C. Because NPA was, by its nature, a centralized, Washington, D.C.-based organization, this change may have been more marked. NPA's style of work was too relaxed for Tony Smith and some of his trustee colleagues. They could see that if the association was to play a leading role in the conservation issues of the day, its leadership needed to be stronger, better organized, and more aggressive. Smith thought he was the man to do this, and enough trustees shared his view to allow him to try.

Lobbying Constraints and a Revised Agenda

Throughout its history, the National Parks Association used lobbying as one of its tactics. (Yard had been a master at soliciting allies for or against specific parks legislation.) All conservation groups used this approach to a degree. The congressional battles over Dinosaur involved extensive lobbying and resulted in public identification of a "conservation lobby." This created problems for the conservation movement and NPA. In 1946, Congress had passed a Federal Regulation of Lobbying Act, which required lobbyists to disclose their activities and the source of their funding. The Supreme Court ruled in 1954 that the act did not violate the First Amendment to the Constitution, and that if persons or organizations engaged in lobbying but did not register, they could be held criminally responsible. If organizations wished to lobby, they must register, and if they registered and were a tax-deductible organization, their tax-deductible status would change.[12]

NPA was reviewed by the Internal Revenue Service and its tax status brought into question in 1958. Was the association publishing editorial material in its magazine that was aimed at promoting or defeating proposed legislation? Was a "substantial part" of its activities "propaganda" or otherwise an attempt "to influence legislation"? Smith reported that the IRS "had taken exception to the reference to working for new parks in the Statement of Objectives on the ground that this implied substantial activity in respect to legislation; and to reference to cooperation with citizen groups on the ground that this might mean cooperation with groups interested in legislation and politics."[13] Smith was able to demonstrate to the IRS that NPA had not been lobbying to such a degree that its tax-deductible status should be revoked, but the IRS thereafter required careful reporting of NPA activity.

The association questioned whether it would defend its tax status by curtailing lobbying, or continue lobbying and likely lose its deductibility. Executive Secretary Smith made the case to the board in November 1959 that NPA simply could not afford to lose its tax-deductible status; it could not expand programs without this incentive for contributions. The board had agreed to expand educational programs and analysis of park policy issues, and to publish its magazine more often and increase membership solicitation.

All of these plans would be jeopardized if donors could not deduct their contributions to the association. "In brief, the opportunities which lie before the Association along educational and scientific lines are great," Smith told the trustees, "but will require financing on a scale which the present and foreseeable possible income of the Association will not provide. If we are to discard our tax-deductible status either intentionally or by taking undue risks, we will necessarily forego these great opportunities and renounce our responsibilities to that extent."[14]

The trustees debated the matter at length, but finally took Smith's advice and decided to protect their tax status. Henceforth, NPA would be careful in how it worked for protection and enlargement of the system of national parks and monuments. It would be a "broad educational operation" that used its magazine as its principal educational vehicle. Descriptive articles would be published about both existing and prospective areas, and judgments rendered as to the national significance and suitability of these areas. Congressional activity would be reported "on a purely informational basis," as it had for years in the magazine's department, "The Parks and Congress." No NPA stance for or against legislation would be stated or advocated. Authors could state positions in articles, with the caveat that they did not reflect the views of the association. The NPA staff would participate in hearings and share their expertise with Congress only when invited. In short, the association would inform and educate but not advocate—at least until the IRS rules changed, which all conservationists hoped would be soon. Significant change would not come, however, until the late 1970s.

Anthony Wayne Smith used this tax crisis to his advantage. The association's loss of its tax deductibility was a severe threat, and a clear understanding of the risk and how it might be countered was essential. Smith used his skills as an attorney to diagnose the problem and propose a solution. Maintaining deductibility would require careful limitation of association activity, a testing of proposed action against a narrow legal litmus. Smith would administer the test, and this gave him the opportunity he needed to centralize authority and exert control. He made the association dependent on his expertise, solidifying his position as its undisputed leader. Some trustees— Wharton, Butcher, Olson, and Woodbury foremost among them—thought the decision to avoid lobbying a mistake. How could the association's long-standing mission of protection be achieved without such activity? Woodbury wrote Wharton that he "was not happy about the situation. If we must be supine and apparently spineless as an organization we lose the respect of those whose respect and cooperation we must have—not to speak of losing our own self-respect as well. I wish I knew how to find a constructive middle ground."[15] Four years later, as he left the board after forty years, Wharton praised Tony Smith for building up the finances and membership, "but I still regret the lack of a sufficiently militant approach to the dangers which confront the National Parks."[16] And Olson, looking back on the situation in 1972, observed that "Underneath the wise and frightened eye of Tony Smith the old outfit has certainly lost its spirit and real objective. What I cannot understand is how the present Board continues to tolerate his ideas."[17] Olson referred not only to the association's politics but to its loss of focus on the parks.

Good-Bye to Diplomacy

While working internally to gain control and reorganize association operations, Smith attempted to establish his presence with the conservation community and the National Park Service in a controversy over a Mission 66 project in Yosemite, the improvement of the road over Tioga Pass. With Yosemite part of its special province, the Sierra Club had long been troubled by this road, which crossed Yosemite from east to west. Since 1933 the club had protested Park Service plans to make the mining road into a modern thoroughfare. Pressure to improve the road was consistently applied by commercial interests at the east end that depended on tourist traffic for their business.[18] Mission 66 provided an opportunity to upgrade Tioga Pass road, and the Park Service set to work.

Tony Smith inspected the project in September of 1958. By the time he arrived, much damage had been done. He found a "hideous gash of new road on the high mountain east of White Wolf" and considered the project a "devastation" and "disaster."[19] The damage could not be repaired, but a portion of the road was yet to be constructed, and Smith suggested an alternative to the engineers' plans. In a long letter to Director Wirth, he offered detailed proposals. He asked rhetorically what the situation demanded of a National Park Service director. "I think the answer is that a man of such quality will reject the advice of subordinates, however competent and well-meaning, made from the limited perspective of their specialties, and apply instead the high standards which have guided men like Mather whose vision and wisdom have been recognized far beyond the boundaries of our own country. Such a man…will decide unerringly upon firm protection to the wealth of natural beauty in Yosemite, as against the inducements of convenience and expediency."[20] In case Wirth did not get the message, Smith closed with a warning shot across the agency's bow. "Unless a solution of this kind be adopted, we see little prospect for any abatement of the rising conflict between conservationists and the Service.…We think that disputes of this kind jeopardize the effectiveness of the Service; as defenders of the Service, as well as of the Parks, we deplore them, and we beg you to take the strong course at this juncture which will solve the current difficulty along the lines that we propose." The new voice of NPA was giving notice that the gloves were coming off. Diplomacy in the style of Olson and Packard was a thing of the past.

Wirth was not impressed. His reply was polite, but made clear the intent of the Park Service, which was to provide for the motoring public. Smith had suggested that parking areas and overlooks be placed so as to minimize their impact, which would require short walks to overlooks. Wirth rejected Smith's suggestion.

> I do not feel that we should build these parking areas off the road and require people to walk in order to obtain a view as superlative as the one we are discussing. We know from many years of park experience

The number of visitors to the National Park System continued to grow in the 1960s. Here, visitors crowd the overlook at Newfound Gap in Great Smoky Mountains National Park.

that although there are many people who would be very glad to walk even a short distance to obtain a superlative view, they are not able to do so.[21]

Although Smith claimed some success in his account of this matter in *National Parks Magazine*, he in fact had little effect. The road was built much as the engineers proposed. Smith concluded that if the Tioga Road project was a manifestation of Mission 66, that program must be changed. He set up an "Action Group" of NPA members to watch the Yosemite situation, and placed road building in the parks on a list of special threats to be monitored.

THE POLITICS OF CONSERVATION
National Policy Initiatives

While NPA struggled with its internal politics, the outside world of conservation and park politics moved on. The Outdoor Recreation Resources Review Commission (ORRRC) was studying the entire province of outdoor recreation. NPA and other conservation groups had supported creation of this commission in part because they saw, in rising recreation visitation to national parks and forests, a growing threat to wilderness. Mission 66 seemed to address the real needs of improving park facilities but paid no attention to the need to protect wilderness. The ORRRC was to inventory outdoor recreation resources, project demand, and prepare a long-range plan to meet that demand.

The National Park Service did not support the ORRRC, fearing that its own status as the lead federal outdoor recreation agency might be threatened. By 1960 the Park Service had, in fact, fallen to third place behind the Army Corps of Engineers and the Forest Service in the number of recreation visits to its lands.[22] The effort to provide statutory protection for wilderness was building momentum. Senator Hubert Humphrey of Minnesota and Representative John Saylor of Pennsylvania had introduced a wilderness bill in the Second Session of the Eighty-fourth Congress in 1957, and that historic debate was raging. The bill would undergo sixty-six drafts, involve nine separate hearings, and collect 6,000 pages of testimony. Preservationists rallied with great strength, but so did the opposition—comprising wood-using industries, oil, grazing, and mining interests, most professional foresters, and mechanized users of the outdoors.[23] These interests had long squared off over public land issues, but the battles were usually regional. On the heels of the Echo Park fight, the wilderness debate was national and strengthened countrywide coalitions for and against preservation.

In national politics, the Eisenhower administration had struggled with natural resource policy under Interior Secretary Douglas McKay. His successor at the Department of the Interior, Fred Seaton, proved a more adept manager of his domain's complex politics. He supported Mission 66 and, in the political jockeying around the Upper Colorado Basin Project and a proposal for a Dinosaur National Park, stated that Interior would not approve any measure that violated the national parks. This assurance somewhat calmed the rising tide of conservation activism. Still, the Eisenhower administration showed little inclination toward the long view in its resource policy. Historian Elmo Richardson summarizes the 1950s:

> ...that decade brought about but slight alteration in the values held by federal policy makers or in the public's understanding of "necessity" in environmental use. The leaders and publicists of those years addressed themselves only to the immediate issues. When Americans were confronted by a total ecological crisis less than a decade after the quarrels chronicled here, they had nothing more to draw upon to cope with that threat than the economic materialism, the bureaucratic inertia, and the political gamesmanship practiced by the men of the Truman-Eisenhower era.[24]

When John F. Kennedy became president, he appointed Arizona's Congressman Stewart. Udall as secretary of the Interior, which changed the politics of conservation and parks dramatically. The Kennedy administration was more interested in conservation than its predecessors, but saw its challenges in new ways. Udall would soon describe a "New Conservation" in *The Quiet Crisis*. He extended conservationist concerns to include cities and minorities, questioned traditional definitions of progress, and redefined the role of recreation in American life. He saw a new, expanded role for the Department of the Interior. Ronald Foresta has argued that Udall's vision of his department moved it "far beyond its traditional concern for natural resources and into an active role in achieving social equality and, in general, improving the quality of American urban life."[25] Winds of change were blowing on the entire conservation movement, with profound effects on the National Parks Association.

At the annual meeting on May 19, 1960, Tony Smith told his trustees that the reorganization of NPA was largely achieved and that he was turning his attention more fully to the "Conservation problems whose solution is a major reason for the existence of the organization."[26] He ran through a list of thirty-six activities. Testimony had been submitted "by invitation" to hearings on the Glacier Peak Wilderness in the North Cascades of Washington, on a proposed Great Basin National Park in Nevada, and on the creation of a Padre Island National Seashore. Various park threats were monitored and efforts made to educate the public about them. These included the threat to Rainbow Bridge from the rising waters behind Glen Canyon dam; increasing use of motorboats on lakes in national parks; opposition to Park Service plans for a Shrine of the Ages in Grand Canyon National Park; plans for building the Fontana road in Great Smokey Mountains National Park, and a road in Zion. Work continued to establish a C&O Canal National Historic Park, and on opposing proposals for dams that would threaten Grand Canyon National Park. The association was assisting in the effort to pass a wilderness bill, as well as following the progress of proposals for Oregon Dunes and Point Reyes national seashores, and for an Arctic Wildlife Range. NPA's plate was full.

Hunting in the Parks

A highly emotional issue moved to the front of the NPA agenda in 1960—hunting in national parks. The long-standing policy of the National Park Service had been to exclude recreational hunting from all units of the National Park System. Managing wildlife populations posed challenges for the national parks from their beginning—bison and elk had been problems in Yellowstone since the 1880s—and the National Park Service Act of 1916 had established the authority of the secretary of the Interior in the matter. "He may also provide in his discretion for the destruction of such animals and of such plant life as may be detrimental to the use of any of said parks, monuments, or reservations."[27] The 1950 act expanding Grand Teton National Park made provisions for "a program to insure the permanent conservation of the elk within Grand Teton National Park with the use of qualified and experienced hunters licensed by the State of Wyoming and deputized as park rangers by the Secretary of the Interior, when it is found necessary for the purpose of proper management and protection of the elk."[28] Fearing a precedent, NPA had not been happy with this, but was unable to defeat it.

Hunters, particularly in the West, resented their exclusion from this prime hunting territory. They thought park wildlife should be managed by state

Wildlife in the National Park System, such as this feral burro flanked by desert bighorns in Death Valley National Monument, drew increasing attention in the 1960s.

game departments and that all lands should be open to controlled recreational hunting. In 1960, a handful of mostly Western state game commissioners passed a resolution through the International Association of Fish and Game Commissioners calling for the opening of national parks and monuments to recreational hunting and opposing any new parks unless the game was managed by the states. The NPA board issued a statement of policy on hunting in the national parks in November 1960:

> This Association is irrevocably opposed to opening any part of the present national park and monument system to hunting. The Association is convinced of the need for the establishment of new primeval national parks or monuments in areas where the scenic, wilderness and wildlife resources are of full national park caliber; it would be completely opposed to permitting hunting in any such parks or monuments....Our national parks and monuments are among the finest of the wildlife refuges, and they ought to stay that way.[29]

After publishing this statement, Smith asked Director Wirth about the Parks Service's position on a recommendation from the Nevada Fish and Game Commissioners that public hunting be permitted in the proposed Great Basin National Park as a measure to control wildlife population. Wirth was rather equivocal, asserting the agency's traditional policy on hunting. Then he revealed a proposal made to the Nevada Commissioners. The law establishing the park might enable the secretary of the Interior "to open a designated percentage of the park to deer management reduction by the public under regulations promulgated by him after consultation with the Nevada Fish and Game Department under which it would aid in such a management reduction program."[30] Smith replied that the approach was completely unacceptable to NPA and that "if the Service were to institute policies of the kind suggested to the Nevada Commissioners and sportsmen, the Association might find it necessary to take the matter to court."[31]

Smith and Wirth exchanged several more letters on the issue, all of which were published in *National Parks Magazine*, and the NPA trustees reiterated and broadened their policy statement at their annual meeting for 1961. They argued for habitat protection and predator restoration. "Insufficient attention has been given to the restoration of predators as a means of controlling wildlife populations. We are not unaware of the limitations of this approach, but recommend its far more extensive use."[32] The NPA position was written by experienced wildlife scientists, including NPA's president, Clarence Cottam. Cottam, who held a doctorate in biology, had enjoyed a twenty-five year career with the Fish and Wildlife Service and had served as the agency's assistant director from 1946 to 1954. He and other biologists among the trustees took a keen interest in this hunting issue.

All of this led Wirth to prepare a statement of objectives and policy on wildlife management in the parks, issued as a resolution of the Advisory Board on National Parks, Historic Sites, Buildings and Monuments, which the association greeted enthusiastically. The statement said that "public hunting is neither the appropriate nor the practical way to accomplish national park and national monument management objectives."[33] The association encouraged Secretary Udall to approve the policy as drafted by the National Park Service.

The issue came up again when legislation to establish Canyonlands National Park was introduced in 1962 by Senator Frank Moss of Utah. Moss's bill, which contained language modeled on the Grand Teton exception to usual park wildlife policy, drew NPA opposition. This prompted Smith to take the case to the original source of the crisis, the fish and game commissioners. He addressed their convention at Jackson Hole on September 13. NPA was not against public hunting, he told them, but it would not compromise on the issue of hunting in the parks. He urged the commissioners to cooperate with park protectors on issues of wildlife conservation.

Interior Secretary Udall appointed an Advisory Board on Wildlife Management in the spring of 1962 to assist him on this festering policy issue. Chaired by wildlife biologist A. Starker Leopold, it was composed of Stanley

Cain, Clarence Cottam, Ira Gabrielson, and Thomas Kimball. Their famous "Leopold Report," submitted on March 4, 1963, was all NPA could have hoped for, and more. It asserted the "no hunting in the parks" principle. The report argued that the primary goal of the national parks and monuments should be "to preserve, or where necessary to recreate, the ecologic scene as viewed by the first European visitors." This would require historical and ecological research, experimentation and testing of management methods, a "set of ecologic skills unknown in this country today." The National Park Service should undertake an expanded research program "oriented to management needs," and the results of this research should guide the management of national parks and monuments. Roadless wilderness in the parks should be "permanently zoned," and nonconforming uses, such as golf courses, ski lifts, and motorboat marinas, should be "liquidated as expeditiously as possible."[34]

Tony Smith, reporting to the board at the annual meeting in May 1963, was delighted with the outcome of the protracted campaign against policy change on hunting in the parks. "One might almost say that if the Association had done nothing else in recent years to justify its existence, its effective defense of the century-old tradition of no hunting in the parks would do so."[35] He thought the Leopold Report would be very helpful in defining legislation for additions to the National Park System, such as national recreation areas and seashores, where carefully regulated hunting might be allowed. NPA was willing to concede that hunting might be appropriate in some of the new "recreation" areas but would support no hunting in the National Park System if it threatened to open the entire system. The Leopold Committee's work made policy clear on this point. Smith also applauded the recommendation that "carrying capacity" be defined and considered in managing wildlife populations in the parks, and he praised the emphasis on "restoration." NPA's task now, he told the trustees, was to help the Park Service analyze its past policies that had led to problems, and to support the recommendations of the report on research and new management policy. The fight over hunting, he thought, had led to more significant policy outcomes than anyone anticipated or intended.

NPA was not the only player in this issue, but this time it was the leader. The issue fell within its traditional province and was viewed as critical to park protection. The very nature of national parks was at stake. With this hunting issue, NPA was defending standards in the classical sense. Ironically, the first issue Robert Sterling Yard had used to gain momentum for the newly formed association in 1920 had involved the Yellowstone elk. Then NPA had advocated rescue of elk herds as a human gesture, even though they had overgrazed their habitat. Wildlife science had progressed far since then, and in the 1960s NPA advocated "biotic management" and defended the Leopold Report, viewed by many as a radical document—as it remains thirty years later. Wildlife management policy in national parks would prove to be one of the most complex and contentious issues in years to come, and NPA took pride in facilitating debate and developing policy.

Tony Smith believed congressional retreat from protection of Rainbow Bridge was a breach of promise.

Rainbow Bridge and Congressional Resolve

The association continued to wrestle with water projects. Its old friends at the Bureau of Reclamation were still hatching grand schemes for the Colorado River, which NPA could not ignore. Conservationists had won the fight over Echo Park, but the battle to protect national parks along the Colorado River was far from over. Grand Canyon National Park and Rainbow Bridge National Monument were new battlegrounds.

Six miles from the Colorado River as it wound through Glen Canyon, up Forbidding Canyon and tiny, twisting Bridge Canyon, Rainbow Bridge arched 290 feet above its foundations in a span of 275 feet. Fifty million years of erosion had shaped the arch of salmon-colored rock and its surroundings. Few knew of this marvelous rock formation until, in the February 1910 issue of *National Geographic*, it was described in an illustrated article by University of Utah archaeologist Byron Cummings. Using the new tool provided by the Antiquities Act, President William Howard Taft created by executive order a 160-acre national monument with Rainbow Bridge at its center. No planning was done for boundary selection—the monument was simply a 160-acre quarter-section judged sufficient to protect the arch from commercial exploitation or other intrusion.[36]

This tiny monument had come late into the struggle over Echo Park Dam on the upper Colorado River; Echo Park opponents had agreed to drop their opposition to the Upper Colorado Storage Project *if* the dam was not built and

Stewart Udall's hand was evident in many national park issues during his tenure as Secretary of the Interior.

there were no other National Park System intrusions as a consequence of the project. The precise language of Public Law 485, Eighty-fourth Congress, was: "It is the intention of Congress that no dam or reservoir constructed under the authorization of this Act shall be within any national park or monument." And of the Glen Canyon project, the act said: "That as a part of the Glen Canyon unit the Secretary of the Interior shall take adequate protective measures to preclude impairment [by Glen Canyon reservoir] of the Rainbow Bridge National Monument."[37]

The Echo Park battle aimed to keep water projects out of the National Park System, and this straightforward language seemed to ensure this outcome. But now, as Glen Canyon Dam neared completion, the prospect of broken congressional promises loomed. The Bureau of Reclamation, some congressmen, and even some Southwestern conservationists were questioning whether the plan to build protective structures for Rainbow Bridge should be implemented.

NPA and most other conservationists took the position that Rainbow Bridge National Monument should not be affected by the reservoir, whatever the cost. Tony Smith wrote that "Far more than Rainbow Bridge is immediately involved in this matter. For some of the same individuals who would prefer to let reservoir waters rise and fall beneath the great stone arch in southern Utah, would also like to see a man-made arch rise within Dinosaur National Monument at the Echo Park site in western Colorado."[38] The issue of keeping congressional promises was foremost. NPA and others refrained from mounting opposition to Glen Canyon Dam to honor the compromise; should Congress not keep its commitment, their trust would be betrayed.

The plan to protect Rainbow Bridge involved constructing a barrier dam downstream from the arch to keep the rising waters of the reservoir out of the monument and building a diversion dam upstream that would keep the Bridge Canyon runoff from creating its own small reservoir behind the barrier dam. A pumping operation was required to keep seepage from accumulating upstream of the barrier dam. Cost estimates for these protective measures varied from $3.5 million to $30 million. New Interior Secretary Udall, a former Arizona congressman who had always supported water projects so necessary (at least politically) for his arid state, including Glen Canyon Dam, inspected the site several times. Though he was troubled by Congress' breach of promise, he concluded that the additional construction around the arch would do more harm than good.

The Interior Department budget for fiscal 1961 submitted by Secretary Seaton included funds for protecting Rainbow Bridge, but Congress refused to appropriate them. Udall's first budget again included a request for funds for the project, but Congress refused, and this time placed a specific restraint on the secretary against using any Glen Canyon project funds for Rainbow Bridge protection. NPA testified for the appropriation in Congress, but congressional intent was clear. It did not consider protecting Rainbow Bridge a priority.

Late in April 1961, Udall, Bureau of Reclamation Director Floyd Dominy, and Conrad Wirth hosted fifty-seven interested visitors at Rainbow Bridge. Among them were David Brower, Tony Smith, and three NPA trustees—Sig Olson, Frank Masland, and Weldon Heald. They inspected the proposed site of the protective dams, and Udall revealed a compromise that he might offer— a greatly expanded Rainbow Bridge National Monument. The land surrounding the monument belonged to the Navajo, but Udall was sure an exchange could be worked out with the tribe if the idea of an expanded monument garnered support in other quarters. Brower, Smith, and the other conservationists liked the idea of an expanded monument, but they remained adamant that the reservoir not invade the monument. Udall's proposal soon met opposition from all sides. Conservationists were suspicious of compromises, and the Navajo tribal chairman did not think exchanging the 274,000 acres would be of any benefit to his tribe. The people of southern Utah also howled that this proposal, coupled with Udall's advocacy of a Canyonlands National Park, would cripple their region's economic prospects.

Tony Smith wrote to Dominy in the summer of 1961 about the schedule for closing the gates of Glen Canyon Dam. The October reply stated that impoundment would begin on January 1, 1963. The reservoir was expected to reach the downstream boundary of the monument in June 1966.[39] Smith informed the trustees that he would prepare to file suit, as they had authorized the previous November, to prevent the closing of the gates of the dam. Once more NPA and others encouraged Congress to appropriate the funds for the protective dams, and again they were frustrated. In August, Smith wrote Udall asking him to postpone closing of the Glen Canyon gates, but he would not. Udall's attempt to compromise had failed. He believed in the Glen Canyon project, and he believed the protective dams would do more harm than good. Politically, he would not take the heat of interfering with the nearly completed Glen Canyon Dam. NPA, joined by the Sierra Club, the Wilderness Society, and the Federation of Western Outdoor Clubs, filed suit against the secretary of the Interior, seeking an injunction against the closing of the Glen Canyon gates until the protective measures for Rainbow Bridge were taken.

The court ruled that conservationists did not have legal standing to sue the government and therefore would not grant the injunction. Although Interior argued that the appropriation acts denying funds for Rainbow Bridge protection had repealed the protective clauses of the Colorado River Storage Project Act, the court ruled that the provisions remained in force. (This ruling provided some consolation as a moral victory for NPA and its allies.) Smith carried the case to the Circuit Court of Appeals and to the U.S. Supreme Court without success. The Glen Canyon gates closed on schedule. Brower traveled to Washington for a last try at personal persuasion with Udall, and Smith continued to send letters to the secretary. He thought Udall had been ill advised by his legal counsel. Udall's response closed the matter:

> While you are not, as you say in your letter of March 20, interested in the opinion of my lawyers, when the issue is a question of law, I am. They are unable to accept your view....First of all, the case was dismissed on jurisdictional grounds. This, alone, makes reliance on the Court's finding legally impossible....The plain fact of the matter is that the Congress after being fully advised by you and by me, and after considering the report of the Geological Survey, concluded that the probability as to the effect of the water which will enter the Canyon beneath the national monument did not warrant the expenditure of funds for the construction of the barrier dams.[40]

Smith told readers of the magazine that "The ability to make and keep a promise is a hallmark of civilized man." There was little doubt of whom he spoke here. He thought the fate of Rainbow Bridge "unpredictable," but "the violation of the national tradition of park protection is undeniable."[40]

The Grand Canyon Threat

Smith was already thinking of new challenges on the Colorado River. On January 21, 1963, with Floyd Dominy at his side, the Interior secretary announced a new project that Bureau of Reclamation planners had dubbed the "Pacific Southwest Water Plan." Since the late 1940s, NPA had been aware of two proposed projects that would affect the Grand Canyon National Park and Monument. One, the Kanab Tunnel, was proposed by the City of Los Angeles. Water would be diverted above the park through a forty-two-mile tunnel to a powerhouse near the confluence of Kanab Creek and the Colorado River below the park. At times, 90 percent of the river's flow through Grand Canyon would be diverted—an obviously severe impact on the national park.

Interior secretaries Krug and Chapman had pledged that no water projects would invade the parks (the Kanab Tunnel would pass under a corner of

Grand Canyon National Park), and the outrageous project had seemed dead. When the Arizona Power Authority applied to build a dam in Marble Canyon upstream from the park, however, Los Angeles expressed its intent to press again in Congress for legislation to construct the Kanab project. The Bureau of Reclamation had also envisioned another dam in lower Granite Gorge near the mouth of Bridge Canyon. This dam would back a reservoir up into the national monument.

NPA formally entered the Grand Canyon struggle on June 26, 1961, when it petitioned to intervene in the Marble Canyon case before the Federal Power Commission. The Arizona Power Authority sought a license from the commission for Marble Canyon Dam. NPA was granted intervenor status and directed its attention to the Kanab project, which would be linked to Marble Canyon Dam. Los Angeles argued that the license should be deferred pending study of the Kanab Tunnel. The city sought legislation to override the provision of the Federal Water Power Act prohibiting the construction of any dam, reservoir, or conduit within a national park or monument. NPA subpoenaed Park Service Director Wirth, who revealed that the agency had considered plans to enlarge the park to include the proposed power development site on Kanab Creek. Matters had even progressed to the point where the Forest Service had agreed to the transfer. In the end, the hearing examiner recommended issuing the license for the dam but rejected Los Angeles's plea for a delay (which it sought while pursuing legislation to approve its Kanab project), effectively killing the Kanab Tunnel.

At this point, Secretary Udall filed to intervene in the case. He argued that the Interior position had been misrepresented, and that no dams should be licensed until larger issues involving the lower Colorado River basin were resolved. Udall awaited a Supreme Court ruling on an eleven-year-old lawsuit filed by Arizona against California. At issue was Arizona's share of Colorado River water. Early in 1963 the court upheld most of Arizona's claims, most important of which was that Arizona was entitled to 2.8 million acre-feet of the Colorado, most of which had been going to California. The time was right: Udall and Dominy presented their Pacific Southwest Water Plan. Arizona's share of the Colorado would be pumped east, and southern California's water shortfall would be replaced with water diverted through a complex system of aqueducts from northern California's Trinity River. Arizona's water would come out of existing Lake Havasu, but power to lift it would come from two new dams—the Bridge Canyon and Marble Canyon dams.

NPA's position was straightforward: no water projects should affect Grand Canyon National Park and Monument. The proposed Bridge Canyon Dam would back water up into the National Park System, and that could not be tolerated. Marble Canyon Dam was tolerable as long as no Kanab Tunnel was attached to it, though protection of the entire stretch of the river, from Lake Mead to Lee's Ferry, should be the ultimate goal. Tony Smith told the NPA trustees at their May meeting that the Grand Canyon struggle would be a long one and he would propose an extensive NPA role in it. In November, he informed them that NPA would conduct economic and engineering studies of the proposals to search for alternatives to the dams, and an educational campaign pursued primarily in *National Parks Magazine*. This approach, he thought, would do the most good while protecting the association's tax status.

As Smith predicted, the Grand Canyon battle was a long one—four years. In 1964 NPA hired a consulting engineer and a consulting economist to study the Grand Canyon proposals. The conservation community, again led by the Sierra Club's executive director and Echo Park veteran David Brower, mounted a national public relations campaign. Defending the Grand Canyon became a popular national cause. NPA advocated presidential proclamation of a national monument for the unprotected reaches of the Colorado between Glen Canyon Dam and Lake Mead.

In August 1965, hearings were held on revised legislation, now called the "Lower Colorado River Basin Project," which contained only one dam, at Marble Canyon. Bridge Canyon Dam had been deferred "pending further study." This proposal met opposition from conservationists, who argued that *no* dams were necessary. The NPA technical studies, said Smith, showed that more than enough power and revenue could be generated to lift the water for the Central Arizona Project. "As far as Brower and his allied conservationists were concerned, the federal government had the blessings of every major conservation organization in the nation to build as many coal-fired or nuclear plants as it needed to pump currently stored water wherever it was lacking, but those organizations would *never*. . .compromise on the issue of the destruction of the Grand Canyon."[41] In his annual report, published in *National Parks Magazine*, Smith told the association that NPA studies demonstrated that "coal power will do the job more cheaply [than the dams], and save the canyons of the Colorado. We have also pointed out that...atomic energy will be available soon in the Pacific Southwest at from 2 to 3 mills a kilowatt hour; thus the hydropower projects will not pay out."[42] In 1966 the legislation became the "Colorado River Basin Project," removing "Lower" and adding five relatively small reclamation projects in western Colorado, in the district of Wayne Aspinall, chairman of the House Interior subcommittee, whose support was necessary to pass any bill. The Central Arizona Project was still in, as was Marble Canyon Dam. This version recommended study and eventual authorization of "Hualapai Dam" to be built, not surprisingly, near the mouth of Bridge Canyon. The conservationists were highly effective at these hearings, their testimony orchestrated by the ever-resourceful Brower. They did so well, in fact, that Brower decided it was time to do more. He mounted a national advertising campaign so effective that it drew a response from the Internal Revenue Service (allegedly prompted by project backers), threatening the Sierra Club with loss of its tax-exempt status. These threats increased the popular opposition. "If the country's citizens hadn't liked the idea of building dams in the Grand Canyon, they <u>really</u> didn't take kindly to the notion of the arrogant IRS flexing its muscles in the face of a little 39,000-member conservation organization."[43]

Finally, on February 1, 1967, Secretary Udall announced that the Johnson administration was withdrawing its support from the latest version of the Colorado River Basin Project. The Central Arizona Project would go forward, but there would be no Colorado River water guarantees to southern California, no water importation from the Northwest, no reclamation projects in western Colorado, and no dams in the Grand Canyon. The Central Arizona Project water would be pumped with energy generated by a massive coal-fired powerplant built near Page, Arizona. On July 31, 1968, congressional conferees agreed on a $1.3 billion Colorado River Basin Project.

The struggle to protect the Grand Canyon was one of the great efforts of the conservation movement. Like the Echo Park struggle, it involved a vast coalition of conservation groups, with David Brower as the field general. NPA played a supporting role, retaining experts to examine the situation and recommend alternatives to the dams. Smith and these experts testified many times during the four years of debate. When Udall announced the final decision, Smith took credit for the association. "More than a year and a half ago this Association advocated precisely the solution which has been adopted," he wrote in *National Parks Magazine* in 1967, "in a statement given on invitation of the Subcommittee on Irrigation and Reclamation of the Committee on Interior and Insular Affairs, House of Representatives."[44]

As early as 1962, the association had questioned whether coal, atomic energy, even solar power would be feasible, less expensive, and less damaging approaches to moving water to Arizona than the proposed dams. It did not question if the water should be moved, but how. As always, it aimed primarily to protect the National Park System from invasion and damaging precedent. In the future, when air pollution occasionally blanketed the Southwest, obscuring views of the Grand Canyon and other parks, NPA would rue its support of coal as an alternative to dams, but in the mid-1960s, it could not anticipate the impending air pollution problem. NPA and its allies had, for the moment, prevented an invasion of the National Park System.

This episode demonstrates a change in NPA's approach to its business. Legal proceedings became a tool, perhaps not surprising, given Tony Smith's background in law. The association also used technical analysis. Smith told the trustees in 1963, "If conservationists are to be more than a shrill voice of protest…they must make the necessary economic and engineering studies, at least in preliminary outline, which may in due course be convincing to the ultimate sovereign, the American people."[45] NPA did so in this case, offering its analyses in legislative hearings and in the pages of the magazine. The motivation was to protect its status as a tax-exempt, nonprofit organization. The trustees made this decision, and Smith adhered to it religiously. The decision clamped strict limits on what the association might do, and directed its activities in new ways. As a consequence of Brower's activities in the Grand Canyon case, the Sierra Club ultimately lost its tax-deductible status; NPA could not afford such an outcome. The record suggests that it found effective ways to work within its limits.

Tony Smith thought the Grand Canyon situation raised basic questions about national values, "whether we intend as a nation to be guided by standards keyed entirely to commodity production, and poorly planned, pressure-influenced production, at that, like uneconomic hydropower and surplus crops, or whether standards of a rational economy and protection of reserves of beauty and knowledge like those at Grand Canyon, as representing key values in our culture, will control."[46] At the end of the struggle, Smith told the trustees in October 1966 that the underlying problem involved the misguided belief that growth is always good. Population growth in the Southwest, a land of little water, could not be sustained without continuous threats to such "key values" as the beauty of the Grand Canyon. Further, rational planning on a large scale was necessary to cope with the problems of population growth and economic expansion. Unless this planning was achieved, Smith saw one natural resource crisis after another looming on the horizon.

Damming the Potomac

While the Bureau of Reclamation was hard at work promoting Colorado River schemes, the Army Corps of Engineers was creating work for itself right on NPA's doorstep—on the Potomac River. Since the mid-1950s, Smith had led the NPA effort to protect the C&O Canal from development. A parkway was defeated and a national monument created, but engineers kept hatching new schemes. Now campaigning to make the monument into the C&O National Historical Park, NPA confronted proposed hydroelectric dams and water storage reservoirs. Not only did the Army Corps oppose the national historical park, so did rural electric cooperatives and local public utilities, all of which coveted the electricity the Potomac dams might provide.

NPA's initial attention to the Potomac region had been drawn by the C&O Canal, but now its concerns were broadening. If the Potomac Valley program proposed by the Corps of Engineers was approved, it would pursue similar programs in the Susquehanna and Ohio river basins. Smith told the trustees in December 1962, "The prospect would be that most of the Appalachian Mountain valleys might be filled with reservoirs with heavy fluctuations and draw-down margins highly destructive to genuine recreation and scenic values."[47] The real need, he continued, was for river basin management that would include soil and forest conservation, eliminating pollution, and small flood control and recreation reservoirs in the upper reaches of the basins. This management should be the responsibility of an agency with a broad planning mandate rather than the Corps of Engineers. Regional planning was of growing interest to Smith. He began to think that park protection would require "protection of the natural environment everywhere,"[47] which would require regional planning.

The C&O Canal Association was housed in the NPA offices, and NPA supported the effort to pass a bill creating the national historic park. Late in 1961, Smith reported strong opposition to the proposed park from the Corps of Engineers, which had recently proffered four alternative multiple-dam schemes. The centerpiece of the schemes was a high dam on the main stem of the Potomac at River Bend. The justification offered for these dams included power production, water supply, and recreation.

During 1962 and 1963, NPA and its allies pressed for the C&O National Historical Park and against the dams. They worked especially hard to make

Tony Smith focused NPA attention on issues involving the Potomac River.

the case that the dams were not needed to supply water to the Washington, D.C., area. The NPA position, developed by a consultant and supported by other scientists, was that water purification was a better way to provide water than impoundment behind dams. The Army Corps kept pushing, and NPA kept up its organizing and educating. Its approach, said Smith, was "founded on de-pollution instead of storage for dilution; on headwater management for flood protection and recreation; and on use of the fresh water estuary as an emergency source of water supply for the metropolitan area."[48]

The struggle for the Potomac finally bore fruit in the late 1960s, although the Corps of Engineers would continue to press its case for years. In 1971, a C&O Canal National Historical Park was established. Smith reported to his *National Parks Magazine* readers that the work of the association and many other groups, coupled with the emergence of a powerful environmentalism in society, had finally defeated the "big reservoir approach to river basin management."[49] A Water Pollution Control Act was passed, the Environmental Protection Agency was created and given a pollution-prevention mission, and prospects were excellent for funding an intake and pumping plant to tap the freshwater estuary of the Potomac for the city's supplemental water supply.

The twenty-year struggle for the Potomac could be counted as another win for conservationists, resulting in a national historical park and a new approach to river basin management. Smith did not claim credit for NPA, but the record shows that the man and the organization were key actors in this long struggle.

The Potomac experience helped define the NPA agenda for the 1970s. When Smith described this agenda, the association had become the National Parks *and Conservation* Association (more on the name change later). Believing that park protection required general environmental protection, Smith suggested extending the NPCA mission to watershed management, ecological forestry ("the governing criterion must be ecological management, not maximum timber production"), pollution prevention and enforcement, and "environmental protection." This latter, he noted, should be defined "in the sense of defense against subdivision, urbanization, over-industrialization, and land speculation."[49] The Potomac experience reinforced Smith's growing conviction that the association should broaden its horizons.

NPA REFOCUSES
Watchdog of the Park Service

Although much of NPA's attention was focused in the early 1960s on the issues described in this chapter, it worked on many other projects. New units were proposed for the National Park System, and the association monitored their progress, describing the legislation in the magazine and doing what it could for and against proposals within the constraints of its tax status. Proposals included a Great Basin National Park in Nevada, a Prairie Park in Kansas, several national seashores (Cape Cod, Padre Island, Point Reyes, and Fire Island), and Canyonlands, North Cascades, and Redwoods national parks. The wilderness bill moved glacially through negotiations, NPA always watchful of national park wilderness interests. Road building threatened Great Smoky Mountains and Mount McKinley national parks and lesser units of the system. The never-ending problem of freshwater supplies for Everglades National Park waxed and waned on the association's active agenda.

NPA had been the watchdog of the National Park Service throughout its history. In the early 1960s, the arguments over wildlife management—particularly hunting—were part of this work. Mission 66 remained a concern. Under Director Wirth, the Park Service seemed bent on developing roads and facilities at the expense of wilderness and other park resources. When the Outdoor Recreation Resources Review Commission issued its report in 1962, the association analyzed it and registered several concerns. The Park Service had opposed even the idea of the commission on the ground that it was the logical agency to study the outdoor recreation situation. In 1936 it had been charged by the Parks, Parkway, and Recreation Act with responsibility for recreation planning, but after a flurry of activity in the late 1930s, the agency had neglected its role as recreation advisor to other levels of government.[22] Fearing the loss of this part of its mission, the agency had opposed the new commission and cooperated with it reluctantly.

This Park Service concern was justified by the commission's findings, which stated that "the National Park Service was not formed to provide recreation in the usual sense but to preserve unique and exceptional areas."[50] This had been NPA's position for decades. The report concluded that the problem for outdoor recreation was not the number of acres available for use but their accessibility to the people, and suggested two solutions. First, the federal government should establish a fund from which it could purchase land and make grants to state and local governments to purchase and develop outdoor recreation areas. Second, a new bureau should be set up to coordinate the recreation activities of federal agencies, oversee aid to other levels of government, and even study the candidates for addition to the National Park System to determine their suitability.

Congress acted quickly on these recommendations, creating a Bureau of Outdoor Recreation in the Department of the Interior in 1962. Its director was recruited from the Forest Service. NPA had mixed feelings about this. On one hand, it had supported the commission. Under Smith's leadership, it agreed there was a need for better planning. Long committed to acquiring park inholdings, the association viewed the fund for "land conservation" as a necessity. It hoped the recreation bureau and its advisory commission might reduce the influence of the Corps of Engineers and Bureau of Reclamation. On the other hand, the association was concerned about the impact of this fund and the new bureaucracy on both the National Park Service and the National Park System; it thought the Park Service should be the agency to rule on all National Park System matters. And what would the consequences be for national park budgets? As the Land and Water Conservation Fund took shape, NPA worried that too much of it was earmarked to the states. It registered objections to proposed allocation percentages, but in the end held ranks with other conservation organizations in support of the fund. Provision of *any* additional funds for acquisition of parkland was a positive step.

Organizational Change

Organizationally, the National Parks Association continued to evolve during this period. Membership grew steadily, increasing from 11,800 in 1958 to 30,800 in 1965. Its financial condition also improved significantly. Relations between the executive and the board settled down after Cahalane resigned. Clarence Cottam became association president, and under his leadership, the trustees seemed willing to allow Smith to operate in the manner he required. Cottam had joined the board in 1956 and was a highly respected conservationist, scientist, and former government official. A native of Utah, he had enjoyed a long and highly successful career with the U.S. Fish and Wildlife Service and commanded respect from his scientific and conservationist colleagues. He first appeared in NPA records as a representative of the Fish and Wildlife Service at the discussion sessions set up by Sigurd Olson in the NPA annual meetings during the early 1950s. After a brief stint as a dean and professor at Brigham Young University, Cottam organized and directed a private wildlife refuge in Texas, a job he held throughout his tenure with the association. Cottam's specialty was wildlife management, but his interests were broad. He seems to have shared Tony Smith's growing view that protection of any part of the natural world required activity in a broad sphere.

The NPA staff also grew and changed. Bruce Kilgore left as editor of the magazine to pursue other interests and was replaced by Paul Tilden. Tilden was a journalist and editor with editorial experience on *Nature Magazine* and *Natural History Magazine*. *National Parks Magazine* changed from a quarterly to a monthly in 1959, and its editing proved a full-time job, although Smith occasionally gave Tilden other assignments. In the early 1960s, with activity on many fronts, Smith tried to set up NPA groups in locations across the country to watch specific parks and issues. He called these groups his "panels of consultants," but found volunteer "consultants" unreliable. Instead he used professionals on retainer rather than expand NPA's professional staff. This approach seems to have worked well during the first eight years of Smith's administration.

The association by-laws changed several times during this period, most significantly at the annual meeting in May 1963. Since 1919, the executive of the association had carried the title "executive secretary." The new title would be "president," and this officer would be appointed by the seven-person executive committee. All association committees except the executive committee would be appointed by the president, and all committees would report through the president and executive committee to the board. These changes merely made official the situation that had evolved since Smith took charge. Committees, which had been an NPA mainstay from the beginning, had all but disappeared under Smith (except for the executive committee). He thought committees should help make business decisions, largely by ratifying the decisions of the executive, but should not have a role in determining policy. Smith was consolidating his power over association affairs. When he reported annually or biannually to the board on NPA activities, he invariably finished his report on a particular activity with the qualifying phrase, "unless otherwise directed by the Board, the President shall…" pursue such and such a course. No evidence suggests that he was otherwise directed by the board. Smith received his new title and soon expanded it, adding "and general counsel." He may have thought this new title would help him to influence decisions in the status-conscious world of government and politics.

A New Perspective

The period described here, from 1958 to 1965, was one of fundamental change for the National Parks Association. Its way of doing business changed. Its focus began to broaden. The business of park defense required attention to many long-standing issues, such as water projects that threatened national parks, but NPA began to see those issues differently. The association was now concerned about population growth, economic expansion and development, a growing technical ability to modify (and assault) nature, and even the philosophies and values that drove these forces. The national parks could not be defended in isolation; they must be viewed as part of regions, as subject to forces in the world surrounding them and extending beyond their boundaries. Even the tools for their defense were changing. Education remained important, but lobbying had to be restrained, and courtrooms became familiar arenas. New terms, such as "ecological" and "environmentalist," crept into discussions; standard terminology such as "conservation" and "recreation" was revised.

Tony Smith told the trustees in 1965 that "A new epoch in national park protection and enjoyment may have been opened by the National Parks Association in its recent public recommendations for comprehensive regional recreation planning for the Yellowstone and Grand Teton Parks region." Recreation resources were part of the plan at all levels of government, including private land and developments. "It may seem surprising," continued Smith, "that this approach had to be suggested to the Government by a private educational and scientific organization; it happened because there has been no place in the Government where a broad perspective could be obtained on these government operations. Thus an autonomous, objective, public-service organization like our Association plays a vitally important part in the democratic process in America."[42] To students of national park history, the NPA leadership in this matter is not surprising. Broad, progressive thinking has seldom characterized government bureaucracies with specific missions, the National Park Service among them. Even thirty years later, Smith's vision of a regional approach to managing the Yellowstone-Teton area has yet to be achieved.

X

The Identity Crisis: 1965-1975

Tony Smith settled into his role as leader of NPA, building his power and influence over association affairs. The board was stocked with his appointees. He recommended policy and program direction with little dissent. He spoke for the association to the exclusion of anyone else, dictated editorial policy in *National Parks* Magazine, wrote editorials in every issue, and personally did as much of the conservation work as he could manage. He did not hire professional staff, preferring to use consultants. Smith dominated the organization from the mid-1960s through the 1970s.

The conservation business changed during this period. Problems with natural resources were seen as the consequence of growing population, lack of an ecological perspective, and an insufficiently broad view of the problems. Smith, ambitious for national and even international leadership in conservation, believed he understood how NPA could grasp leadership by championing ideas that would be central to future conservation and environmental policy debates. One of these ideas was comprehensive regional planning.

Although Smith aspired to extend his ideas to global problems, he found his options in that arena limited. The traditional territory of NPA—national parks—offered the best opportunity. The association had long been engaged with park protection through promotion of a coordinated system of outdoor recreation resources; Smith's comprehensive planning ideas seemed ideally suited to continue that work.

As Smith struggled to establish his personal identity in the field of conservation, NPA struggled with its organizational identity. Because of Smith's

Prescribed burning among the giant trees of Sequoia National Park symbolizes the ecological shift that changed the conservation movement in the 1960s.

dominance of the organization, the two were inextricably linked. As world and national events forced change on all conservation organizations, the challenge for NPA was whether to stay with its traditional mission of national park protection or to embrace emergent environmental issues.

FROM CONSERVATION TO ENVIRONMENTALISM
Issues Old and New

The late 1960s was a pivotal period in the history of efforts to preserve, protect, and conserve the natural environment. Conservationists recognized that the long-term effort to "conserve" land, wildlife, water, forests, and other resources must be broadened to embrace pollution, population growth, and urban decay among other issues traditionally outside their concern. People began to talk about "environment," "environmentalists," and an "environmental movement." For nearly a century, a rather small group of white, mostly male, highly educated American professionals had pursued the goal of conserving natural resources for the future while protecting some unique and scenic parts of the landscape. They had done this through small organizations such as the National Parks Association, the Sierra Club, and the Wilderness Society, each with its special focus on parks, forests, birds, wilderness, or some other part of the natural world. These groups attended primarily to public lands remote from the centers of population. American conservation groups had attended largely to American issues and problems.

All of this now changed. Groups such as the National Parks Association continued to be concerned about the "old" issues and problems, but their attention was also drawn in new directions. Science revealed the weave of the Earth's ecological fabric, and writers such as Rachel Carson brought this new perception to the attention of a widening group. The burst of American industrialization during World War II and economic expansion after the war increased air and water pollution to levels that the general public could no longer ignore. The development boom involving suburbanization, the expansion of the highway system, and industrial growth raised concern about the nature of cities and of American lifestyles. Population growth rates increased worldwide, threatening wildlife, forests, and other resources, and people began to think about the problems facing the global environment. As colonialism collapsed and new countries bent on economic independence and growth entered the international scene, the problem of international "development" began to be seen as a set of environmental as well as social, political, and economic issues. Parts of the Earth previously considered "common" property, such as the seabed and the global airshed, seemed to shrink with growing population and development, and new political issues arose. All of this emerged in popular culture in 1970 when an event called "Earth Day" was organized. Suddenly the environment was no longer the province of a few concerned conservationists. A popular social movement had discovered the cause of protecting the Earth.

Groups such as the National Parks Association were pleased to find more allies and to have their arena of concern discovered by the media and the public, but changes in the nature of the problems and in the institutional environment also posed great challenges. "Environmentalism" spawned new leaders and new organizations, and some of these newcomers were not impressed with the traditional conservation movement. Stephen Fox, in his excellent overview of this period, notes that crusading environmental scientist Barry Commoner disapproved of "a passive admiration of nature" and thought that "people are more important than whooping cranes." Ralph Nader, a prominent consumer rights advocate, thought it unfortunate that "more people were drawn to bird watching than Congress watching," and Charles Wurster of the Environmental Defense Fund criticized conservationists for "talking to themselves in a closed eco-system. They are legally weak, scientifically naive and politically impotent. They lack an offense."[1] The new environmentalism found environmental problems complex and demanded a legal, scientific, and political sophistication some found lacking in the more traditionally focused conservation groups.

Proliferating Competition for Members

The new climate also involved more competition for members. Organizations popped up everywhere. The Environmental Defense Fund emerged in 1967, followed by the Environmental Law Institute in 1969. Greenpeace, the League of Conservation Voters, and the Natural Resources Defense Council appeared in 1970. They were followed by the Trust for Public Land in 1972 and the Cousteau Society in 1973, to name only a few. In addition, issues of specific interest spawned a multitude of grassroots environmental groups. This situation forced traditional conservation groups to reconsider their mission. Could they compete? Could they keep old members and entice new ones? Should they change their mission and their approach?

The National Parks Association began to change in the early 1960s. Its activities reached beyond national parks as never before. Reflecting on the first four years of his administration at the annual meeting in 1962, Tony Smith told the trustees that the conservation program had broadened, that "in addition to a more intense concentration on park management problems, we embarked into the fields of regional planning, metropolitan planning, national forest management, wildlife management, and international conservation activities."[2] Beginning in 1963, Smith examined the issue of population growth, which he thought should be of concern to all conservationists. In his view, all resource problems, including protection of parks, depended on finding a way to control population growth.[3] The population problem became a part of the association's agenda. Although Smith would justify NPA consideration of the problem in the context of park protection, he thought it merited association attention in its own right.

NPA's agenda continued to broaden through the 1960s. Work on the Colorado and Potomac rivers led the association into technical studies. Consultants—economists and engineers—were hired to do this work. NPA

A necklace of clearcuts reaching into pristine North Cascades wilderness spawned a movement in the late 1950s to create a national park there.

became more involved in debates about the best way to combat pollution and provide water during droughts in the Potomac basin and in assessing the economic costs and benefits of alternatives for moving water in the arid Southwest. Among NPA trustees were some of the leading wildlife scientists of the time—Clarence Cottam, Ira Gabrielson, and Durward Allen—and their interest in managing wildlife in parks extended to broader issues of managing wildlife across the landscape. Considering the problem of national park crowding, the association thought that people seeking outdoor recreation should be dispersed through regions surrounding parks, usually on land administered by the Forest Service. Success would require regional interagency planning and forest practices that did not devastate the land surrounding the parks, such as the expanding clearcuts in the redwood country and the Pacific Northwest. The association began to look into "ecological forestry."

By 1967 the association was working intensively on national park wilderness and several critical national park issues. In his report of the year's work to the semi-annual meeting of the trustees, Smith reeled off a long list of these issues: air and water pollution; protection of wild and scenic rivers; protection of estuaries; "socio-ecological forestry"; wildlife protection and restoration (restoring predators in Yellowstone); the abusive use of pesticides; and, of course, the overcrowding of the planet. Smith was aware that he was moving the organization into new territory, but he thought it must be done. "The danger is obvious that we scatter our attention; but most of the conservation organizations have come to realize that their duties are as broad as the environment. We must focus our main obligation on the National Park System, but also relate ourselves productively to these other issues."[4]

New Name—New Allegiance

All of this led, in 1970, to a change in the name of the association and the magazine. The association became the National Parks *and Conservation* Association at the annual meeting; the magazine became *National Parks and Conservation Magazine: The Environmental Journal.* Smith polled the executive committee in January to see if it would support submitting a name-change proposal to the annual meeting. At its March meeting, the committee approved the magazine name change and voted for the association name change, which was then submitted to the entire board in May. After considerable debate, the name change was rejected on the first vote. After lunch and arm twisting, more debate and a second vote resulted in approval. Defending the proposed change, Smith noted that the public often seemed to confuse the National Parks *Association* with the National Park *Service*. The change would, he thought, reduce that problem (though it does not seem to have been a serious one during the previous fifty years, rarely mentioned in the thousands of pages of minutes, letters, articles, and other records of the association). Also, the name would more truly reflect the mission and work of the association, which was concerned about more than parks.

No one could argue that the association's mission had not broadened or that the new name did not more adequately describe it. One could, however, argue that the organization should not be so broad, should focus closely on national parks as it had through most of its history, and therefore should return to its former mission and retain its identifying name. Trustee Ernest Dickerman made this argument in discussing the proposed name change, and reiterated it a year later when asked not to stand for reelection to the board. Dickerman thought there was more than enough work for the association on the national park front. The Wild and Scenic Rivers Act of 1968 specified that a system of national rivers be created, some of which would be administered by the National Park Service. No national organization was soliciting strong public support to ensure that legislation classifying specific rivers, according to the act, passed Congress. NPCA, he thought, should do this. It should also rally to defend interpretation in the national parks, which was falling to the budget axe. Additionally, the Park Service was claiming a backlog of $1.5 billion of new construction to be built within the national parks.

> Does anyone care to envision what sort of places our national parks would be if a billion and a half dollars worth of buildings, roads and utilities were imposed on them? To many people this sort of massive construction program proposed for the parks reveals the most fundamental error in the policies of the National Park Service today. It all too positively shows that the emphasis in Park Service thinking and administration is on the changes which man can make in the parks— rather than on preserving the natural values (the prime motivation in

establishing the national parks) and on increasing the people's under-standing and appreciation of those natural values. The NPCA should be exposing this gross deviation in Park Service policy and rallying public support for a return to the basic policies of preservation and interpretation laid down in the 1916 Act.[5]

Dickerman's words resound with echoes from Robert Sterling Yard and Devereux Butcher. He called for a militant defense of the parks, which should be, as it had been in the past, a full-time job. Dickerman concluded that "the NPCA should be developing the necessary leadership among its members and citizens generally so that they can fight the national park battles effectively. The day the National Parks and Conservation Association embarks on this sort of program in a serious and determined way, it will find that it has a full time job on its hands in protecting and enhancing one of the most precious jewels that we as Americans have, our National Park System."

Dickerman was not renominated to the board and took the floor to share these views at his final meeting in May 1971. He did not change association policy or direction, but his remarks touched a chord that reverberated through association discussions for the next decade. The name had been changed. NPCA was taking new directions, but not everyone among association lead-ers was sure it was on the right track. Everyone (including Dickerman) under-stood that the world was changing and new approaches to environmental pro-tection were necessary. Everyone knew the association had to adapt to its evolving social and political context. And everyone was committed to nation-al park protection. At issue was how to allocate the association's scarce resources to address problems that seemed to grow more complex. Here was the classic specialists-versus-generalists debate, and it would rumble beneath the outwardly placid surface of NPCA for some time.

NPCA Enters the Forests

During the mid-1960s the association's attention was drawn heavily to issues of forest management. In California the decades-long effort to secure a national park in the coastal redwoods crawled on while the big trees fell to the chainsaw at an alarming rate. Should a park ever be achieved, it would be remnant redwood stands surrounded by clearcuts. The task, as NPA saw it, was not only to protect what redwoods were left but to regenerate and prop-erly manage the redwood forest for both economic and "environmental" pur-poses. But ideas of "proper management" were also being challenged by new knowledge. Evidence was accumulating that management of sequoias in the Sierra would require new approaches to fire and root system protection from the impacts of growing numbers of visitors. Ecologists were learning the role of fire in this and other ecosystems, but the fire management policy of the Park Service and other agencies was slow to respond. Long-term management of the giant trees would require application of this new knowledge, and crowd dispersion to reduce visitor impact would require different management of the forests surrounding the park. Meanwhile, in the Pacific Northwest, the

An aerial view of the proposed Redwood National Park reveals how extensively logging had been done in the area.

push for a North Cascades National Park was reaching a climax. There the battle was between two factions: one wanted a park that embraced the ancient Northwest forests; the other wanted to exclude commercial forests from any park. The timber industry had struggled for years to get the big trees in Olympic National Park, and it was determined not to lose more timber to national parks and wilderness. New approaches seemed necessary to soften this either-or mentality. NPA recognized that new approaches to forestry were necessary to create and manage national parks in these forested regions.

In his 1967 report, Tony Smith said these challenges required "environmental forestry," which evolved into "socio-ecological" forestry and finally "ecological forestry." The association had concluded that a feasible way to reduce visitor pressure on national parks was to disperse visitors into surrounding national forests, but people would not seek recreation in clearcuts. Because no one seemed to be developing alternatives to current practices, the association entered the forestry arena. "Ecological forestry," Smith told the trustees in 1970, "implies harvesting methods which protect the soil, watercourses, watersheds, wildlife, scenic and recreational assets, and the timber stand itself against impairment. It means the use of selective cutting, small patch cutting, shelterwood, or at least diameter limits, as against clear cutting. It means a genuine long-term sustained yield." Such forestry may be expensive and thus regarded as uneconomic, "but in long-range terms it is the only socially economic system." Many goals could be addressed by saving forest environments worldwide. The "tide of death which engulfs the planet" could be reversed, endangered species rescued, soils and water conserved, and atmosphere replenished.[6]

NPCA launched its forestry program in 1971. Peter Twight, a forester with a background in the Forest Service, was hired as a forestry specialist. NPCA Vice Chairman Lawrence Merriam, Jr., a veteran forester, would advise and oversee the program, assisted by volunteer consulting forester Edward C. Crafts, Jr., another distinguished alumnus of the Forest Service who had served as first director of the Bureau of Outdoor Recreation. Twight studied examples of ecological forestry and, with the assistance of a forest economist, studied forest types to recommend guidelines for ecological and economic approaches to forestry in those forests. The association sought grants to support this work; in 1966 a substantial bequest to protect the sequoias had been received and would be dedicated largely to a study of that forest.

This initiative started well. The association received a grant from the Culpepper Foundation, and Twight began studying four methods of ecological forestry: individual tree selection, small-patch clearcutting, diameter limits, and shelterwood. Smith testified before the Senate Subcommittee on Public Lands of the Committee on Interior and Insular Affairs in April and laid out his plans for a comprehensive ecological forestry program. The Forest Service would use ecological methods throughout its domain. Large private holdings affected by a federal public interest, such as a valuable salmon stream or navigable waters, would also use the methods. Covenants would be used to establish ecological forestry on small forest ownerships, and federal aid would help all owners make the transition to new practices. Smith attacked the minimum stumpage pricing system used on public lands, arguing that the true value of forests to regions and the nation was being discounted. This, he said, must stop.[7] Although Smith's views were general and short on details, his vision, viewed from the perspective of thirty years, was remarkably prescient. At issue was whether the National Parks and Conservation Association, venturing into this new territory long occupied by others, could make a difference.

Twight set to work, and by 1973 had produced four studies of ecological forestry in the northern hardwood forest, the central hardwood forest, the coast redwoods, and the Douglas fir region. In each case, the studies suggested that some operators were using alternatives to clearcutting effectively and economically. Short-term economic returns might be reduced, but long-term benefits were greater. For all regions, knowledge was available about how to manage forests differently. The NPCA studies were published and widely distributed, and the magazine did summaries. Having completed this basic work, Twight returned to California and was replaced on the NPCA staff by another young forester, Tom Cobb. Cobb attempted to carry on Twight's effort, working on the sequoia and redwood forests and launching a study of ecological forestry in southern pine. Forest-oriented NPCA activity continued to expand, however, and Cobb found himself working on many fronts.

In 1974, Smith described the NPCA organization as having three principal "desks" devoted to parks, wildlife, and forestry. These desks were monitoring activities in the national park, wildlife refuge, and forest systems; thus, Cobb represented NPCA at a meeting convened to discuss the Forest Service's new ten-year "Environmental Program for the Future." A President's Advisory Panel on Timber and the Environment issued a report calling for large increases in annual timber cut on national forests, and Cobb drafted a response to that. The tussock moth was a growing problem in the Pacific Northwest, and the industry and Forest Service called for a lifting of the ban on DDT to combat this pest. NPCA opposed it. For many years, NPCA worked for the survival and recovery of the American chestnut, devastated throughout North America by blight. Cobb continued that effort. Senator Hubert Humphrey called on NPCA to help ensure that the new Forest and Rangeland Renewable Resources Planning Act (P.L. 93-398) "get off on the right foot," and that work also fell to Cobb. So it went, with the ecological forestry thrust fading beneath a growing list of problems handled by the "forestry desk."

In 1975 the forestry program fizzled out, a casualty of difficult budgetary times. The association was highly dependent on membership revenue, and membership recruitment and renewal dropped off precipitously in 1973 and 1974 for various reasons. Cuts in the NPCA operation were necessary, and when Tom Cobb decided to leave the association for an academic appointment, he was not replaced with a forest specialist. He continued to volunteer for a time, but the activity on this front was drastically reduced. Had the effort achieved anything? Another voice was added to the long-term debate about forestry on public lands.

The Forest Service did not change its practices—in fact, logging increased on the national forests under pressure from Congress, and the preferred method was clearcutting. The major reforms called for by Smith did not occur, although two decades later many of the broad concepts he advocated are again under serious study. The association's ecological forestry work was analysis and synthesis of a considerable amount of forestry research; it could not do this research itself. Twight and Cobb reviewed the growing literature, visited field experiments, and suggested the policy implications of what they learned. They told Smith that his direction seemed correct. Fire policy in forest management, dispersal of visitors, regional planning of public lands to include ecological factors, and general forest practices are now changing, but the impact of NPCA activity in this arena is impossible to assess. Latecomer and short-timer that it was in the forest debates, the association seems to have had slight influence.

The Wildlife Desk

At this time the association also made a foray into wildlife management. It had led the struggle against public hunting as a tool for managing wildlife in the national parks. It had embraced the Leopold Report with enthusiasm. Its trustees included reform-minded wildlife biologists. As its programs expanded, wildlife became a "desk." Smith had identified wildlife as an association program area in 1962, but little was done until the late 1960s, when concern about species extinction began to grow among conservationists in the United States and worldwide. NPA had long been involved with the International Union for the Conservation of Nature (IUCN), and this affiliation raised Smith's interest in wildlife. He attended the assembly of the IUCN in Switzerland in 1966, and Harold Coolidge, long-time association trustee, was elected IUCN president at that meeting.

In his golden anniversary year report to members, Smith wrote of the association's concern for wildlife:

> As we have commented often, the national parks are among our finest wildlife refuges. Within the national parks, visitors may see, photograph, approach, and study America's wildlife in its natural environment....One of the main purposes of our Association is to cultivate this sense of community with the animal world. A basic tenet of this relationship is the ecological principle...men have been learning hard lessons about the unbreakable connection between human society and the natural world which surrounds it. More and more we are realizing that we can not tamper with the delicate biological balances in our environment without endangering our own lives.[8]

NPA would attempt to mount a program in ecological wildlife management, as it did in forestry. "The effort is to protect, maintain, and where necessary reestablish natural ecosystems. We therefore favor the protection and reestablishment of predators. We seek protection of all species within complete ecosystems, not merely game animals, nor even those of special aesthetic or scientific interest. The attitude extends to invertebrates as well as vertebrates, and to plants as well as animals." Emphasis would be on pursuing these goals in national parks. Further, the association would work with the IUCN, the World Wildlife Fund, the U.S. Fish and Wildlife Service, and even the United Nations on the problem of species extinction.[6]

An administrative assistant for wildlife was hired in 1971. John W. Grandy was completing his doctorate in wildlife management and assumed the responsibility of putting flesh to the bones of the program about which Smith had so long talked and written. He would explore what the association could do in the complex realm of wildlife management and monitor wildlife-related activity on the legislative front. The magazine would publish one article in each issue on an endangered species, and the association would establish new working relationships with groups specializing in wildlife concerns.

Grandy worked with Smith on ocean mammal protection, conventions and legislation on endangered species, pesticide abuses that affected wildlife, and poisoning programs (predator control) on public lands. He staffed the "wildlife desk," and his responsibilities expanded. He worked to improve funding for the National Wildlife Refuges, analyzed and commented on proposed regulatory rule changes for the management of migratory birds, testified on technical issues in hearings on the Endangered Species Act, and supported creation

In the early 1970s, grizzly bear management in Yellowstone was a subject of controversy.

of new national wildlife refuges. And he found, as did Tom Cobb on the forestry desk, that his territory was growing faster than he could cover it. In 1974, in addition to these concerns, he advocated NPCA's position in the raging controversy about grizzly bear management in Yellowstone and monitored the implementation of the Endangered Species Act. John Grandy was a very busy man.

In 1975, Smith told Grandy that he might be laid off. When another job opportunity appeared, Grandy took it. (He has enjoyed a long and distinguished career with the Humane Society of the United States.) The budget was tight, and Smith decided he could not replace him. Toby Cooper, the heavily burdened administrative assistant for parks, picked up what he could of the wildlife program. Smith told the trustees that work on wildlife issues would continue, particularly involving endangered species. "Many elements of the life environment can survive grave damage but have the potential for recovery; this is true of clearcut forests and former wilderness; time restores areas to their natural condition. Plant and animal species may be depleted and yet recover; but a species once extinct is gone forever."[9] Hunting had appeared again, this time in the debate over proposed parks in Alaska. NPCA had successfully kept hunting from the parks in the 1960s, and its aim was to keep hunting out of the new Alaska parks.

The wildlife program also fell short of Smith's vision. As he repeatedly told the trustees, the association had been drawn into wildlife protection because national parks were critically important wildlife refuges. But protecting wildlife, particularly endangered species, could not be confined to parks. If NPCA were to help ensure survival of precarious species found in the parks, its reach would have to extend well beyond park boundaries. To ensure protection of the grizzly in Yellowstone, for example, regulatory tools were needed for use in the environment outside the park that would complement protection activities within it. Smith envisioned an NPCA that was a pioneer in wildlife management, but the association's limited resources would not allow that level of activity. NPCA contributed to the debate over the Endangered Species Act. Smith claimed it had influenced the legislation by arguing that population segments should be listed, rather than merely total worldwide populations. As with ecological forestry, the association made a contribution, but it was one among many voices in the wildlife management debates.

The third "desk" during this period was national parks. National park wilderness was the principal park effort in the late 1960s. Throughout this period, the association worked on issues involving Everglades National Park. So much was happening on the national park scene, in fact, that the association could not keep up.

THE NATIONAL PARK SCENE: NPCA STRUGGLES TO KEEP UP
A Complicated Mission

Description of the association's national park work from 1965 to 1975 is difficult because it is so diverse. The push was on to establish Redwood and North Cascades national parks during the late 1960s, and the NPA staff worked to support them. The Sierra Club led the effort, and NPA lent its Washington influence to the campaigns, testifying, publishing articles in the magazine, and keeping members informed about developments. A Redwood National Park, insufficient in the NPA view, was established in 1968, and an acceptable North Cascades National Park was finally created the same year. The association continued to work for additions to the redwood park that would provide better protection to the big trees, and it opposed tramways in the North Cascades.

The association took a strong stand on the policy issue of visitor transportation. It believed that park roads should be "stabilized," that is, minimal new miles added to the system and existing roads used primarily, even exclusively in some cases, by small buses. People should park their vehicles outside parks. Further, monorails, railways, tramways, and aerial cable cars should not be built in the parks. Such conveyances would require new rights-of-way. People should use the existing roads. The transit approach and dispersing visitors into surrounding areas should relieve the overcrowding that plagued many parks.

As the federal budget tightened with the escalation of the war in Vietnam, the Park Service was squeezed into reducing personnel, especially in interpretation. The association protested these cuts; it also objected to a proposal to contract national park campgrounds to concessionaires. The Park Service argued that this would save money, but the association countered that it would create more problems than it would solve, and few savings.

Wilderness Review

One major national park push of NPA in the 1960s involved wilderness. Section 3C of the Wilderness Act specified that wilderness review in national parks be done within ten years. Specifically, the act stated:

> Within ten years after the effective date of this Act the Secretary of the Interior shall review every roadless area of five thousand contiguous acres or more in the national parks, monuments and other units of the national park system...and shall report to the President his recommendation as to the suitability or non-suitability of each such area or island for preservation as wilderness.[10]

The act also stipulated that the mandated process of wilderness review be public and involve hearings. A recommendation for or against wilderness might be made by the president to Congress, which would pass authorizing legislation to make recommended wilderness part of the National Wilderness Preservation System. The National Park Service would conduct the reviews.

The Park Service, under George Hartzog, was not enthusiastic about this assignment. It had not supported the Wilderness Act; it did not think the act necessary or helpful, focused more on taking the parks to the people than on wilderness, and did not eagerly carry out the will of Congress as expressed in the act. Because it could not entirely ignore the congressional mandate, it began wilderness reviews. NPA decided to do what it could to influence the outcome of this process, recognizing that when matters reached Congress, its lobbying restrictions would limit its scope of activity.

Deciding that NPA would view wilderness planning as part of regional recreation planning, Smith sought to use the Wilderness Act to promote this regional planning. Wilderness protection would require providing for recreation outside of wilderness areas so that they would not be crushed by overuse. Further, wilderness protection by the public required that people know how their outdoor recreation needs might be met in surrounding areas. These were political and educational challenges made for NPA.

By 1967 the Park Service began its review, and hearings came thick and fast. NPA hired consultants to prepare wilderness studies for the hearings. William J. Hart and later Jonas Morris conducted twenty-four wilderness studies between 1966 and 1971. The association took the position that wilderness protection should be provided in nearly all roadless areas: The test for wilderness designation was not remoteness but whether the area needed protection from development. Streams and lakes in parks should be wilderness.

NPA and other wilderness advocates objected to many aspects of the Park Service's studies. Much, they thought, was left out of proposed park wilder-

North Cascades National Park, established in 1968, met the standard for national parks as wild and inspiring places so long championed by the NPA.

ness because the Park Service insisted on pulling boundaries back from roads and the exterior boundaries of the parks. NPA had earlier advocated buffers around parks, but the Park Service used them in a way unacceptable to the association. Buffers would be *inside* park boundaries, and wide road corridors would buffer wilderness users from the sound of automobiles. The association believed that only quiet mass transit vehicles should be allowed in parks with wilderness, and because these would not be noisy, wilderness boundaries need not be drawn far from the roads as the Park Service proposed. NPA also objected to the Park Service's omitting areas where it had plans for future roads and other developments. It accused the service of minimizing wilderness protection to allow for future development well beyond what was reasonable. The Park Service even left enclaves in the remote backcountry where it might build shelters and huts, as in the new North Cascades National Park.[11]

By 1968 Smith reported to his trustees that NPA's work on wilderness was having little effect. "In view of the indifference accorded to our studies by the National Park Service, it is possible that we should now retrench and prepare materials elucidating our general approach," he told them. The association would do occasional studies when individual park planning required it, but it would curtail its level of effort on this front.[4] By 1969 Smith was thinking that no wilderness designation would be better than what the Park Service was promoting. "Within the parks, the danger is that hasty wilderness designation, without reference to broader plans for the dispersion of visitation, will result in unduly small wilderness areas, with large potential facility areas."[12] In 1971 the association published *Preserving Wilderness in Our Parks*, which presented its twenty-four wilderness studies. After that, NPCA largely retired from wilderness work. The Park Service was not budging, and there was much else to do.

As if preparing for these wilderness hearings were not enough, master planning was also under way in many parks, and the association attempted to review and comment on all these plans. Much legislation was introduced into Congress that affected parks, such as the Wild and Scenic Rivers Act and the National Trails Act. Staff members monitored these bills, testifying when invited and describing the legislation in the magazine. They also followed all new park legislation.

Smith was reluctant to expand the staff. During the 1960s, he did much of this park work himself, assisted by consultants. By the early 1970s the association had growth and assets sufficient to expand the staff, so it hired administrative assistants for forestry, wildlife, and the national park system. Jonas Morris became the national park system specialist in 1971. He served ably in this role for two years, then Clay Peters replaced him, only to leave before he really started. Toby Cooper was hired to work on parks in 1974. He worked closely with T. Destry Jarvis, who joined the group in 1972 and whose initial job description had him working on legislative and international affairs. When Cooper left during the budget crisis of 1975, Jarvis became administrative assistant for parks and conservation.

Regional Recreation Planning

This national park work had three principal dimensions: responding to crises and proposals of the moment; tending to long-term problems such as water for the Everglades, national park wilderness, and national parks in Alaska; and working on large issues of principle, such as promotion of comprehensive regional recreation planning. Regional recreation planning consumed considerable association thought, time, and resources during the 1960s and 1970s. As early as 1962, Tony Smith told the trustees that regional planning was needed to protect parks. In 1964 he reported that the association would mount a major campaign to promote comprehensive regional recreation planning as a way to reduce pressure on national parks. The Wilderness Act added impetus to pursue such planning.

Smith thought some of the visitation crushing national parks could be diverted to other lands around the parks. Activities not appropriate in national parks, such as motorboating on Yellowstone Lake, could be enjoyed on the reservoirs of the Bureau of Reclamation. He saw the recently created Land and Water Conservation Fund as a wonderful opportunity to identify recreation needs in regions and respond to them with local, federal, and state sites and facilities, all developed to be complementary. The national parks could be protected and enjoyed, but they must find a way to limit visitors. Planning could solve problems in Yellowstone, the Everglades, the Potomac Basin, even the

Lower Colorado. This initiative would start with a demonstration of the concept in the Yellowstone and Grand Teton region.[13]

A Yellowstone plan would involve the two parks, Reclamation Bureau lands and waters, Bureau of Land Management lands, Indian lands, state and local public lands, and private holdings in a large area between the 42nd and 46th parallels and the 108th and 112th meridians. This region encompassed much of Montana, Wyoming, and Idaho and in 1965 included eight national forests with four wilderness areas and six primitive areas. Thousands of campsites were either built or contemplated in the forests and areas administered by the Bureau of Land Management and the Bureau of Reclamation. Surveys revealed that most motorboat visitation to Yellowstone came from within a 400-mile radius and that within that radius were forty-six reservoirs, most of which were open to motorboats. If launch facilities were installed in those reservoirs and drawdowns timed around peak recreation use, the need for motorboat recreation in the region could be met while protecting the wildlife and other natural values of Yellowstone Lake. Nonmotorized boaters would have a few lakes in the region to themselves, and a conflict between recreation users would be reduced or eliminated. Visitors could get comprehensive information about the options in the region, and a reservation system for the national parks could be used to control visitation. Those without reservations would be advised on their options.[14]

Planning for national parks was not a new idea. Master planning had long been used throughout the system by the National Park Service. What was new was the idea of interagency planning over a large region, and plans that included state and local government and private providers of outdoor recreation services. Federal agencies were wrestling with new classification systems offered by the Park Service and the Bureau of Outdoor Recreation. Smith explained his ideas in the context of these systems, noting that there were uncertainties in another round of Yellowstone master planning complicated by the recently passed Wilderness Act. The matter seemed very complex, but the basic idea was simple: Get together, cooperate, think big, and all would be well. He was convinced that the time was right for this approach. The statutory authority was in place—the Land and Water Conservation Fund Act bestowed it on the Bureau of Outdoor Recreation. All that was needed was for the secretary of the Interior to tell the Bureau of Outdoor Recreation director to "embark on this course."[15]

This planning concept tied in with a number of other NPA projects. Ecological forestry would allow timber production in national forests while retaining recreational values. Transportation systems could be developed that would keep the noisy internal combustion engine out of the parks while providing for it elsewhere; the sounds of nature in the parks would not be drowned out by the electric motors of the small buses transporting people in the parks. Areas managed as wilderness could come close to park roads, for quiet vehicles would not intrude far into the backcountry.

Smith met with Director Hartzog to discuss his planning ideas. He reported that he was finding considerable support among Park Service people, although the agency did not seem to be altering its approach. In 1967 the President's Council on Recreation and Natural Beauty invited the association to recommend government policy on comprehensive recreational regional planning, and it did. The recommendations, "Providing for Ample Outdoor Recreation Facilities of all Kinds and at the Same Time Protecting Trail-type Recreation Country by Means of Comprehensive Recreational Regional Planning and the Dispersion of Visitation," included all precepts guiding the association's wilderness planning effort. Noting the strong emphasis on protection in the National Park Service Act and most of the statutes establishing individual parks, the report said:

> This fundamental policy of protection is not in conflict with the national policy of providing abundant outdoor recreational resources for all the people; the two purposes can be achieved simultaneously if planning for protection and recreation is done on a sufficiently broad regional basis.[16]

The report described recommended roles for all federal outdoor recreation agencies and specific proposals for action. Buffer zones on public lands surrounding the parks should be established in which management "would be conducted with a high priority for scenic and ecological protection. Wilderness should begin at the roadside in the parks. The Bureau of Outdoor Recreation will be the coordinator of all of this."

The president's council did not change the face of outdoor recreation planning. The secretary of the Interior did not instruct the Bureau of Outdoor Recreation director to take this new approach. Why not? For one thing, the agencies jealously guarded their territories. If Smith had studied the history of federal interagency rivalry, he might have been less optimistic in his rhetoric. The Forest Service was not about to change its management policies to benefit the parks. After all, it had lost 4.5 million acres of its domain to the national parks and in 1967 was facing the prospect of losing more in the North Cascades, where it was fighting tenaciously against a national park. Although it was involved in recreation, its principal mission was "commodity production," as it euphemistically called its timber operations, and it had little incentive to cooperate with a recreation agency at the expense of timber production. There was also, as Clarke and McCool have described, a power differential among the various agencies that would have to cooperate to make the regional planning approach succeed.[17] In addition, the Bureau of Outdoor Recreation was not in a position to direct its older and more powerful brethren among federal natural resource agencies to do anything. Its relations with the Park Service, which viewed it as an interloper on its territory, were not good. Furthermore, its priorities were elsewhere, as Foresta has noted:

> Although coordinating the outdoor recreation activities of other federal agencies was part of BOR's official charge, the new bureau, a small one with support in the administration but with no power base

in Congress and no articulate body of supporters in society at large, was no match for the older agencies like the Forest Service or the Park Service, agencies whose policies it was supposed to coordinate. BOR quickly recognized the obvious, that is to say, since it could not impose its will on them, it would have little real say in recreation policies for federal lands. Accordingly, it abandoned efforts in this direction under a rhetoric of compromise and conciliation and put its energies into its responsibilities for federal liaison with lower levels of government.[18]

The idea of coordinated and consistent interagency planning for outdoor recreation was attractive and compelling but, at least in the 1960s and 1970s, not practical. Smith's views to the contrary, the time was not right for this approach. NPA could be visionary at times, and this was such a time, but its efforts did not yield significant consequences. Some of the ideas it espoused, such as dispersing visitors and using transit in parks, began to be explored in the 1970s (and Smith claimed credit for the association), but the radical new regional approach did not catch on in the late 1960s.

The Everglades Coalition

NPCA continued its traditional role as watchdog of the National Park Service, scrutinizing its activities and voicing its concerns. It also continued to watch for threats to the parks. George B. Hartzog succeeded Connie Wirth as National Park Service director, and with a Democratic administration in the White House and an activist secretary of the Interior, this was a busy period in national park history. The Mission 66 program pumped nearly a billion dollars into park development, but the system remained stressed by growing visitation. This ten-year project did "comparatively little for the plants and animals," and "nothing at all for the ecological maintenance of a park."[19]

As 1966 approached, the use-versus-preservation dilemma that had plagued the Park Service from its beginning loomed larger, and in 1964 Hartzog and Udall attempted to address it with a new administrative policy. They divided parklands into three categories: natural, historical, and recreational, an approach that allowed the Park Service to adopt preservation as the leading goal on some but not all of its lands. At first NPA liked this new approach, but as the National Park System expanded by fifty-two new areas between 1964 and 1968, Smith and others voiced concerns. Planning for the natural parks proceeded at a snail's pace and did not provide enough protection for resources. The largest park defense during this period was mounted for Everglades National Park. The association had been involved in Everglades-related issues since the early 1930s. In the 1940s, it promoted land acquisition for the park, and throughout the 1950s and 1960s worked on ensuring a supply of fresh water to the Everglades in the face of rapid and massive development in south Florida. The ecological threat of an inadequate water supply remains even now. A proposal to construct a vast jetport for south Florida galvanized conservationists in the late 1960s and engaged NPA and other conservationists with the Everglades as never before.

George Hartzog, Jr., succeeded Wirth as NPS director and turned the agency's priorities in new directions.

Routine monitoring of Everglades National Park problems in the mid-1960s revealed that the Army Corps of Engineers was working at the behest of Congress on another water engineering plan for south Florida. The association did not think the Army Corps was the proper agency to plan for south Florida, the Everglades, or any other region and said so in hearings. It promoted its comprehensive regional planning approach and threatened suit if Army Corps plans did not provide essential water to the national park. Late in 1968, it detected a new threat to the Everglades—a jetport with connecting highways proposed by the Dade County Port Authority and backed by the U.S. Department of Transportation. The project was well along before NPA and other Everglades protectionists learned of it. The airport would be constructed in Big Cypress Swamp west of Everglades Conservation Area Number 3 and north of the park.

Smith wrote editorials about this new threat and told the trustees that the Everglades problem now had three parts—the airport, the inadequate water engineering plan of the Army Corps of Engineers, and the need to protect Big Cypress Swamp. In March Elvis Stahr, president of the National Audubon Society, called Smith and suggested a meeting on the Everglades situation, which brought various conservationists together on April 13. The group agreed to protest the airport, and Smith was drafted to prepare a letter to Transportation Secretary John Volpe, to be signed by as many conservation organizations as could be recruited. The letter was signed by twenty-two organizations, which became the Everglades Coalition. Stahr and Smith co-chaired the coalition.

The letter went to Volpe and was copied to other officials on April 17. Six days later, the coalition met at NPCA headquarters and agreed to accept no

NPA began working to protect the Everglades in the early 1930s.

less than complete abandonment of the airport project. Coalition members shared their view in a press conference after the meeting. Soon the secretaries of Interior and Transportation issued statements that they could not support the jetport, although they thought protective measures could be arranged to allow the use of a training strip, already almost completed. They appointed an interdepartmental committee to study the situation and recommend how or if this goal could be achieved. The Dade County Port Authority, meanwhile, issued its own statement that it would continue with the project regardless of the federal government's decision.

The coalition considered legal action, contending that in providing funds to support the project, the federal government had violated the Transportation Act, which explicitly stated that transportation facilities would not encroach on park and conservation areas unless and until all other siting possibilities had been explored and plans drawn to minimize ecological damage. The interdepartmental committee reported that the jetport project seriously threatened the Everglades. An agreement was reached among the departments of Transportation and Interior, the state of Florida, and Dade County to find another site for the airport and remove the existing training runway. The conservationists, now called "environmentalists" in the early 1970s, had won a major fight. Tony Smith claimed a central role in building the coalition and leading the struggle with his colleagues from the Audubon Society. The struggle for the Everglades was not over, but another battle had been won.

Smith told NPCA trustees in 1970 that final success in the Everglades would require a multi-faceted program, with regional planning at its core. A firm commitment must be wrung from the Army Corps to provide a reliable

A huge jetport was proposed and construction begun just seven miles north of Everglades National Park. The Everglades Coalition stopped the project in 1970.

supply of fresh water to the east side of the park, even in periods of drought like the current one. To stop the development that threatened the park's ecological integrity, Big Cypress Swamp and inholdings in the park had to be publicly owned. The Land and Water Conservation Fund should provide resources for acquisition, which he estimated would cost about $80 million. Water flow through Big Cypress into the west side of the park had to be ensured. (Audubon and NPCA jointly acquired land in Big Cypress and successfully filed suit to stop the creation of a drainage district in Gum Slough, which they believed would lead to development.) In the long run, Everglades National Park could be protected only through comprehensive planning for federal administration of the region that should include protecting wilderness, dispersing visitors, and preventing development in the park. NPCA had developed a regional plan that might provide a beginning.[20] Water supply and threat of development would continue to plague Everglades National Park.

GLOBAL CONSERVATION
An Enduring International Concern

While he focused on wilderness, regional planning, and park defense, Smith's attention was also drawn to international conservation affairs, an arena that had long been one of his personal interests and an interest of the association. Robert Sterling Yard and Fred Packard had both nurtured interest in this international scene. In 1969 Smith saw a world in severe crisis. Great environmental problems loomed on every front. The United Nations was planning a world conference on the environment for 1972, and Smith thought

NPA should be a part of that historic meeting. Working with the IUCN, the World Wildlife Fund, and others, NPA advocated a U.N. Environment and Population Organization comparable to the Food and Agriculture Organization. Smith also hired Robert Cook, former president of the Population Reference Bureau, as a consultant to help the association monitor government activity on population while it pushed for this new U.N. organization. Arguing for attention to population, Smith told the trustees:

> Most of the environmental problems must be attacked on their merits, but at the same time most of them are undergirded by over-population. This is not merely a matter of environmental niceties, of parks and scenery; nor even a matter of the survival of other species; it is a question of human survival in terms of water and air pollution, food supply, and the psychological pressures which result from congestion and which may be leading segments of society toward suicidal and murderous decisions.[6]

The association set up a Special Committee on Population Policy, which studied the situation and made policy recommendations. This sparked a debate that continued for several years.

In 1971 the association sponsored a meeting about endangered cats of the world and increased its efforts to be involved in the U.N. Conference on the Environment at Stockholm in 1972 and the U.N. Conference on Law of the Sea in 1973. Some trustees questioned this international effort, but Smith told them that "The major conservation and environmental organizations are moving strongly into the international field. It is important on the merits that NPCA involve itself in such work. Without regard to the merits, NPCA would not be able to maintain its competitive business position and make necessary appeals for new and renewal membership unless it were to do so."[21] Support was sufficient to forge ahead, and Smith hired a specialist in ocean law, Robert Eisenbud, who would be his "personal representative in United Nations Matters," especially in Law of the Sea work. Smith worked with others to set up a structure of nongovernmental organizations for the Stockholm meeting, and NPCA attained accredited observer status for the meeting.

Smith attended the Stockholm conference, launching several years of furious activity on the international front. NPCA submitted a twenty-three-point program to the plenary meeting and tried to revise the official recommendations of the preparatory commission on wildlife management, arguing for ecological valuation of wildlife to complement the economic values emphasized by the commission. With a tiny delegation of two, NPCA had minimal effect. Smith was pleased the meeting had occurred but not happy with the U.S. positions on key issues. He thought private organizations such as NPCA should work to influence the official U.S. government position on issues, moving the country to take greater responsibility for solving international environmental problems. The Stockholm experience firmed his resolve to make the association a player on the world environmental stage.

In 1972 Smith served on the Advisory Committee to the Secretary of State on the U.N. Law of the Sea Conference. Gilbert Stucker, a trustee from New York, was the official NPCA representative on environmental matters at the United Nations. Smith traveled to Banff, in Alberta, Canada, to participate in the 1972 IUCN meeting, then to Grand Teton National Park for the Second World Conference on National Parks. The following year found him at Morges, Switzerland, with the IUCN, and in Geneva attending the World Assembly of Nongovernmental Organizations. He attended two weeks of the Law of the Sea Conference in New York and prepared to be part of the official U.S. delegation to the meeting of the Seabed Committee, also in Geneva.

Questions of Survival

Throughout this period, Smith explained to the trustees the importance of NPCA's "World Program." On the one hand, this work was a matter of survival. "The issues with which we are dealing in our World Programs are basic to the survival of life on earth," he told the trustees in October of 1973. "It cannot be repeated too often that the present ecological crisis does indeed involve the survival of life on earth. This is not irresponsible alarmism. On the contrary, irresponsibility attaches to those who would play down the danger and have us believe that all will turn out well, even if we do not do much about it." On the other hand, Smith was convinced there were pragmatic reasons for this work.

> The competitive position of the NPCA in the environmental movement in the United States compels us to participate in these activities. The Sierra Club and the National Audubon Society maintain full-time international representatives in New York. The National Wildlife Federation publishes a separate magazine, *International Wildlife,* and maintains an international department. The Friends of the Earth maintain chapters in quite a number of other countries. The NPCA has a special position to maintain in relation to the international movement for international parks.[22]

All major environmental groups grappled with their mission at this time, and most, like NPCA, expanded to embrace global concerns. "Environmentalism," in contrast to traditional conservation, seemed to require this. Anything less, many thought, might result in winning a battle here and there for a park or a species, only to lose the war to global ecological disasters.

Administrative duties for the association demanded more of Smith's attention in 1973 and 1974. Although he managed to attend a Law of the Sea Meeting in Caracas, Venezuela, and an IUCN meeting in Kinshasa, Zaire, he was forced to reduce his travel and cut back on his ambitious "World Program." With his law background, Smith was especially interested in the Law of the Sea challenges, which he continued to pursue as much as possible. He did so, in fact, until he left the association in 1980. In his view, the environmental community was not adequately involved in this work. In 1976 he told the trustees that "I find myself in the same position as always, being the

lone environmentalist engaged with these problems."[23] Decline in membership and fiscal difficulties required the reduction of international activity, and Smith admitted in 1976 that "We have fallen far short of what could have been done."[24] Still, he carried on, and in 1977, when the fiscal situation stabilized, he stepped up his global work, focusing on population growth and illegal immigration along with Law of the Sea. In 1978 he hired an administrative assistant for immigration and population and organized an Environmental Working Group on the Law of the Sea.

What did immigration have to do with environmental concerns, with national parks? Someone must have asked this; in 1978 two editorials in *National Parks and Conservation Magazine*, one by Tony Smith and the other by Gerda Bikales, the new NPCA population expert, attempted answers. Underlying all environmental problems, they argued, was human population growth. In the United States, that growth was fueled by illegal immigration. "The natural increase of our population is now about 1.2 million a year," Smith told his readers. This "natural" growth will dwindle, but the 1.2 million annual population increase by immigration, two-thirds illegal, would continue.

> Conservation organizations which say that illegal immigration and a rising population are none of their concern, should think the problem through.... If they are concerned with the preservation of parks, wilderness, forests, wildlife, they must be blind indeed, if they think they can cope with the shock-waves of an uncontained population explosion, of which illegal immigration is now the most important component.[25]

Environmental organizations, said Smith, should join the AFL-CIO in its strong stand against illegal immigration, and he offered a detailed program to combat the problem. His suggestions ranged from a forgery-proof social security card to economic aid to help poor countries, such as Mexico, create jobs that would keep people at home.

A careful reading of Tony Smith's editorials in virtually every issue of *National Parks and Conservation Magazine* leaves no doubt that he was strongly and genuinely convinced that NPCA's mission was to pursue a wide agenda on both national and international fronts. Smith led the association for nearly twenty-two years, and each year his agenda became broader and more ambitious. To the job, he brought a long-standing interest in international affairs, which was fed by events in the late 1960s and the 1970s. The years would show that his views were often on the mark. Illegal immigration would create problems. Nations unable to agree on how to manage the "commons" of the oceans would seriously deplete fish stocks, a critical source of protein for a growing human population. But should NPCA deal with such issues? Time after time during his tenure as association president, Smith addressed this

Growing understanding of ecology revealed new challenges to those who sought to protect rich natural systems like the Everglades that were tied inextricably to areas outside park boundaries.

question directly and answered in the affirmative. The association throughout its history had limited means with which to address its agenda. As long as membership growth continued, it could take on larger challenges, but when growth lagged, as it did in the mid-1970s and again in the late 1970s, difficult choices had to be made. What were the priorities? Some trustees still believed the national parks should be the association's primary concern.

Vision and Reality

The Everglades episode illustrates that NPCA continued to be active on the park protection front, even as its resources were drawn into forestry, wildlife protection, and international affairs. It lent its hand in many park issues during these years, seldom playing a leading or central role, but always doing what it could when development threatened or a substandard area was proposed for inclusion in the system. Revisions of the tax code in the mid-1970s allowed the association to be more active in Congress. Since 1960 it had testified only "by invitation" and studiously avoided expressing views on legislation. Late in 1976 Smith told trustees that NPCA could now express its views on legislation in *National Parks and Conservation Magazine* without threatening its tax-deductible status. Environmental groups could propose legislation, draft it, and even lobby. Staff members T. Destry Jarvis and William Lienesch led a more active parks program on Capitol Hill.

The financial crisis of 1974 and 1975 led to the closing of the forestry and wildlife "desks." Smith did not stop his international work or completely abandon his interest in forestry, wildlife, population, and other parts of the broad agenda, but NPCA refocused its attention on national parks. This renewed interest went in three main directions: Jarvis concentrated on issues involving the National Park System, such as the long struggle for new parks in Alaska, park planning, and reorganizing of the National Park Service; Lienesch, formerly an Environmental Protection Agency official, was hired in 1977 to oversee efforts on the urban and regional park initiatives under way in Congress; and Helen B. Byrd oversaw association efforts on the historic preservation part of the National Park System.

After the stormy initial years of Tony Smith's administration, NPCA's organizational "command structure" settled down. Smith ran the show, the board was supportive, the association grew, and there was no dearth of work. The decade from 1965 to 1975 was a difficult time to lead a conservation organization, as Smith's executive colleagues David Brower of the Sierra Club and Stuart Brandborg of the Wilderness Society learned. They were removed from their executive positions in painful organizational upheavals. Frank Graham, Jr., in his history of the National Audubon Society, summed up the period in a rather understated way: "Amid the conflicting claims of science, politics, and public relations, conservation organizations of the 1970's found it increasingly difficult to set policy."[26]

In the mid-1970s, the National Parks and Conservation Association seemed to have survived its identity crisis, but trying times were ahead. The association's focus was returning to national parks, but it could not revert to old ways of doing business. The National Park Service seemed to be in a perpetual state of crisis and change. People flocked to the parks in growing numbers. The problem of reconciling protection with use loomed large. A bewildering array of environmental and park issues faced those who, like Anthony Wayne Smith, believed in progress, reason, hard work, and the potential of government to solve problems.

XI

A Crisis of Leadership

As NPCA entered its sixth decade, its leadership was dominated by a single personality, and its organizational success or failure depended on him. The "old" leaders—Wharton, Butcher, Olson—were gone from association circles, and with them, the tradition of shared responsibility in which the executive and the board carried part of the burden of leadership. Tony Smith had set the organization on new paths, and he directed virtually every detail of its affairs. He believed that the new direction was the correct one, and that the association would rise to new levels of prestige and accomplishment as a consequence of its redirected effort. The association bylaws had been amended to give the executive control, and he wielded that control in an aggressive and autocratic fashion.

In following Smith's career in the 1970s, one wonders whether he noticed what had happened to his colleague David Brower in the Sierra Club. Brower had taken bold action in fighting Grand Canyon dams, which had cost his organization its tax-deductible status, and he had not heeded the direction of his board. As good as he was at leading and winning conservation battles, when his actions led to financial strain on his organization, his directors removed him from power. While Brower was struggling, Smith was riding high, his organization growing, its finances improving, and his programs expanding. He saw himself guiding NPCA indefinitely into the future.

Smith was trying to redefine the nature of the association. The change in name to add "conservation" was more than symbolic. Smith intended, as the last chapter reveals, to alter its mission, and he seemed to have the support he needed to achieve the changes. Over the years his personal interest in national park matters waned. He knew the association should continue to attend its traditional mission, but he was drawn elsewhere. Because he was so much the

The struggle over national parklands in Alaska raised many issues of concern to NPCA. One of them involved the level of protection for wildlife and prohibition of hunting in national parks and preserves there.

leadership of the association, his waning interest in parks reduced the association's influence in that arena, and when his plans to move into global environmental affairs, regional planning, forestry, and wildlife protection were thwarted by obstacles including a drop in membership and financial resources, both he and the association were in trouble.

NPCA FALTERS
Eroding Finances and Administrative Disarray

Anthony Wayne Smith reported to the executive committee early in 1969 that the financial condition of the association, although good, was fragile. A building had been purchased in 1967, and stocks were liquidated in 1968 to retire the debt. The association depended on income from memberships and contributions for 95 percent of its operation. He warned that should returns on membership solicitation decline, there would be trouble. Programs were expanding, and staff members were hired to cover new responsibilities. Until recently, the policy had been to hire consultants to do additional work, but the time had come to expand the in-house professional staff. Smith was confident that with adequate investment in member solicitation, the prospects for financial strength were good. He offered his plan—the key to which would be upward revision of the dues structure—to expand the magazine, enlarge the staff, increase salaries, develop advertising, and maintain a broad environmental program.[1]

Smith's approach was approved, and he began implementing it. Dues were increased in 1969 and again in 1972. He began to hire staff, bringing on assistants for his international work (Robert Eisenbud), administration (Larry LaFranchi), wildlife (John Grandy), and the National Park System (Jonas Morris). The association's assets grew to nearly $600,000 in 1969 and 1970, excluding the building. Operating costs increased steadily at 5 percent per year, and Smith told the board in 1971 that he expected to hire two more professionals, one as press officer and the other as administrative assistant for communications. Late in 1971, however, he began to see warning signs. He had said in 1969 that a recession could hurt membership recruitment, which seemed to be happening. The country dropped into a recession, and membership followed the downward track. Total association memberships, including schools and libraries, reached an all-time high in 1971 of nearly 53,000, then started slowly downward in 1972 and dropped precipitously in 1973. The December 1973 membership report counted slightly more than 40,000, and the following year it stood at 39,000.

As Smith predicted, the financial pain from the decline was severe. To meet the payroll and other obligations, capital assets were drawn. By 1975 these assets had dropped to $250,000, not counting the building, and the situation was serious. As staff resigned for one reason or another, they were not replaced. The size of the magazine was reduced. Savings were sought everywhere. Smith could not travel to his international meetings (at least not as frequently), and the association significantly reduced its programs.

Extension of the National Park System into urban settings might meet recreation needs but was, in the NPCA view, stretching the ideals of the system to and beyond its limits.

What happened? Smith's prediction that a recession might severely hurt the organization proved correct. He also discovered that staff had erred in several of the new-member solicitation campaigns; mailings were ill timed. Mailing lists were in disarray, and membership service declined. People had paid their dues and not received *National Parks and Conservation Magazine* for months, if at all. Blame could be laid on various links in the chain. In some cases, staff members were responsible for the breakdowns and were replaced. The Postal Service was cited as part of the problem.

The time had come to move to a computerized system, and that was accomplished, though not without difficulty and considerable expense. Smith reported to the trustees that NPCA was not the only suffering environmental group. Many organizations had seen a drop in membership after the burst of growth in 1970 with Earth Day and the emergence of environmentalism as a popular cause. The "falling away of public interest in environmental questions during recent months may well be due to the intense concern which began in the United States several years ago...," Smith told the annual meeting in May 1973. He thought this concern had degenerated into alarmism and produced "a tendency toward complacency."[2] The only recourse for the association was to cut costs, redouble recruitment efforts, and work through the difficulties.

By late 1974, Smith could report modest success in dealing with the crisis. Membership had climbed back to 46,000. Significant capital assets remained. Dues had been increased again (planned increases were postponed during the crisis so as not to drive members away), and dues categories were restructured. More cuts were made throughout the operation. By late 1975, membership seemed to stabilize at about 45,000, assets climbed, and the professional staff was down to two: T. Destry Jarvis covered "parks and conservation," and Rita Molyneaux was assigned information and legislation. The crisis curtailed some of Smith's activity, forcing him to focus on the business affairs of the association.

Clarence Cottam stepped down as chairman of the board at the 1973 annual meeting. He had served for a dozen years and was forced by his age and health to relinquish the chair. Spencer Smith succeeded him, and then he, too, stepped down in November of 1975, resigning from the board after serving for twenty years. He had taken a position as an economist for the Democratic Steering and Policy Committee and wished to avoid any conflict of interest. With the departure of Cottam and Smith, Tony Smith lost two of his staunchest supporters. They had led—some might say controlled—the board for nearly fifteen years. The chairmanship was assumed by an archaeologist from New York, Gilbert Stucker.

Although he admitted no responsibility, Tony Smith was not entirely blameless in the fiscal crisis of the mid-1970s. In October 1974 he sent a confidential letter to the trustees in which he described the internal problems of the organization as he saw them and asked for their help in identifying potential wealthy donors. In this letter, he described how he had changed the director of membership services twice in fifteen months, changed business managers three times in the same period, and been through three computer companies.[3] Some trustees believed that this turnover, a "comedy of errors" one called it, revealed administrative ineptitude. One of the dismissed staff members wrote a detailed letter to the trustees explaining that he had suggested measures to help in the crisis, but that Smith had rejected them all. The president would not take the staff's suggestions, wrote O. J. Neslage:

> He has been insulated too long and too much the master of the Association. He will not listen to anyone but himself nor will he brook interference on any matter whether it be a major decision or something simple…The Association has gone through entirely too many Business Managers and Membership Directors over the last several years. If I were the Board, I would wonder why and then I would do something about it.[4]

The board took no action against Smith, though when he polled them on taking out a line of credit, he received a strong negative reaction from several trustees.

Back to the Parks

This association upheaval seems to have brought a modest refocusing on national park work. Smith and the trustees realized that there might be problems to solve on many fronts, but NPCA simply did not have the resources to cover all of the territory. If it could not do everything, perhaps it should cover the parks territory on which it had the strongest claim. Furthermore, there was much work to be done on parks, even more than five years earlier when Ernie Dickerman attempted to direct their attention back to more traditional con-

A stern NPCA Board, led by Clarence Cottam, faced serious challenge to the viability of the association in the early 1970s. Tony Smith sits to Cottam's right.

cerns. Tony Smith, though, could not bring himself to focus *totally* on the parks. He deeply believed that at least some effort on the international scene was essential. As he told the board repeatedly, the association needed breadth to compete with other environmental groups for members. He claimed that many people thought globally and would not be content to join an organization concerned only with what they might think of as provincial problems. Although the focus might return in large measure to the parks, work on Law of the Sea and population would continue.

The situation had improved enough by 1977 that Smith could once again talk about building the professional staff. He told the board that his first priority was to "round out" the parks staff by adding an administrative assistant to cover historic parks and preservation. He hoped to add specialists in forestry and wildlife to reopen those "desks," but these new hires required increased grant support. He would write foundation grant proposals to dedicate funding to these staff additions. Because he was still restoring the association's capital reserves, he could not expect to fund these staff positions from membership revenues for a while.

Bill Lienesch was hired as administrative assistant for "Parks and Land" in 1977. A foundation grant received early in 1978 allowed Smith to hire a staff specialist in historic preservation. Helen B. Byrd became special assistant for the "Historic Parks, Heritage, and Preservation" program in 1978. With yet another grant, Gerda V. Bikales was brought on as special assistant for "Immigration and Population." These new staff members joined T. Destry Jarvis and magazine editor Eugenia Horstman Connally (who had replaced

Paul Tilden in 1969) as the association's "professional" staff. They presented reports every six months, which were submitted to the trustees, and these reports indicate that most of the substantive work was done in the national park arena, with Jarvis and Lienesch contributing significantly.

Staff reports for April 1978, for instance, find Jarvis working on a huge menu of projects: he coordinated "unity meetings" in which environmentalists met with National Park Service leadership to explore mutual concerns; he monitored reorganization of the National Park Service and its preparation of a new approach to planning; new oil and gas regulations were prepared at Interior, and he watched how they might affect parks; he monitored and commented on developments in concessions policy; he prepared comments for general management plans submitted by the Park Service for Acadia and Mammoth Cave national parks and Assateague, Canaveral, and Fire Island national seashores, among other units; and he was recording NPCA opposition to extending the runway at the Teton airport. In the legislative arena, Jarvis represented the association's views on bills to expand Redwood National Park, purchase lands to protect the Appalachian Trail, protect the Boundary Waters Canoe Area, create a New River Gorge National River, and pass an Omnibus National Parks Bill. In the massive campaign to pass protective legislation for significant portions of Alaskan public land, NPCA's contribution was the Alaska Coalition's effort in the Senate, which he coordinated.

Lienesch also had a full plate. His chief responsibility was to work on national park initiatives involving "units of the urban-regional type." He supervised a survey of threats to parks on adjacent lands, monitoring National Park Service appropriations and the Service's Urban Recreation Study. He represented the association's views in opposition to alternatives presented in the report of this study, which recommended that Land and Water Conservation Fund monies be allocated to meet "indoor recreation" needs in urban areas. Lienesch was also working on the "Greenline Park System" idea, watching work by Interior on a Nationwide Outdoor Recreation Plan, commenting on general management plans proposed for Cuyahoga Valley, Gateway, and Golden Gate national recreation areas and Indiana Dunes National Lakeshore and supporting proposals for a Chattahoochee River National Recreation Area, a Lowell National Historic Park, a Santa Monica Mountains National Recreation Area, and a Channel Islands National Park. He registered opposition to a Jean Lafitte National Historic Park (NPCA opposed the "park" on the grounds that the area was not of sufficient national significance but supported a "national reserve.") Lienesch coordinated environmentalists' efforts in the Senate on the National Parks and Recreation Act, which passed.

These two industrious staff members did all this in the six-month period preceding March 1978. Smith assigned Helen Byrd the daunting task of reviewing the management plans of all 171 historic parks in the National Park System. Further, she was to work with Jarvis and Lienesch on the aspects of their programs that touched on areas of historic interest, such as the National Parks and Recreation Act, among other legislation moving in Congress. As part of her work on the "National Heritage Program," she reported attendance

Bill Lienesch played a central role in NPCA park initiatives from the early 1970s through the 1980s.

at a meeting of the Advisory Council for Historic Preservation to discuss the new Heritage, Conservation, and Recreation Service. President Jimmy Carter had recently renamed the Bureau of Outdoor Recreation and transferred to it the historic preservation responsibility of the National Park Service. NPCA had opposed this move, but it was done, and the task now was to determine how the interest of historic preservation could be served. Byrd was to decide how this might be accomplished. She also reported visits to several historic sites involved in legislation or management planning, including Lowell, Boston, Morristown, Thomas Edison, Manassas, and Chesapeake and Ohio National Historical Park.

Gerda Bikales and Tony Smith worked on immigration and population issues. Bikales met with Immigration and Naturalization Service officials to discuss immigration policy. She studied identification systems for immigrants. The thrust of the Immigration and Policy Program was to educate the public about the connections among environment, resources, and population. Bikales found herself clarifying these connections on television and radio and in Congress and *National Parks and Conservation Magazine*. A large part of the education effort involved explaining to colleagues in the environmental movement why they should support and join the NPCA effort in the immigration and population field. Education was even needed internally. Staff and some trustees needed to be convinced that these fields concerned the association.[5]

The Board Reasserts Authority

Even as the association staff grew again and worked in its traditional national park territory, a new crisis brewed. Membership and financial resources began another precipitous decline in 1978. As reported at the annual meeting in May, membership stood at 37,888. With expenses going up and membership down, operating deficits reappeared. Smith and his staff tried to find ways to counter the downward trends. They thought they might be too exclusively a Washington, D.C., organization. In response, they tried, with little success, to set up a national contact network, a system of area representatives, a system of affiliated organizations, and a system of conservation education committees. Smith reported to the trustees that NPCA was not alone in its difficulties:

> The organizational and financial chaos within the environmental movement is incredible. New groups are popping up all the time with purposes which overlap those of the existing organizations. Each group seeks separate funding from special donors or foundations; none of them achieves adequate financial stability. Often they disappear after one vigorous effort and cannot be found when the battle recurs.[6]

Membership had dropped to 35,000 in September when Chairman Stucker called the executive committee together. The financial crisis had deepened, and Smith had arranged a $250,000 line of credit to draw on in the emergency. Some of the trustees were growing concerned about the association's situation, and the act of arranging the credit finally elicited a response from the executive committee. It issued explicit instructions to the president:

1. That written approval by a majority of the Executive Committee be obtained before any borrowing under the new loan commitment.

2. That a development director be hired by joint efforts of the President and the Executive Committee.

3. That there shall be no staff personnel changes without prior consultation with the Executive Committee.

4. That there be no promotional materials relating to population, immigration and energy.

5. That a 1980 budget be submitted to the Executive Committee on or before November 15, 1979.[7]

Smith disagreed with the action taken by the executive committee but said he was "obliged to comply." He told the committee "that in his opinion the executive committee was not empowered to forbid action on the immigration issue, nor indeed was the Board of Trustees, without amending the Articles of Incorporation."[7] The committee did not agree with his interpretation.

The executive committee met on November 29, before a meeting of the full board of trustees. It received the budget requested from Smith and noted that he had hired a development director as instructed. Committee Secretary Bernard Meyer said Smith was wrong in stating that neither the executive committee nor the board could order him not to issue promotional materials relating to the immigration issue. The board, in Meyer's view, was the seat of authority in deciding "what policies would be adopted and programs executed" by the National Parks and Conservation Association.[8] The assertion of board authority to make policy that some trustees knew must be made was finally in the open in full view of Tony Smith. After discussing this matter, with Smith holding firmly to his position on his authority to make policy decisions, the committee went into executive session and emerged with a startling announcement:

> The Executive Committee, with the abstention of William Zimmerman, is asking for the resignation of the President.

Reasons for this action were the continuing deterioration of the association's financial condition, low staff morale, the president's failure to work with the executive committee, and continuing decline in membership. Smith was asked to resign "effective at the earliest practical date" and with a "reasonable settlement of his contract."[8] The executive committee's recommendation would be taken to the board the next day. The president stated that he would not resign.

Fashioned in Smith's Image

Why was Smith asked to resign? Was it a capricious action, as Smith suggested? To understand the decision of the trustees in November 1979, one needs to look back a decade and beyond. Tony Smith was smart, autocratic, and ambitious, as he demonstrated at the beginning of his tenure as president of the National Parks Association. He was confident in himself and his ideas and believed firmly that he knew best what needed to be done and how NPA should conduct business. He would tolerate little or no dissidence. In 1963 he engineered a revision of the bylaws that gave him nearly complete control of association affairs. If trustees disagreed with him, they would not be asked to continue on the board. Ernie Dickerman's experience was an example. In notes made at a 1972 meeting of the trustees, Richard A. "Red" Watson observed that "The Board of Trustees is an instrument on which the President can play.... Who chooses and prunes Trustees? The President; he picks those who pick him.... The Board is advisory; the President has made the organization in his image, he has the rights of immense ability and a lifetime's work."[9] Most trustees, including Watson, were willing to go along because during Smith's first fifteen years at the helm, the association seemed to prosper as never before. Membership, resources, and influence all were increasing. When problems in the organization appeared, however, some of the trustees were forced to reconsider their position on their autocratic executive. Even so, they knew that any effort to defy the will of the crafty Smith would be met with swift and effective response, and the organization's survival might be seriously threatened. The situation seemed to be one of "we can't live with him and we can't live without him."

Watson identified another problem in those same notes. The trustees supported Smith's expansion of the association's mission, but many had doubts about its wisdom. Again, having Dickerman's experience, they were not vocal about those doubts. Watson wrote that "Our present President aspires to (and in large part succeeds to) the role of senior American Conservationist; he has shaped his organization into an instrument to work on the central problems of our time. Would it have been better if he would have moved on to control of a less specialized, more central organization?"[9] Although Watson shared Smith's belief that broad environmental problems needed attention, he was convinced that NPCA was not well served to do that work.

In 1973 trustee Carl Reidel told a board meeting that he thought the association should focus on park work. He was the first to take this position openly since Dickerman, and he received little overt support from fellow trustees. Watson wrote Reidel that he sympathized with his view. Watson was especially concerned about what might happen should Smith become incapacitated or die. In many ways, Tony Smith *was* the association. Watson thought the board should change the way the organization conducted business. At the least, it ought to plan for a transition, perhaps groom a younger man—Smith was getting up in years. The trustees ought to change the power structure slowly to achieve better balance between executive and board. Watson was corresponding with Vice Chairman Lawrence Merriam, who shared his views but cautioned that they would have to work for change gradually and carefully so as not to jeopardize the stability of the organization.[10]

Reidel responded that he had hoped to discover other trustees who felt as he did. He was weary of NPCA's being "simply an extension of Tony's personal interests and activities." If trustees were there merely to give their stamp of approval, he had better things to do with his time. NPCA, he thought, needed change:

I, too, have great respect for Tony's past accomplishments, but am not at all convinced that he is very effective on the international or national level in as broad a range of activities as he is now engaged. By far the great majority of key leaders in other conservation related organizations with whom I have talked do not take Tony's activities very seriously. They seem to feel that he is involved in so many activities covering so many major topics, that he has become a sort of conservation gadfly. And as a result, the NPCA does not seem to have a very substantial reputation as either a defender of the National Park System or of anything else.[10]

Reidel and Watson agreed that any significant change would have to wait until Smith stepped down. Reidel moved on to other arenas, and Watson hung on.

At the board meeting in May, Treasurer Donald M. McCormack, a trustee who had served on the executive committee for a decade, resigned from the committee and from the board. He tried to gain executive committee agreement on four points: that NPCA should stay within its budget; that policy should be conceived and directed by the executive committee and not by editorials in the magazine (virtually all of which were written by Smith); that NPCA should have some plan for management succession; and that NPCA should return to primary concern with the parks and pull away from diverse involvements.[11] When he could not achieve agreement on these points, he left the association. Two other NPCA leaders soon followed—Francis Young of the executive committee and long-time vice chairman, Lawrence Merriam. Merriam, a distinguished professor of forestry and grandson of John Merriam, had finally decided that he would seek Tony Smith's resignation. Outmaneuvered by the redoubtable president, Merriam resigned from the executive committee while remaining on the board. Dissidence was rising, but Smith was still firmly in control.

The association emerged from the financial crisis in 1975, and Smith responded to concerns about the association's mission, refocusing to a degree on national parks. The executive committee renewed Smith's contract for another five years in 1977, but the issues continued to bother the trustees. Smith's program to address illegal immigration, launched in 1978, seems to have been the catalyst for the final confrontation between Smith and the executive committee. Some thought Smith had again strayed too far from the national parks, that his proposals flirted with racism, and that the association was losing its effectiveness. Someone needed to tend to the growing crises in the national parks, and the National Parks and Conservation Association should do that job. Smith forged ahead on his own path. When membership again began to decline in 1978 and 1979, the executive committee finally decided it had to take action.

No one expected Tony Smith to go without a fight, and he did not. At the board meeting that followed the executive committee request for his resignation, he told the trustees that he did not believe he was responsible for the association's problems, nor did he believe the problems were as serious as some seemed to think. He proposed fiscal measures that were reasonable and had worked in the earlier crisis. Soliciting members by direct mail was a tricky business that caused problems for other groups, and given the chance, he was sure he could resolve the difficulties. He loved a fight (as some trustees observed to each other in correspondence about the confrontation), and he would fight. He seemed confident that he would win, continue as association president, and carry on to new heights of achievement.

The President Fights Back

During the weeks after the board meeting, there was furious activity from Smith and his opponents. A stream of memoranda flowed from Smith to the trustees. He sent them copies of a letter from an assistant to President Carter, along with a photo of Carter with Smith and others at a recent White House meeting. There were copies of proposed budgets requested by the executive committee but not acted on, notice of a new grant from the Rockefeller Family Fund for work on the Heritage Program. In a letter to Chairman Stucker, he wrote that under the circumstances, the proposed contribution of the Smith family farm to the association would not be possible, and he copied in all of the trustees. This was a campaign to demonstrate that under Smith's leadership, the association was doing well, that it was reaching to the centers of power, that its financial prospects were brightening, and that removing him from office would be costly to NPCA in many ways. He reminded the trustees that he had a contract with the association and he would demand that it be honored. Doing so, he told them, would be the right and prudent thing to do. Smith also probed the trustees to determine what support he had, and he lobbied to build that support.

This only hardened his opponents' resolve. They knew the resourceful Smith had fought off earlier attempts to remove him or reduce his power and might do so again. Richard Watson advised Chairman Stucker to take the initiative, to move steadily forward and not allow Smith to snow the trustees with the avalanche of material he sent them. Watson and executive committee member April Young were working on proposed changes to the bylaws.[12] Merriam analyzed the budgetary situation, writing Stucker that the projected budget would likely result in a loss for 1980 of $258,000 and "the NPCA may very well face bankruptcy at the end of 1980 or 1981 at the latest." He thought the mortgaging power of NPCA "overrated" and the "NPCA under Tony Smith doesn't enjoy the confidence in conservation circles it once did in terms of outside support." Smith tried to convince the trustees that there were excellent prospects for grant funds; Merriam did not think this view realistic.[13]

The executive committee met on January 9 and decided to negotiate a revised contract with Smith. It would reduce his time and compensation and assign him to work on specific programs such as Law of the Sea. It would propose revisions of the bylaws and, on reaching agreement with Smith, hire an executive director to oversee the management of the organization. Trustees Carl Bucheister, John Quarles, and Michael Brewer would negotiate with Smith, and trustees Bernie Meyer and Betty Phillips would oversee the staff at headquarters during this difficult period. The staff, of course, was under great stress. Further restrictions were placed on Smith's financial and programmatic discretion.

Stucker informed Smith of the decision and expressed his hopes for an amicable resolution. Offers and counteroffers were made. Smith continued to lobby the trustees, and several sent letters to all trustees arguing that Smith's contract should be honored without alteration. A special meeting of the board was called for February 22, 1980; a resolution passed at that meeting set financial and temporal limits on negotiations. If no agreement could be reached by the end of the month (one week later), Smith was through as of March 31. Smith's opponents clearly had the upper hand, and on February 28 Stucker accepted Smith's resignation. The retiring president received a severance package and would become "special counsel" to the association, though he was not obligated to perform any duties. In essence, Smith had refused to accept a reduced role in the association.

One factor that may have encouraged the trustees to take a hard line with Smith was a letter from seven members of the association's professional staff.

The staff did not believe the president functioned well and thought a change in leadership was essential. Current leadership could not solve NPCA's "severe problems."

> In our opinion, NPCA needs: (1) a reorientation to and principal focus on the National Park System, particularly in communication with our members and prospective members; (2) dynamic new leadership that understands and actively involves itself in the political process, in modern fund-raising techniques, and in development of its constituency, particularly "grassroots" constituency; (3) a new leader who has the respect of and ability to work with other conservation organizations; (4) a leader with the desire and ability to actively pursue and obtain funds for the Association from foundations and wealthy contributors.[14]

Unless such a prescription is followed, continued the staff, "it is only a matter of time before NPCA will first cease to be able to influence national park policy, and then will quickly cease to be able to sustain itself."[14] These comments could not be ignored. Staff would not lightly offer such remarks, for they knew they would not be able to work with Smith after writing the letter. They knew Smith had friends on the board who would share the letter with him. The trustees now saw that they had to decide between the staff and the president. One had to go.

The board also adopted new bylaws at the February 22 meeting. The revisions were made to remove some of the power of the executive that, in the view of many trustees, had led to the current unfortunate circumstance. Standing committees on finance and nominating were established, with members to be appointed by the chairman and approved by the board. The executive committee would be the source of association policy; it would make policy recommendations to the board for its approval at the annual meeting. The executive title would be "executive director." The key change was in relation to policy making. The new bylaws returned this function to the board, where it had been until Smith's revisions had been adopted back in 1963. Richard Watson described the situation that needed correcting in a letter to April Young:

> The Board and Executive Committee can paralyze the Association by refusing approval of all moves made by the President. But the Board and Executive Committee cannot run the Association except by approving the moves made by the President...because the Board and Executive Committee cannot hire, appoint, or make on its own; the Bylaws explicitly give these to the President...(subject to approval, etc., and so on, of course). Incredible.[12]

In his letter Watson speculates about how the earlier board had allowed the association to fall into this situation:

My guess is at the time Tony took over, he was viewed as something of a volunteer. And he viewed the NPA as his creation, much as volunteer autocratic leaders of voluntary associations often do. Probably the Board at that time saw it all that way. And of course these kinds of organizations often work best with autocratic leaders who have a lot of authority and power. And Tony was very successful with the Association. So it worked ok for a long time. But he designed a real stinger with those Bylaws.[12]

The Smith years bear out much of Watson's analysis. He came from inside the board to the executive role and moved quickly to change the nature of the association's governance. He joined with younger members of the board to wrest control from a group that had controlled association affairs for decades. He offered new approaches and great ambition, and he enjoyed initial success. The complete control he exercised, however, jeopardized the existence of an organization that should transcend him. He did not create it, and the association in the end proved to be bigger and more important than he.

Acrimonious Departure

Tony Smith did not leave quietly or generously. He felt he had been wronged and said so in his final report. "The NPCA is headed toward financial disaster. The outgoing President is not to blame. The Executive Committee's actions and failure to act are to blame." Having absolved himself of any responsibility for the straits of the organization he had led for twenty-two years, he observed that "The record of my administration has been exemplary, and increasingly successful in recent years, in respect to contribution, grants and renewals." The problem was recruiting of new members with direct mail, and that could have been solved if he had been given a chance. He went on for sixteen pages in what was, essentially, a rant. He cast blame on a number of individuals and expressed no regret or apology for any of his moves. His report leaves no doubt that he thought the association was doomed, that it would not survive without him. His strategies were good, he said, and had it not been for the "hatred" of the executive committee and a subversive staff, all would be well.[15]

Study of Smith's entire career with the association, which reached back nearly thirty years to his initial appointment to the board, reveals a man driven by ambition. Blessed with prodigious energy and stamina, with intellect and broad knowledge, he thought he could be "Senior American Conservationist," as Red Watson called him. He thought he could save the world, that he could be a national and even global leader in conservation and, later, environmentalism. Others might see limits to his ability and influence. They might see him as hard, mean, manipulative, vindictive, but he did not see these qualities in himself. A man of massive ego, he always thought he was right, and he demanded unquestioning allegiance and support from his staff, from colleagues, and from board members. Anyone who did not agree with him was simply wrong.

As the account of his twenty-two years as leader of the association reveals, there is no question that Tony Smith did much good work. But in the end, those who were on the association journey with him, and some trustees who had been around from nearly the beginning of his "reign," must have wished he could have retired with pride in his accomplishments and support for his successors. His work was far from done, and that work could have been pursued in other roles than that of NPCA president. Tony Smith, however, could only be who he was, and he left his NPCA office an angry and bitter man. Henceforth he would have nothing to do with the National Parks and Conservation Association.

Revising the Agenda

With Tony Smith out of the picture, the board met in May to select a new leader and correct the association's course. When the Smith matter was resolved, the trustees formed a search committee to seek an executive director. Mearl Gallup was named acting manager to oversee association business until the new executive was hired. The senior staff—Destry Jarvis, Bill Lienesch, Mearl Gallup, and Eugenia Connally—prepared a briefing paper for the executive committee that explained their views on measures to revive and redirect NPCA. The strongest advice in this lengthy document was to *focus on the national parks*. Staff members agreed that this single measure would probably improve membership recruitment and retention, increase the chances of foundation support, achieve the most significant outcomes, and result in the association's survival. They took one last shot at Tony Smith's controversial agenda:

> For more than a decade, NPCA has purported to carry out the broadest possible agenda of programs, issues and action, ranging afield from our traditional role in national parks, to include wildlife, forestry, historic preservation, farmland protection, air and water pollution, water resource projects, and population. More often than not for at least the past five years these "programs," other than national park system issues, have in fact been only illusory, consisting primarily of articles in the Magazine." [16]

Despite this, in their view, NPCA had maintained its significance in national park work. "No other single conservation organization monitors all phases of park-related conservation activities as closely or as effectively as does NPCA." Jarvis and Lienesch could take credit for this, although they did not. The association had worked on projects specific to one or another unit of the National Park System, and it emphasized nine general park policy areas: new areas, land acquisition, budget and personnel, concessions, incompatible uses, transportation, management planning, wilderness, and adjacent land uses. The staff thought that park-specific projects, in addition to these nine policy areas, should constitute the "program for parks" that would be the core of association activity.

The staff made another major recommendation: that NPCA find a way to develop a field network. "Outside of the Washington political arena, and the immediate conservation community of organizations' staffs and activists, NPCA suffers from an identity problem with the great mass of people who are potential members."[16] Throughout its history, leaders of the association recognized this problem and sought ways to combat it: Yard had reached out with his network; Butcher had traveled, acting as a one-man field corps; a Western Office was established in the 1950s; and Smith had tried at least twice to set up a national system of contacts and activists. Despite these attempts, the association presence remained primarily in Washington, D.C. The staff thought that at this crossroads in NPCA's history, a renewed outreach and communications effort was essential. "We must take decisive action to give greater visibility to NPCA's role in the conservation movement, and particularly to our supremacy in the parks area." Other conservation groups might do less for parks than NPCA, but they did a better job of publicizing their work. They sold themselves better than NPCA. The staff believed solutions to this lay in a field coordinator, field representatives, and associated organizations. Admittedly, development of an adequate outreach system would be expensive and would have to be done carefully, but it must happen.

> NPCA has a role, a niche, vital to the continued viability of the conservation community, and we are performing services in the public interest which no other organization, at this time, is willing or able to do. We know that we are performing these valuable services to the country, and particularly to future generations, but we need to make sure that others know it too." [16]

Mearl Gallup's report on financial objectives revealed just how much concerted effort was necessary to move to a position where growth, like that involved in a field network, would be possible. The level of operation projected by the 1980 budget, with no expansion, required a sustaining membership of 46,000. There were currently 24,000 members. Returns on direct mail promotion were low and could be raised by focusing on parks and targeting those who visited parks and other "outdoor people." The association would also have to retain more members. Better organized development and grant-seeking programs were essential.

Eugenia Connally, editor of the magazine, described a development plan that she thought would make *National Parks and Conservation Magazine* a more effective tool for communication and membership. She reported that she had worked under severe restrictions imposed by Tony Smith: editorials were written only by Smith or by others he invited; Smith would allow none of his numerous contributions to be edited; each issue was to contain a park article, an endangered species article, and a broad conservation article; special issues, book reviews, and staff reports were forbidden. Connally had asked for a member survey, to determine who the readers were and what they wanted to read, but Smith would not allow it.

Connally suggested that a member survey should be done immediately, and that the magazine editorial content be diversified. Editorial comment

Paul Pritchard became executive director of NPCA in the summer of 1980.

should be reduced; more political reporting and human interest stories should be featured. "The Environmental Journal" should be dropped from the magazine's name, and perhaps it ought to be retitled *National Parks Magazine* or just *National Parks*. Above all, the focus of the magazine should return to national parks. She would be happy to have articles exhibiting the expertise of the NPCA staff. Who, after all, should know the issues better than NPCA experts?[17]

The board had much to do at the May meeting. (Its "briefing book" for the meeting ran to 148 pages.) The most important decision was selecting a new executive. Trustees agreed that this should be done as quickly as possible. To head off further financial crisis, new directions needed to be set immediately, and a new executive should be in on those decisions. The search committee recommended an executive director, Paul C. Pritchard. The NPCA trustees looked for a bright, young, energetic, politically savvy person who was comfortable in the charged political environment of Washington, D.C. They hoped to find someone who could lead, yet who would be comfortable with a board that set policy and direction. They needed someone confident but humble, independent and creative, yet cooperative and consultative. The new executive director would need skill in managing people, for one of the first tasks was to heal the wounds and build the morale of a staff that had suffered in the recent power struggle. They needed someone who enjoyed a challenge, who would not be daunted by NPCA's difficult financial straits. The new leader must be able to make difficult decisions, for reduction of staff would be one of the first tasks. If these qualities could be found in someone who was knowledgeable about national parks and the public land bureaucracy and who also understood the complexities of the competitive direct-mail world of

Gil Stucker, as Chairman of the NPCA Board, presided over the association during the difficult period leading to Anthony Wayne Smith's resignation as president.

conservation and environmentalism, the trustees could be confident that their organization would pull through.

Paul Pritchard, trained in natural resource management and planning, seemed to fit this extensive bill. His career had developed rapidly: staff assistant to Georgia Governor Jimmy Carter (1971-1974); staff member at the National Oceanic and Atmospheric Administration (1974-1975); executive director of the Appalachian Trail Conference (1975-1977); deputy director of the Bureau of Outdoor Recreation (1977-1980). The Bureau of Outdoor Recreation had changed to the Heritage Conservation and Recreation Service during his tenure, and he had played a key role in the transition. He had chaired a task force to develop a national heritage trust. At the age of thirty-five, he was widely experienced in the ways of the federal bureaucracy and in conservation politics. His service with the Heritage Conservation and Recreation Service had given him extensive contacts with politicians and conservationists. As executive director of the Appalachian Trail Conference, he had effectively managed a nonprofit organization and developed direct-mail promotion and fund-raising skills. He knew and loved the parks. NPCA's new direction seemed, to him, the correct one. The match seemed good, and agreement was reached that Pritchard would be the new executive director of NPCA beginning on July 1, 1980.

Back in 1973, trustee Richard Watson reflected in a letter to Lawrence Merriam that when Tony Smith left the NPCA presidency, the basis of association business should "change from operations based on personality to operations based on law." More dependence would be placed on legal or administrative methods. Watson continued, "It is interesting that Ralph Nader—himself a charismatic personality of impressive dimensions—stresses the administrative, legal approach. When one depends on a personality, the strength of the organization (and its longevity) is based on the survival of that personality."[18] In 1980 Watson was still a trustee and had helped draft the new bylaws as one means to achieve new association management. Watson worried over the years whether the association had depended so much on Smith that a collaborative and trusting partnership between a new executive and the board would be impossible. Now he would learn whether Paul Pritchard and the board, under the leadership of Gil Stucker, could work together to revive the association.

Reaffirming the Mission

T he removal of Tony Smith from his executive position was the greatest trauma in the history of NPCA. The painful and prolonged confrontation shook the organization to its roots. Pressures built toward the upheaval for a decade, years during which the association lost its focus and its identification with the national parks. Smith was a complex man of great intelligence, drive, and ambition, but he lacked perspective on himself and his organization. Ambition led to self-delusion about his importance and that of NPCA. His aspirations for himself and the organization were unrealistic and exceeded his grasp. The result was a severe organizational crisis.

Smith's successor at NPCA's helm faced a daunting array of challenges. Foremost among them was placing the association back on its organizational feet. This meant rebuilding a functional relationship among executive, board, and staff. The financial situation must be stabilized, which required building membership. Reasons for membership decline must be identified and antidotes administered. The new executive must exercise power and authority carefully and strategically to restore confidence in NPCA's structure or, alternatively, restructure it. This latter course would be time and resource consumptive, making the association even more vulnerable. Everyone agreed that repair of the existing structure should be attempted.

Other challenges involved the national parks and politics. The times were not good for the parks, nor were they propitious for advocates of expanded government programs. The national parks were suffering the consequences of

NPCA was part of the Alaska Coalition of environmental groups that successfully fought for protection of areas such as the Wrangell-St. Elias mountain region.

Chairman Bruce Vento, House Subcommittee on Parks, and NPCA President Paul Pritchard presented the NPCA's National Park System Plan *in 1988.*

ever-increasing visitor pressure, of budgetary restraints caused by war and recession, and of unstable leadership of the National Park Service. The Park Service itself was struggling with its mission, trying to adapt to an expanding array of responsibilities and social pressures. Just as NPCA suffered from a prolonged identity crisis in the 1970s, so did the National Park Service as it tried to meet the needs of urban populations, plan parks in remote Alaska, and maintain the troubling balance between use and protection required of it in 1916. All of this was happening in a dramatically changing political climate with the election of Ronald Reagan in 1980. The new administration would not support increased expenditures or expanded programs for national parks.

From his first day on the job, Paul Pritchard knew that the association must return to its original focus on national parks, but how to do that was not clear. He knew most of the players in the national park game, and he had a feel for how the association could work its way back in. He could waste no time in demonstrating to NPCA's faltering membership and its environmentalist colleagues that the association could return to the game with energy and effect.

THREATS TO PARKS FROM ADJACENT LANDS

During the 1970s NPCA studied the damage done to national parks by activities on adjacent lands. This had been a problem for decades, but the extent and scale were increasing. Loggers were cutting right to the boundary of Redwoods National Park, streams were silting up, and root systems of ancient trees were eroding. Air quality was degrading in the Southwest, affecting park vistas there. Oil, gas, and geothermal drilling were increasing adjacent to Yellowstone, Glacier, and other parks. NPCA hosted a meeting of conservationists, congressional staff, and Interior officials in 1977 on these growing threats. All agreed that these problems would continue to grow unless something was done, and NPCA decided the first step must be to document the extent of threatening activity adjacent to parks. Park superintendents would be surveyed.

New staffer Bill Lienesch was assigned the project, and by May of 1978 he was able to report that the survey substantiated a serious and growing threat to parks from adjacent activity. Two successive articles in *National Parks and Conservation Magazine* in 1979 published the survey results and NPCA recommendations on addressing the problems.[1] The report (including information gathered by The Conservation Foundation in a survey of its own) claimed a return rate from the park system of approximately 90 percent. The results were impressive:

> Nearly two-thirds of the 203 respondents stated that their units suffer from a wide variety of incompatible activities on adjacent lands that affect the parks in every conceivable manner—everything from trespassing livestock that trample vegetation, to industrial dyes that regularly change the color of a park creek…nearly 50 percent of the superintendents believe that they do not have sufficient authority or appropriate policy directives to respond to problems emanating from lands outside the boundaries of their units.[2]

The list of incompatible activities was long: residential, industrial, and commercial development; road building; grazing; logging; agriculture; energy production and extraction; mining; reservoirs; hunting; military facilities; flood control; dredging and sewage treatment. These activities and their effects on water, air, silence, wildlife, and general biotic conditions, among other resources, were presented in ten tables.

The second of the two articles described the actions that superintendents were taking to address these problems and NPCA's recommendations for improving the situation. The Park Service needed more specialized personnel trained in urban planning, pollution control, law, and other fields. The shortage of Park Service scientists was especially critical. "More scientists assigned to the field," said the report, "could provide more definitive and quicker identification of the problems, sources and solutions."[3] Cooperative ties with other agencies, especially the Environmental Protection Agency and the Heritage Conservation and Recreation Service, should be strengthened. The Park Service planning efforts should reach beyond park boundaries, and land acquisitions should be made with consideration of external threats. In the design of new units, the regional and potentially troublesome external activities and developments should be considered. This would take congressional help; legislation giving the Park Service authority to deal with external threats was essential. This was a large order, but NPCA concluded that unless something was done, "resource management within park boundaries eventually would be rendered meaningless by external forces."[4]

The parks subcommittee of the House reviewed the NPCA survey results and requested that the Park Service carry out its own survey. Published under the title *State of the Parks 1980: A Report to Congress*, this report revealed extensive internal and external threats to park resources. It substantiated NPCA conclusions about adjacent lands problems and went further.[5] The report identified no fewer than 4,345 specific threats to park natural resources and mentioned, but did not cover in detail, threats to cultural resources. *State of the Parks 1980* also revealed an abysmal lack of research to describe and document the problems accurately. Fewer than 100 scientists were among the agency's 9,000 employees, and only 200 were trained in resource management. NPCA work in this arena had brought the scale of national park problems to the attention of interested parties; the challenge of protecting the parks was greater than ever. Journalist Robert Cahn wrote that these "threats to the national park resources and values are the most serious they have ever been."[6]

PRITCHARD GOES TO WORK

Paul Pritchard faced major challenges when he became NPCA executive director on July 1, 1980. His organization was deep in crisis, its membership down, its finances precarious, and its staff reeling from prolonged turbulence. Yet NPCA, weak as it was, had stimulated a historic examination of the national park situation. If ever there was a need for a strong citizen organization for national park protection, that time was the early 1980s. In monitoring the national park situation, the NPCA staff had found a deepening crisis. It had identified an agenda that might provide the impetus for NPCA's revival if the organization could respond.

Pritchard went to work. In their briefing paper for the board in April, staff members described a program they thought could revive the association. Pritchard examined these recommendations closely, conferred with staff and trustees, and formulated a program that followed many, though not all, of the staff's suggestions. The most important task was to restore the organization's financial health, and this meant more members. When Pritchard began work on July 1, 1980, the association counted 26,000 members. Analysis indicated that at least 46,000 dues-paying members were necessary to check the association's drain on its reserves. Return rates from direct-mail solicitations must improve dramatically. Ways must be found to retain more members, especially new ones. Services to members must be enhanced to make people see their association membership as a source of pride. In the long run, financial reserves must be restored.

The immediate challenge was to increase revenues and reduce spending. Pritchard cut staff and stepped up membership recruitment efforts. By November he reported to the trustees that the operating budget had been reduced by $150,000. The downward slide of membership had stopped, with growth recorded for four months. Two grants were received, and an aggressive pursuit of additional grants was under way. There was an operating deficit, but it was not as severe as had been feared earlier in the year. The organization was not out of its crisis, but prospects were improving, and it seemed NPCA would survive.

National Parks and Conservation Magazine was a critical ingredient of NPCA operations. As Yard had recognized in the association's early days, members needed a tangible return for their dues. They needed to know that the organization was doing good work, but they also needed something they could hold in their hands that reminded them of their membership, gave them pleasure, and told them about the activities of the organization they were supporting. The magazine had been quite good under the leadership of a series of fine editors, but according to its current editor Eugenia Connally, its recent editorial policy had been inflexible. It did not keep up with some of its competitors, and its content reflected no knowledge of what members were interested in reading. No reader surveys had been done—the magazine simply contained what President Smith thought it should.

Connally recommended a number of changes. The name should drop "The Environmental Journal" subtitle; the board did that in May 1980. Ultimately it was shortened to simply *National Parks*. Editorial comment must be reduced to a single page, readers consulted, color used in the magazine's interior (as soon as budget allowed), and membership service elements, such as news and calendars of events, should be printed. All of these suggestions were eventually adopted. Editor Connally cautioned against cutting back the magazine, arguing that it was the association's most visible and effective communication device, but Pritchard and the board agreed that the monthly publication should become bimonthly. Total pages would remain the same, but mailing and handling costs would be significantly less.

The most important advice accepted by the new executive was to concentrate on the national parks. The need for this was as clear to him as it was to most of the board and staff. He did not agree with the staff's opinion that NPCA, despite its problems, had recently been an effective advocate of national parks. Pritchard's work in the Heritage Conservation and Recreation Service involved him in the lengthy and complex negotiations over Alaska lands considered for addition to the National Park System. Although the association was represented in the Alaska Coalition, in his view, it had not been a major player. This Alaska initiative was the most important park issue in a half-century because it would double the size of the system; Pritchard thought that NPCA, associated as it had been with parks, should be a leader. He plunged into the last phases of the Alaska negotiations. The focus of the association would be national parks.

Revitalizing the Board of Trustees

The national parks certainly needed the full attention of the National Parks and Conservation Association, but for two more years the work of rebuilding strength continued. Pritchard worked to revitalize the board, develop a mission statement and corporate plan, and bring the organization

NPCA Board Chair Gil Stucker (third from left) urged NPCA to refocus on national parks, a process that began with a conference in Jackson Hole in September 1981.

out of debt. His analysis of how the association had gotten into trouble in the 1970s led him to conclude that major responsibility lay with the board of trustees. The board had not done its job of providing policy direction and oversight of the executive. Many members of the board over the years perceived this problem but were unable or unwilling to act on their insight. The problem went back to the late 1950s and early 1960s when the board, over the protests of Wharton, Butcher, Cahalane, and others, allowed Smith to take nearly total control of association affairs. No board since had regained control.

Pritchard placed revitalizing the board at the top of his agenda. The crisis of 1980 led to the resignation of some trustees, and those remaining were determined that they would play a more central role in association affairs. Pritchard's challenge was to strengthen the board without asserting unhealthy influence over the group that had hired and could fire him if he did not do the job. He tested his situation by arguing that the title of "president" should be restored to NPCA's chief executive. His logic was that this title was more appropriate in a modern environmental organization than "executive director," the title favored by the board. Trustees were willing to concede the point, which encouraged him.

Leadership of the board fell to Chairman Gil Stucker, and he usually agreed with Pritchard's assessment of what needed to be done. The two worked together to make necessary changes. The greatest need was to bring new blood into NPCA leadership, and Pritchard found the board not as committed to doing this as he expected under the circumstances. When the nominating committee invited all incumbent trustees to continue, whether or not they had been active, a showdown ensued. Pritchard challenged this, arguing that the association needed an active, working board, people who would attend meetings and take their responsibilities as trustees seriously. He prevailed, and a new slate of nominees was drawn. New members with better geographic distribution joined the board, and they understood they were expected to attend meetings and contribute. Committees, an important part of early NPCA activity, once again were formed to complement staff functions. The board members objected to establishing a system of NPCA regional representatives, but Pritchard and the staff convinced them that establishing an active NPCA presence outside of Washington, D.C., was critical. Regional representatives were appointed as the budget allowed. Pritchard knew he had achieved his goal of an active, committed board when, in the mid-1980s, the trustees decided to establish the National Park Trust. A divided board barely approved this program, but when it did, trustees rallied behind it and raised $250,000 to boost the effort. The board, its chairman, and the association executive all worked together, with the board setting policy on the staff's recommendation.[7]

Setting the Finances in Order

When Pritchard took the reins, NPCA's financial crisis was acute. Tony Smith had borrowed against the building, and its mortgage carried a high interest rate. The operating budget was running at a loss. Membership services and recruitment were in shambles. To balance the budget, Pritchard first cut costs by reducing staff. The mortgage on the building was renegotiated. He slashed magazine costs. The goal was to operate on a pay-as-you-go basis, which was quickly achieved. Grant proposals were written and several received for special needs, such as color magazine covers and special meetings and publications. The flow of red ink stopped in 1981.

Two ingredients in NPCA's financial recovery were improved membership recruitment and retention, and renewed confidence in the association among potential donors. Effective recruitment involved using direct mail more strategically. Membership mailings targeted specific markets that promised higher yields, and slowly recruitment improved. Former association members were targeted. While staff worked on the technical aspects of membership recruitment, Pritchard and his parks staff worked to raise NPCA's profile as incentive for former members to return. People had been members because they cared about parks and, as an incentive to return, needed demonstrations that the association was doing something significant for parks. Pritchard understood this. With the magazine and other publications, he made it clear that national parks were the association's central concern. Membership services also improved, and slowly the membership increased. The worst of the financial crisis was over.

The Jackson Hole Conference

At the November 1982 board meeting, Chairman Stucker told his colleagues that the time had come to focus less on internal organization and more on issues in the national parks.[8] As it reorganized itself, NPCA worked to understand the state of the National Park System and how its problems might be addressed. It had convened a conference for these purposes in Jackson Hole, Wyoming, in September 1981. In Pritchard's words, the conference aimed "to bring together concerned citizen experts to provide a framework for the development of a comprehensive National Park System plan; and, second, to provide a basis for alerting the public about the current problems facing the national parks."[9] Former Interior Secretary Cecil Andrus and Park Service Director Russ Dickenson attended, along with forty others. Ric Davidge, special assistant to Assistant Secretary of the Interior for Parks and Wildlife G. Ray Arnett, represented the Reagan administration. The association published *National Parks in Crisis,* a volume of talks, related papers, and recommendations from the meeting.

Conferees at the Jackson Hole meeting agreed that the national parks faced massive problems, although they disagreed on which were most severe and deserved highest priority and on what methods would best solve them. Politics weighed heavily on the group. The Reagan administration was in office, and James Watt was secretary of the Interior. Watt had recently called the condition of the parks "shameful" and offered a plan that called for a five-year moratorium on acquiring new parks, returning some of the urban parks to cities, more development near parks, and greater freedom for concessionaires.[10] Park Service Director Dickenson attempted to put the best face on a difficult situation. He offered the administration line, stating that "The national parks are alive and thriving." He was, he said, most optimistic about the future.[11] Michael Frome, a journalist who had critically scrutinized the national parks for many years, offered the prevailing view of most conferees. "The national parks are endangered—all of them," he told the group. "If it isn't one thing, it's another." The only hope for the parks was to build a stronger constituency, one that reached out to groups not previously interested, such as urban people. Frome blasted Watt and the Reagan administration's approach to natural resources. "If the nation accepts the course he advocated, ultimately fish and wildlife and parks will be sacrificed, while the natural resources will be used up and lost, long before their time."[12]

The conferees talked and argued, lounged on the deck with its spectacular view of the Tetons, and generated many ideas for dealing with the situation. They described twelve "general principles" along with forty specific recommendations for addressing the current crisis and achieving effective planning. The principles emphasized resource protection, and they rang with traditional association rhetoric. The "rich natural and cultural heritage of our nation" was preserved in the parks. The system should be protected and enhanced; "appropriate uses of the resources are for inspiration, education, and appreciation." All Americans should be "actively committed to preserving the heritage represented in the system."[13] The meeting closed on a high note with an inspiring (at least to attending conservationists) talk by Dee Frankfourth, a leading Alaskan conservationist. "Every step along the way there will be dilution of the ideals and aspirations we strive for in our nation-

The conference at Jackson Hole in 1981 set the association on its path toward a major planning project for the National Park System.

al parks. We have a sense of what those ideals are. We should not settle for anything less."[14]

The Jackson Hole meeting was useful to NPCA, helping to propel it back into the thick of national park debate and politics, but it had little effect on the immediate course of national park events. Secretary Watt continued to promote his agenda of improving park facilities while allowing, some would say encouraging, the degradation of park resources. A vision for the National Park System was needed, an alternative to Watt's view that repair of sewers, roads, and buildings would solve the national park crisis. Studies by NPCA, The Conservation Foundation, and even the Park Service itself refuted Watt's approach.

Exploring New Approaches to Park Protection

Paul Pritchard and Gil Stucker, discussing the straits of the parks and the Park Service, had concluded that NPCA should press for a major park system planning effort. At the Jackson Hole meeting, Destry Jarvis asked if the time was right for a comprehensive plan for the system. He traced its planning history and concluded that a plan was needed:

> I conclude that even though there have been a variety of long-range plans for the parks; despite the fact that most of these have been prostituted into construction programs; and because no comprehensive plan has ever been produced which deals with all aspects of the

Park Service's mandate, even in one hundred and nine years, a comprehensive National Park System Plan is needed. Desperately needed. It should not, however, be prepared on a crisis basis. It should especially not be prepared with built-in constraints such as those being imposed by the present Administration with its budget-cutting frenzy and bias against new additions to the system.[15]

The plan needed was not, as Secretary Watt might think, a plan for facility construction and improvements, but rather "a plan to save the parks." It should "have as its core the preservation and interpretation of the natural and cultural resources of the system."[16]

The National Park Service was in no position to do the type of work called for by NPCA; Reagan administration approval for such an undertaking was out of the question. NPCA, then, would itself do the plan. Jarvis was assigned to oversee this massive undertaking. Pritchard, with the help of Laurence Rockefeller, approached the Mellon Foundation for support. A major grant came from Mellon, and work began in 1983.

Countering the Watt Agenda

While NPCA reestablished itself among the leaders addressing park issues, the political crisis of the parks intensified. The Service followed its *State of the Parks 1980* report with a second report to Congress titled *State of the Parks: Strategy for Prevention and Mitigation of Natural and Cultural Resources Management Problems*. Released in January 1981, it proposed an agenda for improving stewardship of cultural resources.

These reports were a start, but little was actually done. The Reagan administration was not serious about improving the situation. Paul Pritchard met with James Watt and offered eight points on which he thought NPCA and Interior could work together to improve the park situation. Watt was cordial but not interested in cooperating. A disappointed Pritchard, who had recently argued against NPCA's joining a "dump Watt" campaign being mounted by other environmental groups, concluded that trying to work with Watt was futile. Henceforth the association would work with Interior and the Park Service when possible—the Service was willing, indeed eager, to cooperate in an unofficial and sub-rosa way—and focus more of its energy on Congress.

Watt brought an anti-government philosophy to Interior and favored development and privatization. He did his best to impose his agenda on the Park Service, which offered, at most, passive resistance. Watt's appointments of his assistants revealed his priorities. William Horn, who as a congressional aide had fought the Alaska Coalition on Alaska park and wilderness matters in the late 1970s, was appointed undersecretary for Alaska issues. Ray Arnett, an advocate of sport hunting, commercial fishing, and off-road vehicle use in some parks, was appointed undersecretary for fish and wildlife and parks. Ric Davidge, Arnett's special assistant on park matters, had been a lobbyist for the National Inholders Association, which led the opposition to federal acquisition of private lands for parks. And so it went, through the ranks of the Interior

James Watt, Ronald Reagan's first Secretary of the Interior, offered challenge and opportunity to NPCA and other national park advocates.

bureaucracy.[17] Watt and this cast of characters demanded the constant attention of the association and other park advocates who lobbied Congress with determination and considerable success. As one NPCA staff member describes the period, the strategy was to "blow the cover" off Watt's plans, bringing them to the attention of Congress and the public, and to fight attacks on the parks on legal and regulatory grounds when necessary.[18]

Paul Pritchard, the first association executive to bring government experience to his role, insisted that issues needed to be addressed in new and more aggressive ways. His predecessors were handicapped by tax laws, but when the strictures of these laws were relaxed, NPCA had been slow to respond. No longer should testimony before congressional committees be the primary approach. Staff were expected to "work" members of Congress and their staffs, to act as lobbyists and aggressively counter the lobbying of special interest groups pursuing anti-park and anti-regulation agendas. NPCA continued its tradition of working with allies, even when those allies seemed unlikely. In the struggle against Ray Arnett's efforts to relax hunting regulations in the parks, for instance, the association recruited the handgun control advocates into its successful campaign. It went to court more frequently than ever before, often joining other environmental groups in their suits. It joined the National Wildlife Federation, for instance, in a case supporting National Park Service efforts to control burros in Grand Canyon National Park and Bandelier National Monument.[19] The association successfully reestablished its presence in Washington as a defender of the national parks.

The National Park Trust

NPCA's search for ways to accomplish its park protection mission led it in a new direction in 1983. At its November 1982 meeting, the board decided to establish a land trust program, although it was split on whether to take this new path. After agreement that the trust would not draw funds from other work of the association, the proposal passed by a single vote.[20] Pritchard reminded the board that the association had a long history of assisting the Park Service in its efforts to acquire park inholdings. Progress had been made on the acquisition program, but as the system expanded, so did the acreage of inholdings. The threat from inholdings was that they might be developed or used in ways that were inappropriate within a park. There was urgent need to acquire these lands, and a national park-oriented land trust could help.

When the decision was made to create the National Park Trust as an NPCA program, few had illusions about the scale of the enterprise. The trust would be small but would educate about the inholding problem, and would assist in land acquisition to whatever level allowed by its funds. The need for the program was dramatized by James Watt's oft-stated opposition to land acquisition. The eighty-million-acre park system had more than two million acres of private inholdings in 1983. When the Reagan administration took office, the Park Service had plans to acquire 423,000 acres at a cost of $800 million, with the money coming from the Land and Water Conservation Fund (LWCF). Since its creation in 1965, the LWCF had provided money to purchase 1.7 million acres of land for parks. Secretary Watt simply placed a moratorium on all parkland acquisition, despite the $1 billion in the LWCF. He attempted to divert LWCF monies to park facilities improvement, even though the law stipulated the fund could be used only for land acquisition. Robert Cahn reported in 1983 that Watt "was enforcing several kinds of de facto moratoria to avoid buying new parkland" and had effectively stopped all acquisition.[21]

This situation motivated NPCA to create the National Park Trust. Other land trusts were not working on national parks, and Pritchard saw an opportunity to do important "hands on" park protection work that would yield immediate, demonstrable results. The goals of the new program were fivefold: 1) to proceed as funds allowed to acquire inholdings, which would then be held until they could be purchased by the National Park Service; 2) to assist other local land trust groups that might help with inholding acquisitions; 3) to seek out and acquire property that contained physiographic elements not yet represented in the National Park System; 4) to help complete the Gates of the Arctic National Park acquisition project; and 5) to help resolve clouded title and foreclosure issues when they might occur.[22] The first project completed was the acquisition of a five-acre wilderness inholding in Gates of the Arctic National Park and Preserve in Alaska. Other projects quickly followed involving Acadia National Park and Big Cypress National Preserve in Florida. The trust remained an NPCA project until the trustees, in 1990, fearing personal liability associated with acquiring and holding land, decided to make it an independent, nonprofit land trust.

Terri Martin and Russ Butcher proved the value of NPCA field representatives in the early 1980s.

These initiatives placed NPCA solidly back on its feet. Its reputation was slowly restored, and membership grew at a modest but steady rate. By 1982 there were 36,000 members. Grants came in steadily, and finances were gradually improved. The organization remained on a pay-as-you-go basis. A regional organization was developing. Devereux Butcher sponsored his son as the association's first regional representative in the West. Russ Butcher knew the parks well; he was a writer and grassroots activist who was as committed to the national parks as his father. Dev Butcher soon withdrew his support over a disagreement with the association's stand on hunting in Alaska parks (he thought it too soft on hunting in national reserves), but Russ had by then established himself, and Pritchard convinced a reluctant board that the time had come to set up the regional system. It would be another ingredient in establishing an NPCA presence outside of Washington. A young intern named Terri Martin so impressed Pritchard that when she completed her internship, he sought external support to keep her on staff. She soon was the second regional representative, her territory spanning the entire Rocky Mountain region, and she proved to be highly effective. A Midwest regional representative, Stephen Burr, was also added. Although Pritchard, like most Washington-based executives, had qualms about creating staff beyond his immediate control, he recognized that all of the park protection work could not be done from Washington and systematically extended the regional system as funding allowed. He cultivated board support for this expansion by holding one board meeting annually in a park. The regional staff could explain the work they were doing, and trustees often examined the tangible results first-hand.

Gil Stucker stepped down as chairman of the board at the November 1983 meeting. Stephen Mather McPherson, grandson of Stephen Mather, assumed the chair. He was pleased, he said, to carry on the family tradition. Stucker could leave the chairmanship knowing he had helped the association survive its most severe internal crisis. "The crisis we went through was no beer-and-skittles affair in any man's language; it was a knock-down and drag-out. We were working against time."[23] When Stucker left the board in 1985, rotating off as was by then the practice to ensure a constant supply of "new blood" among trustees, he wrote to McPherson:

Those years of association with all of you at NPCA span the most meaningful and gratifying period of my professional life—sharing in your high purposes, and your equally high expectations and accomplishments. Looking back, I am mindful of what a signal distinction it has been to be one of you, and how special our organization is.[24]

Stucker, a modest, low-key person, credited trustees Bernard R. Meyer and John R. Quarles with much of the labor that carried the association through the 1980 crisis. Paul Pritchard, he said, had brought the association back to health. Although Stucker was criticized by some for not pursuing the removal of Tony Smith more aggressively and quickly, most of his NPCA colleagues credited him with patiently, diplomatically, and effectively guiding the organization through a most difficult period.

The effort to build a sound financial base continued. In 1980 Pritchard hired a director of development, Karen Raible Kress, with whom he found he could work effectively. She soon was promoted to vice president for operations. Her work resulted in significant increases in contributions. Pritchard thought the association's building was an albatross that drained funds from its central mission. It was old and in need of considerable renovation and repair.

NPCA leaders (left to right) Karen Kress, Board Chair Stephen McPherson, Paul Pritchard and Destry Jarvis marked the association's anniversary at the 1984 Annual Dinner.

In 1985 he sold the building and realized a significant profit. The organization's net worth and reserves were growing nicely, ensuring financial security and modest expansion of staff and programs.

Long-Range Planning at NPCA

By 1984 Pritchard and the board believed NPCA's credibility had been reestablished. The organization had been rebuilt, and basic systems were in place for steady organizational growth and program expansion. The future seemed secure enough that long-range corporate planning was necessary. The atmosphere of crisis had passed, and the first strategic plan in the history of the organization was written. NPCA counted 47,000 members and an annual income of nearly $2 million. Its strength was its focused program and commitment to national park protection; its greatest resource was in-depth knowledge of national park issues. On a continuum of environmental organizations, NPCA found itself "positioned within the range of conservative to moderate actions such as research, education and cooperation, and acquisition, lobbying, litigation and pressure." Its strategic position was similar to that of the National Wildlife Federation and The Nature Conservancy.[25]

Goals set in the five-year plan included increased direct action for park protection; strengthened programs to influence legislation, federal policy, and regulations; increased regionalization; and continued improvement of finances. The National Park Trust would be the vehicle to increase direct action, and a $1 million revolving acquisition fund was the goal to firmly establish that program. The federal activities program would expand to increase its influence in that sector, and expansion of a grassroots network, outreach programs, and a regional office system would advance regionalization. Specific development targets were an 80 percent increase in total income, a $500,000 endowment, and a 75,000 membership base by 1990. This plan was specific throughout, establishing quantifiable goals that would allow evaluation of progress. The mission statement in the plan clearly spelled out the aim of the renewed National Parks and Conservation Association. The mission would be:

> To protect and improve the quality of our national parklands and to promote an understanding, appreciation and a sense of personal commitment to national parklands on the part of the American people.[26]

When Pritchard took the challenge of reviving NPCA in 1980, he believed that if he could survive the current crisis, he would see the organization through three stages. First would be rebuilding and establishing basic operational systems (membership, development, financial management, personnel, and technical systems, principally computers) while building expertise in park affairs. The second phase would involve laying the foundation for growth of membership, programs, and influence, and the third would be expansion to an organizational level of maximum effectiveness. Although he had no idea how long each phase might take or how it would be achieved, he envisioned where he thought the organization needed to go. The five-year plan adopted by the board in 1984 would be the second phase.[27]

The National Park System Plan

Strategic planning can provide insight into the condition of an organization as well as focus for future action. The process cannot foresee all forces that will act on an organization, and in 1984, NPCA did not appreciate the effect of the National Park System Plan. After Destry Jarvis's challenge in 1982 was refused by the National Park Service (because the political climate would make meaningful planning by that agency impossible), NPCA took up its own challenge. It would attempt something never done before—a nongovernmental organization would write a plan for a government agency. There were great risks in this, for a private organization would not have access to the inside information available to the agency, nor would there be any assurance that the agency would pay attention to the outsider's work after it was done. But in the tradition of Robert Sterling Yard, who was not daunted by the prospect of his tiny organization attacking Albert Fall's Department of Interior, Pritchard, Jarvis, and the NPCA board resolved to take on the work shirked by Watt and his underlings. At the very least, the undertaking would advance NPCA's expertise in national park matters. As it turned out, the mammoth task placed NPCA under considerable strain.

The project's emphasis would be on park resources. Little attention would be paid to the maintenance and administrative dimensions of National Park Service responsibilities, although they would be factored into analysis where essential. The intended audiences for the plan were the Park Service, Congress, and the public. An executive summary of the project report would be prepared for the general reader.

The project planning team, headed by Jarvis, divided up the task. The first step must be an analysis of precisely what was involved in protecting park resources. What are the resources, and what is known about them? What are the threats to those resources, and how well are they understood? What is the National Park Service doing to address these threats, and what should it be doing? Is the Park Service approach to resource protection based on current knowledge and adequate vision?

A second part of the task, related to the first, involved examining the role of science in decision making about park resources. Any effective resource management program must have a scientific basis, and how was the Park Service doing in this regard? If the scientific foundations of resource protection in national parks were inadequate—and NPCA expected to find that they were—then why were they so? What was necessary to raise the level of scientific research and the quality of the information used to make park resource management decisions?

A third element to be examined was visitors and their park experience. The constant growth of visitation was common knowledge, but how much was known of the visitors, what they came for, and what they did in the parks? How was the Park Service assessing visitor needs and wants, and how was it responding? What kind of experience did the Park Service try to provide for visitors, and how did that experience affect resources? These questions were of special interest to NPCA, given its long history of involvement with issues of standards. Here the issue of the purpose of national parks would come in. Were visitors being provided education and inspiration, or merely recreation? Were appropriate uses being made of parks? Were visitor services, often provided by concessioners, compatible with the goal of protecting resources? What role did interpretation play in visitor services and management?

The adjacent lands survey had revealed many problems related to activity outside park boundaries; therefore, another part of any analysis of resource protection must examine boundary issues. Where do boundaries adequately protect park resources, and where do they not? Boundaries were historically drawn for political and economic reasons rather than to protect resources. As parks become islands amid seas of development and as science identifies new park values such as preserving biological diversity, should boundaries be adjusted, and if so, how? Does park planning include boundary questions? Should it? What significant natural and cultural areas not already part of the National Park System should be added? What criteria should be used to identify new areas—a central NPCA question for so long?

These were some of the questions Destry Jarvis and his team had to address to prepare a National Park System plan. Consultants and project staff were hired. Jean McKendry, Dave Simon, Terry Kilpatrick, and Kirsten Artmen were recruited to work on the project. As work progressed and the scale of the undertaking became evident, Washington staff members were drawn into the work. Susan Buffone, Steve Whitney, Laura Beaty, Robert Pierce, Bill Lienesch, Laura Loomis, Frances Kennedy, Brien Culhane, Kathy Sferra, and Bruce Craig all labored on parts of the project. At one point, according to Bill Lienesch, all the staff usually involved in conservation and legislative affairs except him were working on the plan, and he was stretched to the limit. Everyone was stretched. The work was demanding, requiring exceptionally long hours. Tempers sometimes flared, and schedules were pushed back.[28]

The plan eventually sorted itself into nine volumes that were released in April 1988, well beyond the project's anticipated deadline. Titled *Investing in Park Futures: A Blueprint for Tomorrow* and called by everyone "The National Park System Plan," the report was impressive in its scale and its scope. The plan offered 147 recommendations. NPCA recommended that forty-six natural areas and forty cultural areas be added to the system. It recommended specific boundary adjustments for many units. Internal debate raged about the wisdom of releasing the specifics of these boundary recommendations, and two versions of the boundary study were prepared, one with specifics and one without them. The fear was that identification of suggested boundary adjustments would give ammunition to the anti-park forces such as the National Inholders Association and the emerging Wise Use Movement. Land prices might be inflated in areas suggested for addition to existing units. Each volume was packed with data, much of it acquired from the National Park Service, which cooperated throughout the effort. The rationale for the recommendations was provided at some length. Appendices provided supporting data and extensive bibliographic information. The report was a compendium of information about the state of resource protection in the National Park System as of the mid-1980s.

NPCA reported to Congress, the Department of the Interior, and the National Park Service that resource protection was not adequate, which was not news to anyone except some members of the Reagan administration. Further, natural and cultural resource protection was the key to the National Park System's future. Resource management must be elevated to central importance in the National Park Service agenda and organization. Science and research must be given higher priority and funding and serve as the foundation for all policy decisions involving resource protection. Visitor numbers and their impacts must be carefully and scientifically monitored and controlled, and activities incompatible with resource protection curtailed. Concessions policy must be reformed and regional planning conducted to disperse use and control adjacent land activities that degrade the park resources. (The recommendations on regional recreation planning demonstrate Tony Smith's foresight many years earlier.) Boundaries should be adjusted, inholdings and mineral rights acquired, and new natural and cultural areas added to the system.

This weekend haul of "poached" vegetation at Joshua Tree National Monument illustrates the need for visitor impact management.

Personnel policies should be changed to improve Park Service career paths, provide better training in resource management, and reward people for good work in resource management and research.

A reading of the 1988 National Park System Plan suggests that NPCA had been preparing for this exercise for seventy years. The plan contains new ideas and perspectives, but many of its conclusions and recommendations echo with the ideas of generations of NPCA leaders. The ideas of Yard, John Merriam, Devereux Butcher, and Tony Smith ring through the recommendations. The recurrence of problems, issues, and recommendations is striking. The world may have changed—there were many times more units in the park system than earlier, and new kinds of units—but the basic questions remained the same. What uses should be paramount? What standards should be used to judge appropriate uses of parks, and what criteria should govern system expansion? What should constitute the national park experience? How can threats from commercial development both within and outside the parks be reduced? What role should education play in park management? How can adequate funds be acquired for park protection?

The answers to these questions are familiar: education, inspiration, and resource protection, rather than recreation, should be paramount; resource

NPCA alerted the public in the late 1970s to the problem of external threats to national parks, such as air pollution from nearby power plants.

quality, national significance, and protection of America's natural and cultural heritage should govern what goes into the system; science and resource protection should drive Park Service decisions and dictate budgets as much as possible. New ingredients appear. "Ecosystem management" is introduced, a scientifically informed approach to regional resource management. New tools for resource protection derived from advancing science and technology should be applied, and new planning processes embraced. Yet the National Park System Plan seems mostly to update earlier recommendations. New data had been analyzed, using new tools, and a truly comprehensive analysis done, yet the results were not surprising.

The National Park System Plan was not met with unbridled enthusiasm in the national park world, though no one expected it would be. The report's message was, after all, that the job of protecting park resources was not being done adequately. The Park Service had admitted this early in the decade, but the planning study indicated that little had been done by Congress or the Park Service to improve the situation. This was a message that everyone needed to hear but did not necessarily want to hear. The report, consequently, went out to mixed and rather subdued reviews. No immediate revolutionary changes occurred. The study did not claim to address all of the problems facing the parks, and received criticism for not dealing with maintenance and infrastructure problems. The political climate remained relatively conservative and not amenable to dramatic changes in policy direction.

Preparing and releasing such a report would be only the first step in the effort to implement its recommendations. NPCA revised its strategic plan in 1988 and incorporated in it strategies for implementing the National Park System Plan. An effort to educate Congress and the National Park Service about the plan and its recommendations was foremost in the program for park protection; however, internal stresses generated by the massive effort to produce the plan interfered in the followup necessary to its promotion and implementation. A blowup resulted in the departure of Destry Jarvis from NPCA after sixteen years of service.

As is often the case in organizational upheaval, the problem involved money and executive control. The project proved much more demanding of NPCA resources than anticipated. Once it began, there was no turning back. The association's reputation rode on this project's successful completion. As the work progressed, more staff time was drawn into it. Budget shortfalls required that grant funds be supplemented from the operating budget, and at the peak of the effort, NPCA's operating budget ran into red ink for the first and only time during the Pritchard administration. Pritchard supported the project as a vehicle to provide coherence and systematic structure for the association's park protection work, but he was not willing to let it bankrupt the enterprise. Although there was no real danger of that happening, the stresses of the effort strained the relationship of the two men. The final straw for Pritchard came when he thought he and Jarvis had agreed they would not release the detailed version of the boundary study. When Jarvis released the study, Pritchard decided that Jarvis should move on.[29]

Precisely when the political work to sell the National Park System Plan was needed, its principal architect and advocate left the organization. Bill Lienesch shifted into Jarvis's role, but he did not have the broad knowledge of the plan and could not lead the educational effort that would have been possible under Jarvis's direction. As a result, followup was less than needed to achieve the plan's full impact in the policy arena. The National Park System Plan was in circulation, but NPCA could not aggressively promote it. The organization moved on to other priorities. The plan was a solid body of work and contained many reasonable recommendations. The association worked on the Park Service and Congress to implement some of those recommendations, but promotion for the plan as a whole—as a comprehensive long-range plan— was less than it might have been if its architect had remained with NPCA.

Secretary Watt departed Interior late in 1983 after three years and was succeeded by William Clark. Watt's undersecretary, Donald Hodel, succeeded Clark after fourteen months and forced Park Service Director Dickenson to retire. NPCA saw the opportunity to promote one of its own trustees, seventy-five-year-old William Penn Mott, Jr., a veteran park administrator who began his fifty-four-year parks career with the National Park Service. Mott had directed California State Parks and Recreation under Governor Ronald Reagan. Pritchard supported Mott for the director post. Mott was appointed.

This should have been cause for rejoicing at NPCA, but there were few illusions about how much Mott would be able to improve matters for the

*With the support of NPCA,
William Penn Mott was appointed
NPS Director in 1985.*

national parks under Secretary Hodel, who was pursuing Watt's agenda less stridently and with more success than had Watt himself. Mott repeatedly clashed with William Horn, whom Hodel had appointed as assistant secretary for fish and wildlife and parks when Ray Arnett resigned to become executive director of the National Rifle Association. Horn repeatedly interfered in Mott's efforts to run his agency. NPCA published a series of articles in *National Parks* documenting this interference, and Congress held hearings on the matter.[30] In short, Mott and NPCA spent much of their time and energy fighting to maintain the status quo in the face of administration attempts to encourage commercial uses of resources with little regard for park values.

ORGANIZATIONAL DEVELOPMENT
Building Grassroots Participation

One goal of the 1984 NPCA corporate plan was to expand grassroots activity, and work proceeded on that front despite the demands of the National Park System Plan and political battles in Washington. Russ Butcher and Terri Martin worked industriously and effectively in the Southwest and Rocky Mountain regions. Martin proved her mettle in a struggle to protect Canyonlands National Park from encroachment by a nuclear waste dump. She mobilized grassroots forces in her region, placing NPCA at the forefront of the "Don't Waste Utah" campaign. Davis Canyon, a mere quarter-mile from the boundary of the national park, had been selected as one of five finalists for a high-level nuclear waste depository. This was an undeniable external threat that required new levels of scientific and legal expertise and grassroots organizing in the campaign against it. With the astute behind-the-scenes legal help of her law professor husband, Bill Lockhart, Martin successfully led the fight against this threat to Canyonlands.

In Washington, D.C., Laura Loomis, whose love for national parks had originally brought her to the association in a secretarial role in 1976, was charged with developing a national grassroots system. Her title was "director of grassroots." The aim was to mobilize a constituency, people who lived near parks and cared enough about them to act in their defense. Local park activists, to be called "park watchers," were encouraged to participate in park planning processes, monitor inappropriate uses of parklands, identify threats from activity on adjacent lands, review park budgets to identify deficiencies, publicize threats to the parks, and encourage other citizens to support the parks. NPCA supported these efforts by alerting Park Watchers to developments on the national scene with a newsletter and provided the resources of the Washington office to work on specific issues that involved direct contact with Congress or the National Park Service.

This approach was inspired in part by the example of local groups all over the nation. "Friends of the parks" groups watched many units of the system. The North Cascades Conservation Council and Olympic Park Associates, for example, were actively involved in issues at North Cascades and Olympic national parks. More local watchdogs were necessary, and NPCA tried to recruit them to the cause. (The National Park Service also encouraged "friends" groups at this time.) The National Park Action Program was launched in 1982. Although it never became a major undertaking, the program has slowly evolved for more than a decade.[31]

These goals were achieved by a steady improvement in *National Parks* magazine. Eugenia Connally, editor since 1969, was given greater editorial freedom. Many of her recommendations were adopted, such as reduced editorial comment (confined to a single page), color inside the magazine as well as on the cover, more attention to national park news and events, description of member services, and more feedback from readers. Although the magazine came out less often (bimonthly rather than monthly), its graphic and literary quality steadily improved. Readers could learn what was transpiring on the national park front and what role NPCA was playing in those events. Photographic quality, which had been outstanding during Devereaux Butcher's years as editor, had lagged as magazine funding was pinched, and this quality was restored.

Connally's last issue as editor was the first of 1984. She was succeeded by her assistant editors, Marjorie Corbett and Michele Strutin, who shared editorial duties throughout that year. Strutin became senior editor in 1985, editor in 1988, and was succeeded by Sue Dodge in 1990. All of these editors succeeded in making the magazine increasingly attractive and informative, just the vehicle for promotion of the association needed at this period in its history. Yard's vision of the magazine as a recruitment and retention tool was again achieved.

With a dramatic increase in NPCA membership and resources in the late 1980s and early 1990s, the regional system expanded. NPCA established regional programs in the Northeast, Southeast, heartland (Midwest—the earlier regional office there had been short-lived), Pacific Northwest, and Alaska to complement those already established in the Pacific Southwest and Rocky Mountain regions. The presence of NPCA regional directors with whom Park Watchers could interact directly has boosted the grassroots effort; after seventy years of hoping and scheming, NPCA claims a national presence. The organization is not truly grassroots—that is, derived from local-level activism. Most of its activity is generated from Washington, D.C., and done there, but in the 1990s, its communications with people near the parks has improved, and these people have become an integral and important part of NPCA's approach to its park protection and advocacy work.[32]

Developing Financial Security

The growth of association activity on all fronts depended on increased revenue, and a revenue base required membership. Pritchard struggled to increase membership throughout the 1980s. With millions of people visiting the parks, the potential for membership growth seemed limitless, yet growth was slow. One reason for slow growth early in Pritchard's administration was the precarious state of the association. Prospective members wondered whether it would survive its crisis. Pritchard demonstrated that the organization's troubles were manageable and laid to rest concerns that NPCA might vanish from the scene.

Competition for members was intense. If prospective joiners chose one environmental organization over another, they had to be convinced that their choice would help achieve their goals. Some of NPCA's competitors in the membership recruitment game also claimed the national parks as beneficiaries of their efforts, the Sierra Club and Wilderness Society foremost among them. Also, many prospective members wanted immediate, tangible benefits for paying their dues. The association's magazine competed with *Sierra*, *The Living Wilderness*, and *Audubon*; all were fine publications. One of NPCA's failings under Tony Smith had been its public relations, and public relations was still an important factor. NPCA had believed its cause and its message so obvious that they needed no selling. Smith thought of the association as an "educational and scientific" organization that should be above marketing itself.[19] Now the competition for members was stiff—all groups were practicing the fine art of direct-mail solicitation.

Direct mail as a way of reaching a diffuse constituency with a membership appeal appeared in the 1960s, and many of its techniques were not developed until the 1970s.[33] The direct-mail appeal usually focuses on the need for action to combat a particular threat. The proposed action by the group seeking support is described, and a donation is requested. A material incentive, such as a magazine subscription, is offered, and a convenient self-addressed envelope is provided. When many groups appeal to a limited pool of supporters, it is necessary to craft the appeal carefully.

Direct mail resulted in the dramatic growth of environmental groups between 1960 and 1983. During this period, the number of these groups almost doubled, and the number of people contributing to all groups expanded rapidly. Mitchell describes three growth spurts: one in the 1960s, attributable in part to the concerns raised by Rachel Carson's *Silent Spring*; a second around Earth Day 1970; and a third stimulated by the extreme anti-environmental rhetoric of James Watt. The third spurt led to the most dramatic increases, with the Wilderness Society growing by 144 percent between 1980 and 1983, the Sierra Club, by 90 percent, and Defenders of Wildlife and Friends of the Earth, by about 40 percent.[34]

NPCA grew during the first two periods at approximately the same rate as the Wilderness Society but well below the pace of the Sierra Club and the Audubon Society. The association did not yet have the direct-mail system in place to take advantage of the opportunity for membership growth that Watt provided. It was still recovering, building its operational systems and reestablishing its credibility. Large-scale recruitment that paid such dividends for many environmental groups was costly. NPCA was not yet in a position, financially or philosophically, to make such a financial commitment to recruitment. The staff member who oversaw this effort left in 1983, and a consultant was hired. Market analysis showed much potential, but also revealed that the effort would be costly. Membership in 1984 again approached 50,000. The consultant conservatively estimated that the association could grow to 80,000 members, and that was the goal set in the first corporate plan. The association's goal at this stage was to gain as much as possible from its recruitment investment—a quality rather than quantity approach, with a high retention rate. Pritchard and the board were committed to an aggressive effort to increase membership, but in the Reagan years, they held to this relatively conservative strategy.[35]

The Strategy Changes in the Late 1980s

In 1988 the association began participating in the Publisher's Clearinghouse sweepstakes, which increased membership in leaps and bounds through the gimmick of offering an inexpensive magazine subscription to register for the sweepstakes drawing. Retention of members recruited in this fashion was very low, but the effort did expose many new people to NPCA. Membership, excluding Publisher's Clearinghouse "members," stood at 95,000 in 1989. At this point a new consultant, Dave Dawson, was hired, and he further broadened the approach to recruitment. Dawson had compiled an impressive record with other environmental groups, and his analysis showed that NPCA should be able to expand rapidly to 300,000 if it would make the commitment and investment. His strategy involved offering heavily discounted, first-year memberships. The National Park System Plan had just appeared, and he counseled that it should be used as the centerpiece of the membership drive. The scale of direct-mail appeals must increase dramatically, and there must be considerable investment in developing long-term members from the large number of people who took advantage of the discounted first-year offer.

Pritchard liked Dawson's approach, and the board decided to make the investment. Mailings went out in huge quantities—6 million pieces in the first year, as many as 1.5 million at once. Advertisements were run in mass-market publications, and a public relations campaign to raise the association's general profile was launched. Operational systems were changed to cope with this huge increase in activity, the principal change being a switch from handling membership services with an in-house computer to contracting out these services. Immediate and consistent provision of services to a rapidly increasing membership was essential to the long-term success of the campaign. By the end of 1990, the membership had grown to 200,000, and reached 350,000 in 1993. Dawson's strategy yielded remarkable growth that significantly increased NPCA's resources and allowed program expansion. The association had moved well beyond survival to new heights of prosperity and activity.

STUDIES OF PARK PROBLEMS

NPCA turned its attention in its National Park System Plan to the status of research in and for the parks. The perception of most national parks observers had long been that science and research played an inadequate role in park management. Not since the Leopold and Robbins reports of 1963 had there been any major policy statements on park science. NPCA's work on the plan confirmed the need for a congressional mandate for scientific research to aid decision making on natural and cultural resource issues in the parks.[36] As work on the plan progressed, new National Park Service Director Mott drew up a "twelve-point plan," aimed at revitalizing national park management. This Park Service plan would involve citizens more in park management. One part of the plan proposed convening a panel of experts outside the Park Service to assess the role of research in the parks. Interior Secretary Hodel was not excited about Mott's proposals; he would not support a panel on research.

NPCA decided to supplement its efforts in the National Park System Plan on research by reaching outside its own ranks of experts and consultants. It sought and received a grant to set up the commission on research called for by Mott. Thus was born the NPCA-sponsored Commission on Research and Resource Management Policy in the National Park System. John Gordon, dean of the Yale School of Forestry and Environmental Studies, agreed to chair the commission. Sixteen scientists and national parks experts joined Gordon on this panel, which convened on April 18, 1988. Its work began by examining the Leopold Report and its contention that "natural regulation" should dominate most national park ecosystems and Park Service management policy. The crucial question of whether the Park Service should continue this approach, and what research of the past twenty-five years suggested on the question, lay at the center of discussion. Even as the commission began its work, conditions moved inexorably toward the Yellowstone conflagration of the summer of 1988 and the spotlight it cast on this very issue.

In his welcome to the commission, Paul Pritchard told the group that "The conclusions of our National Park System Plan clearly show the urgent need for greater emphasis on scientific management of the National Park System. But this commission is unique; it will deliberate independently of NPCA and the National Park Service and is encouraged to reach its own conclusions."[37]

Controversial approaches to managing bears in Yellowstone were one impetus for examining the role of science in national park management.

The commission's work took it to fifty park areas and through in-depth discussions with National Park Service staff at all levels. NPCA's Dave Simon and Destry Jarvis, who became an independent consultant in the middle of the project, provided staff support to the effort. As Gordon summarized the project in *National Parks*, the group "set out to identify elements that are found in all National Park System areas and to develop an integrated approach for managing park resources. We recognized that among those resources are remnant ecosystems, rich gene pools, cultural benchmarks, and places for recreational and spiritual renewal."[38] The commission went beyond the Leopold Committee in its analysis, examining the status of research not only on natural resources but on cultural resources and park management itself. The implications of its findings for National Park Service professional standards and educational mission were also explored. The commission worked at its task for a year.

The "Gordon Commission," as it has been called (in the tradition of the "Leopold Committee"), concluded that "The National Park System is threatened by many things, but by none more than ignorance and inattention by the American people." If the system was to be rescued from the many threats described and documented in the previous decade, the "best concepts and tools" would be required.[39] Protecting cultural and natural resources was the most critical obligation of the Park Service. The philosophy underlying management of park resources had evolved since the establishment of Yellowstone, and the latest step had to be "the integration of park resource management into larger regional, and even global, environmental protection activities."[40] The centerpiece of this new stage must be "ecosystem management," an approach guided by the understanding that any ecosystem is not a collection of discrete units but an interacting whole that includes cultural as well as natural elements.[41] Further, because ecosystems do not correspond to any political or administrative boundaries, management must, at the minimum, include interagency efforts.

NPCA convened a one-day conference in March 1989 to allow the research commission to present and discuss its report. This meeting was timed to precede the opening of the Fifty-fourth North American Wildlife and Natural Resources Conference and was titled "National Park Research and Management Policy into the Next Century: the Successor to the Leopold Report." An extensive list of specific recommendations was unveiled. The commission's report made a concise, convincing case for why it was necessary to put new emphasis on research, professional development, and education to achieve goals for cultural and natural resource protection.

Visitor Impact Management

The association also sponsored a study of visitor impact to complement the National Park System Plan and Gordon Commission. Congress in 1978 had required the National Park Service to establish a visitor carrying capacity for each unit of the National Park System. Although the agency had established trail and river capacities in many units, NPCA judged its efforts inadequate; a more systematic and comprehensive response to the congressional mandate was needed. Collaborating with the University of Maryland and Pennsylvania State University, the association set out to develop a management process that incorporated the physical and social aspects of recreational carrying capacity, was based on scientific principles and natural resource management concepts, would be applicable throughout the system (in cultural as well as natural units), and would be usable by field managers and planners. Because much work had been done in this field, the project would involve analyzing and synthesizing a considerable body of literature.

The project was carried out under contract by three scientists and advised by a panel of Park Service and Forest Service people familiar with the issue of carrying capacity. NPCA staff provided assistance, and the undertaking was funded with a grant from the Laurel Foundation. A thorough research review yielded a set of principles and corollaries that should govern design of a process. The undertaking resulted in VIM, the Visitor Impact Management process, an eight-step program to identify impact problems, their causes, and effective management strategies to reduce impacts.[42] NPCA published the results as a contribution to examination of a complex problem and in the spirit of scientific inquiry that had been called for by both the National Park System Plan and the Gordon Commission. Paul Pritchard, in his introduction to the publication of the study results, noted that VIM was not intended to be the final word on the subject but "a condensation of what is currently understood about recreation impacts. We view them as a comprehensive foundation to further research and understanding of the effect man has upon not only park environments, but the world as a whole."[43]

The VIM project was aimed directly at assisting the Park Service. Its technical reports were for researchers, managers in the field, and the Park Service at the Washington office level. As NPCA resources increased, more of these carefully focused projects were conducted. NPCA, for instance, hosted a symposium in October 1989 on the National Park System Advisory Board. The objective was to discuss the role of this important citizen board, examine its strengths and weaknesses, and formulate ideas about changes that might be recommended when the board came up for congressional reauthorization. Other such symposia have examined wilderness in national parks, the future of the National Park Service, the special problems of the Everglades, and various cultural resource issues. Dave Simon edited a collection of essays by legal scholars on use of environmental laws to protect parks, which NPCA published as *Our Common Lands: Defending the National Parks*. NPCA studied the acid rain threat to national parks and published the results. In each of these projects, the scope of discussion was carefully delineated to allow for useful, specific outcomes.

These projects—the plan, the Gordon Commission, and the VIM study— illustrate how the association is attempting to move beyond its traditional methods of education and lobbying into substantive work aimed at assisting the Park Service with specific problems. When the means or political will to do critical work is lacking in government, NPCA has tried to step into the breach. It has marshalled expertise, compiled and analyzed information, and offered what it hoped were useful recommendations. The impact of these recent efforts cannot yet be assessed, but during the Pritchard years NPCA has invested resources in realizing a vision, traceable to the founders, of a nongovernmental organization doing necessary work for parks that the Park Service, often for political reasons, is unable to do.

CULTURAL RESOURCES

During the Pritchard administration, NPCA has paid much attention to cultural resource issues. Since the early 1930s, the association has been concerned with protecting historical and cultural areas. John Merriam chaired an NPA committee that described the importance of preserving the cultural heritage of the United States and raised questions about who should do this. The association long harbored concern about standards in selecting cultural areas for the National Park System, and worried that the proliferating number of cultural units would draw resources from management of the natural parks.

Despite NPA concerns, the cultural dimension of the National Park System expanded until two-thirds of the units were cultural areas. The first

NPCA staff member hired to work especially on cultural issues was Helen Byrd in the late 1970s. For many years she had worked at the National Trust for Historic Preservation, which was engaged in historic preservation of private property. She observed that historic preservation was rather low on the agendas of the National Park Service and NPCA and convinced Tony Smith that NPCA should do more in this arena. With a grant from Mrs. DeWitt Wallace, Byrd came to NPCA to launch the Historic Park Program. Her agenda included efforts to raise the status of the Cultural Resource Division within the National Park Service. By her account, the division was at the bottom of the pecking order when the Office of Archaeology and Historic Preservation moved out of the Park Service into the new Heritage Conservation and Recreation Service in 1978. Reporting to the board in 1980, she said that NPCA could take much credit for recent elevation of the status of the division. "Today the Cultural Resource Division is moved up to an Assistant Directorship with an office on the Director's corridor, which is a status symbol in the Park Service."[44] She worked to build a constituency within NPCA to support cultural resource efforts by creating a "Contact" program, which identified people willing to rally to the support of individual historic parks. She analyzed proposals to add and expand cultural areas in the system. Smith assigned her the mammoth task of reviewing the plans of every cultural unit in the system, but he left the association before she could begin.

Byrd, too, left the association in late 1980 and was replaced by Laura Beaty. Beaty's title was "administrative assistant, historic heritage," and she brought a strong background in history and archaeology to the position. She increased the activity in cultural resources. NPCA joined seventy other groups to form the American Heritage Alliance and began coordinating efforts to pass a National Heritage Act. This became the National Historic Preservation Act Amendments of 1980, which were a major modification of the landmark 1966 Historic Preservation Act. The amendments required more local participation in the National Register designation process, extended the Historic Preservation Fund through 1987, and authorized the National Historic Landmark program. This success significantly boosted historic preservation efforts, and would provide plenty of work for NPCA's cultural resources program in years to come.

By the time NPCA prepared its first corporate plan in 1984, cultural resource considerations were a part of all association programs. The association's general park protection and education efforts embraced cultural issues whenever they arose. Beaty was especially interested in Southwest archaeological sites and paid particular attention to issues involving them.

When Beaty moved on in 1986, Bruce Craig came to NPCA to oversee cultural resource efforts. Craig had been with the National Park Service for a decade and had won NPCA's Freeman Tilden Award for outstanding accomplishments in the field of national park interpretation. A historian by training, Craig joined the NPCA staff because he saw its influence growing. He felt he could make a greater impact on national park affairs from NPCA than from within the Park Service, which struggled with the constraints of a conservative

NPCA was instrumental in the campaign for Great Basin National Park, which was established in 1986.

administration and extensive political meddling. He broadened NPCA's focus on cultural resource activities, reducing somewhat its emphasis on the Southwest and refocusing on issues in the East, especially protection of Civil War sites. He continued Beaty's successful effort to integrate cultural resource issues into most NPCA programs. Major association projects, such as the National Park System Plan and the work of the Gordon Commission, considered cultural resources throughout their analyses. The natural and cultural aspects of the National Park System had finally achieved equal status in NPCA affairs.

GREAT BASIN NATIONAL PARK

One NPCA project during the 1980s was the proposed Great Basin National Park. The association's efforts to establish this park reached back to 1956, when trustee Weldon Heald and NPA Western representative C. Edward Graves first visited Ely, Nevada, to meet with park advocates. Writer and editor Darwin Lambert, an NPA/NPCA trustee from 1958 through 1983, was a long-time leader of the movement to create this park. The NPCA board first officially endorsed a Great Basin Park proposal in 1958.

Many proposals for the park came and went over the years, usually becoming casualties to opposition from Nevada congressmen, but in the early 1980s the political situation in Nevada and in Congress seemed to offer usually good potential to realize the national park dream. In the fall of 1983, Russ

Butcher organized an "exploratory meeting" at Lehman Caves National Monument. Destry Jarvis and Laura Loomis traveled to the meeting from Washington. Other park advocates and key congressional staff attended. No agreement was reached on the scale of the next proposal, but many ideas were floated.[45] Butcher testified for NPCA at a congressional hearing in November 1985 in support of a bill that would create a 174,000-acre Great Basin National Park.

The battle over this park took interesting turns. In 1986 anti-park forces brought Charles Cushman of the National Inholders Association to Nevada to whip up the opposition. The White Pine Chamber of Commerce, which supported the park for its economic boost to the community, invited NPCA to counter Cushman. Bill Lienesch and Russ Butcher went to Ely, where they found that Cushman had been characterizing Lienesch in a radio spot as the "hired hatchet man from Washington." On the advice of Ferrel Hansen of the chamber, Lienesch and Butcher took a low-key approach, ignoring Cushman's accusations.[28] Lambert describes what happened:

> Lienesch and Butcher, speaking quietly, described the NPCA as a nationwide, nonprofit organization that had supported high standards for national parks ever since 1919 and that believed the Wheeler area would make an excellent park. Butcher emphasized the bristlecone pines, saying they were overdue for national park protection. "The scenic environment in which those groves are located is the finest there can be." There was discussion of economic benefits. Nobody could know for sure in advance, but the NPCA men gave recent figures showing that national monuments upgraded to national parks averaged a 35 percent increase in visitation per year. Nevertheless, the Ely headline the next day read, "Park proponents get rough reception," but the accompanying photograph showed Lienesch and Butcher looking calm and reasonable—a consistent attitude that carried them undefeated through two harrowing hours.[46]

Several Great Basin park bills were introduced in 1986, with proposed size ranging from 44,000 to 109,000 acres. NPCA sent *National Parks* editor Michele Strutin to Nevada to prepare a feature on the proposed park, which appeared in the September/October issue. Secretary of the Interior Don Hodel, true to form, said that his department could not afford another national park, prompting a response from Congressman Harry Reid of Nevada that Hodel's comment "reflects the arrogant view of administration bureaucrats that Nevada is nothing more than a federal colony," good enough for a nuclear dump and for bombing ranges but "not good enough for a national park."[47]

The debate heated up. Hearings were held in July, with NPCA recruiting and coordinating pro-park witnesses. After maneuvering and compromising that involved Nevada senators Paul Laxalt and Chic Hecht and representatives Barbara Vucanovich and Harry Reid along with House natural resource leaders Bruce Vento and Morris Udall, a 77,000-acre national park passed

Congress and was signed into law by President Reagan on October 27, 1986. A sixty-two-year effort to create the park, in which NPCA had played a role for the past thirty years, successfully concluded.

The Great Basin story is one example of NPCA's carrying on normal protection and advocacy work in the 1980s. It is only one such story among many. In the Great Basin case, the association pursued its traditional role of park advocate. Other new areas for which it worked included Petroglyph National Monument, the National Park of American Samoa, Jimmy Carter National Historic Site, Zuni-Cibola National Historic Park, and Weir Farm National Historic Site. In other instances, it pursued its park *protection* role. When adjacent shopping mall development threatened the Manassas National Battlefield Park, NPCA joined the National Trust for Historic Preservation to form the National Heritage Coalition, a group dedicated to better protection of all historic park areas. The fight to protect the Manassas battlefield was successful.

The association also revived the Everglades Coalition to readdress the threats to that park. It successfully led opposition to a proposed marketing of state-owned land within the boundaries of Utah national parks, and joined efforts to draft a Clean Air Act to protect and enhance air quality in parks and other primitive areas. It renewed efforts to reform concession policy in the national parks; in a number of instances, lawsuits were filed to protect specific parks.

NEW LEADERS AND NEW IDEAS

Stephen Mather McPherson chaired the NPCA board until 1989. McPherson had worked very effectively with Paul Pritchard to build the organization's strength. He presided over successful board participation in the first and second corporate planning efforts. McPherson was succeeded in 1989 by Norman Cohen, a board member since 1984 who had played a major role in developing the National Park Trust. A business executive, Cohen was to lead the board through decisions that resulted in massive expansion of the member-recruitment effort and dramatically increased association membership. Businessman Gordon Beaham succeeded Cohen as chairman in 1992.

The association launched new programs in the 1980s to protect parks. Annual awards were established to recognize outstanding service to the national park cause and publicize and encourage the important work being done. A March for Parks was initiated in 1990. One of the organizers of the original Earth Day, Denis Hayes, convened a group in 1988 to plan a celebration for the twentieth anniversary of the popular Earth Day event. The goal was to renew the event and create new, ongoing activities. Paul Pritchard was a member of this group. As he was pondering how NPCA might participate, he had a conversation with an organizer of the program that raised funds for cystic fibrosis projects by staging marches. Why not, thought Pritchard, stage marches to build support for parks? This seed grew into March for Parks, which would allow local groups to organize activities coordinated and publicized nationally by NPCA. Local groups could promote their issues and parks while NPCA focused attention on park issues in the national media. The pro-

NPCA Board Chair Stephen Mather McPherson presents the Conservationist of the Year award to Florida Senator Bob Graham.

gram continues, providing an annual media event and fund raiser for the park protection cause.

A FULL RECOVERY

When Paul Pritchard came to NPCA in 1980, the organization's survival was in question. It not only survived, but moved to new levels of achievement and prosperity in the 1980s and 1990s. This period in the association's history has not been without problems. As it begins to celebrate its seventy-fifth year, some observers wonder if it can sustain its exceptional membership growth. Some wonder if its size will become burdensome and interfere with its effectiveness. Some raise the question of whether NPCA's approach maintains a proper balance between legislative activity on Capitol Hill and monitoring the National Park Service. But all agree that NPCA has become *the* expert on national park affairs outside government. They agree that the organization is stronger than ever before, and that its influence is at a new level.

The central feature of this most recent NPCA episode is the return to its focus on national parks. Its traditional core mission has been restored. During this period it has dealt with overriding issues such as planning and the role of science in park policy, without neglecting specific park battles. The search for ways to extend its presence beyond Washington, D.C., has led to a system of regional representatives, with six regional offices operating in 1993; more grassroots activity; and "media events" such as March for Parks. The association has struggled to define and limit its agenda to ensure significant success, but new challenges to national parks appear weekly. In 1993 it labors on concessions reform, external threats, improving the situation of National Park Service personnel, and increasing the agency's standards of professionalism. It promotes legislation to protect the fragile California desert. Reports such as the National Park System Plan and the Gordon Report provide a huge inventory of work to be done.

Gil Stucker, reflecting on NPCA's revival in the 1980s, calls Paul Pritchard a "miracle worker."[48] Pritchard has presided over NPCA's recovery. He has made controversial moves during his years at the helm, but his leadership has undeniably resulted in new levels of achievement for the association. Anthony Wayne Smith had allowed the organization to become too dominated by his personality and his personal agenda. His dominance was intentional, and he failed to see where this was leading him and his organization. With the Smith example and a board mandate to guide him, Pritchard used a much more collaborative approach, working closely with board leadership in setting direction. Pritchard's style was to lead but not to dominate, to consult rather than dictate. Most of the time his approach has worked, bringing the organization back to its focus on national parks, reestablishing its identity as a protector of those parks, and guiding it to unprecedented size and stability in the highly competitive world of environmental nonprofit organizations.

NPCA: Seventy-five and Working

A NATIONAL PARK SYSTEM?

When Robert Sterling Yard and associates gathered in May 1919 and formed the National Parks Association, the National Park System was new and fresh, still being defined. The National Park Service, recently organized, was finding its way while administering seventeen national parks and twenty-three national monuments. The Park Service was created largely because concerned people realized there was no consistency and organization in park management or even in park selection. There was no system of national parks, and they hoped the new bureau would build one.

Argument continues today over whether it succeeded, whether the diverse units under Park Service management constitute the "rational system" that early NPA leaders knew could be achieved. No one disagrees, however, that the scale of the national park enterprise has increased dramatically during NPCA's seventy-five years. The nature of national parks and the business of managing them has changed. In 1991 the "system," according to the *National Park Index*, consisted of fifty national parks, thirteen national preserves, seventy-eight national monuments, sixty-nine national historic sites, thirty-one national historic parks, twenty-six national memorials, eighteen national recreation areas, eleven national battlefields, ten national seashores, four national lakeshores, nine national military parks, ten national wild and scenic rivers and riverways, six national rivers, four national parkways, three national battlefield parks, three national scenic trails, and miscellaneous other units

Massive water projects near Everglades National Park have severely affected the natural systems of the park.

including the White House, the National Capital Parks and National Mall, and the Wolf Trap Farm Park for the Performing Arts, among others. The total of these numbers from the *Index* shows that the National Park Service administered 357 units that include slightly more than eighty million acres. This diverse array would bewilder Yard, John Merriam, and perhaps even Stephen Mather himself.

Seventy-five years have seen remarkable growth of the national park concept—a growth that has been painful, tumultuous, sometimes planned and sometimes not, and always contentious. Nineteenth-century visionaries conceived of a system of parks where the greatest scenic and natural wonders of the United States might be preserved for all time. Part of that vision has been achieved. Yet the world has changed in the century since the national park "movement" really gained momentum, changing the nature of these parks. The NPCA story is one of this changing world and evolving concept.

The association's attention drifted occasionally during the years, especially during the Anthony Wayne Smith period, but history shows that its missions have always been to protect the national park concept and serve as citizen guardian of the parks. Its mission has been *conservation*, and the organization has been *conservative* in its approach. Scholars argue over terms in their attempts to describe the process of protecting nature and natural resources that began in the last quarter of the nineteenth century. They identify a "conservation" school of utilitarians bent on ensuring long-term supplies of commodities from nature, and "preservationists" who aimed for complete protection of parts of the natural world. Throughout its history, NPCA has called itself a "conservation" group, although its business has been preservationist to a fault.

Whatever it is called in the conservation-versus-preservation debate, NPCA has been "conservative" in defining and carrying out its mission. Until the 1960s, it fought doggedly to define national parks as places of exceptional scenery and natural value. *National* parks should be landscapes that inspire and educate, not merely places for enjoyment and recreation. They should be sites of national significance—they should be special. During two periods of social unrest in the United States, the 1930s and 1960s, when social pressures forced the National Park Service to create new categories in the National Park System, the National Parks Association stood strongly for traditional standards of park selection and management. In this sense, it was conservative. It made its strongest stand in the 1930s, leading the organization to positions that seem surprising fifty years later. How could the pro-park NPA be *against* Olympic National Park, Kings Canyon National Park, and expanding Grand Teton National Park? The answer lies in the association's unyielding defense of standards, the central theme of its seventy-five-year history.

Detractors of NPA have called it "purist," and justifiably so throughout most of this history. NPA leaders such as Yard, Wharton, Butcher, and even Olson bore the epithet proudly. They tried to save and protect the best of the nation's wild, scenic, natural, and historic places—and *only* the best, at least within the *National* Park System. They raised no objection to preserving sites

NPCA has argued with the Park Service and fellow conservationists over places such as Olympic National Park, and continues today its long defense of standards.

of regional and local value—they actively encouraged such protection—but state and local government and private groups should do that work. If every local park were to be part of the National Park System, it would be too large, too diverse, and too expensive for the kind of management that national treasures deserve. Early on, NPA saw that Congress could make national parks into valuable political capital. Sullys Hill, Platt, and other areas of questionable or no national significance cheapened the system. The Park Service never had enough money to do its job properly, and if all members of Congress and local promoters (even the Interior secretary) could bring their personal parks into the system, the whole enterprise would slowly sink into mediocrity. Purist NPA would proudly be.

Association leaders may have erred occasionally in their enthusiasm. At times they may have carried their aggressive support of standards too far. Had they worked with Secretary Ickes in the latter half of the 1930s to promote national park wilderness, might the Park Service have embraced wilderness and provided more protection during the past sixty years? Might they have unwittingly aided the anti-park movement against Grand Teton National Park expansion in the 1930s by too strictly adhering to standards? Had they naively allowed themselves to be manipulated by the wily timber industry in the battle to create Olympic National Park? Might they have used more aggressive tactics in the 1950s to convince Park Service Director Connie Wirth to protect natural values in his Mission 66 program? Hindsight allows these questions and tempts a response, but NPA leaders pursued the course they thought was right at the time; they were idealists, and they followed conscience wherever it might lead. Occasionally it led to political stumbles and unanticipated outcomes, but these men could do nothing else. They were mostly amateurs when it came to politics, and their ideals and uncompromising stances sometimes worked against them and their cause.

ADAPTING TO NEW IDEAS

Robert Sterling Yard was an association leader for twenty-five years. William P. Wharton served as a trustee for forty years; Calvin Coolidge was in the White House when Wharton became a trustee, and Lyndon Johnson was there when he retired. Trustee Huston Thompson was active in the organization for nearly a half-century. Susan Schrepfer, in her history of efforts to save the redwoods, calls early park advocates "moderate reformers."[1] They were accomplished men, mostly scientists and educators, who believed the bureaucratic system worked and that they could influence it by employing their persuasive powers and the political skills that had allowed them to be successful in their professional lives. Uncomfortable with confrontation, they sought ways to accommodate their interests without conflict. The literature of the early NPA rings with martial rhetoric ("defense," "battle," the "war on our national parks"), but the park defenders preferred a quiet and orderly process to achieve their ends.

The world in which the association's conservation work was done changed constantly during the long terms of the early stalwarts. While Yard and his colleagues wrestled with the daily challenges of protecting their parks from senators and other "invaders," scientists built new theories and models and derived new understanding of nature. As the science changed, so did conservation, and in the 1950s conservationists redefined their missions. "Ecology" became the watchword, highlighting new threats to nature, as illustrated by Rachel Carson's description of pesticide problems in *Silent Spring*. Such threats were often complex in origin, difficult to detect, and threatening to human health as well as natural resources.[2] All conservation groups were forced to examine themselves—their purposes and their methods—which was painful for the National Parks Association. Anthony Wayne Smith was a difficult man, but the association would have struggled whether or not he was president.

As traditional conservation evolved into environmentalism, the nature of the work changed. The culture of the 1960s forced more activist methods on staid organizations like the National Parks Association. Pressing environmental problems such as population growth and pollution demanded attention; people could not continue saving scenery and protecting wilderness and ignore potentially life-threatening issues. The public, for many reasons, became more concerned about quality of life and demanded action on a broad front. Sociologists Dunlap and Mertig have written that "Many of the existing conservation organizations broadened their focus to encompass a wide range of environmental issues and attracted substantial support from foundations, enabling them to mobilize increased support for environmental causes. In the process they transformed themselves into environmental organizations."[3] NPA nearly failed in its attempt at this transformation. When it tried to place its work in new ecological frameworks and address "second-generation issues" such as pollution and population, it discovered that it had sold itself too well as the defender of national parks. People wanted it to stick with that mission. When it did not, they wandered away.

The association adapted, changed, and survived. Its travails of the 1970s and refocusing in the 1980s resulted in new perspectives on its mission and new methods. As they struggled to refocus in the 1970s, NPCA leaders did not see that they should redefine *park protection* in light of new scientific perspectives and social priorities. They *did* broaden their focus on national parks, incorporate new ideas and methods into park work, and make parks part of the effort to solve problems such as loss of biological diversity. This, it turned out, was what association members and potential members wanted them to do. When they refocused in this way, they began to attract supporters. NPCA leaders in the 1980s did not completely abandon the ideas that had consumed (and jeopardized) the organization in the 1970s, but they realized that threats to the parks are ecological in nature. These threats often come from outside park boundaries and require regional effort, as Tony Smith knew they must.

NPCA reaffirmed the importance of parks. National parks are becoming isolated as ecological islands, and if the values in them—particularly the wildlife, the diversity, the gene pool—are to be protected, creative ways of protecting them outside these islands must be found. Thus, NPCA has returned to its traditional territory, but it is not the old National Parks Association. It is a thoroughly modern organization, competing in the fast-paced world of direct-mail recruiting and public relations.

THE WORK WILL NEVER END

The NPCA story demonstrates that the work of protecting national parks will never end. Parks will never be protected to the degree that people can sit back and say, "Now, that's done. What next?" Yard and his colleagues may have believed their task finite, that they could help create boundaries, write policy, and define good management, and when that was done, the parks would be safe and their work complete. They learned immediately that they were wrong, that boundaries gave little protection, policies changed with political winds, and definitions of "good" management changed with the context. Having fought off the threats of Albert Fall, Yard breathed a sigh of relief in 1923 and said that now the association could get on with its real work, education. But soon the entrepreneurs and developers were ready with new challenges, and Yard was mounting new defenses in the "war on the national parks." Twenty years later, at the age of eighty-three, Yard was still at work, inspiring young acolytes such as Devereux Butcher to carry on the work that he knew by then would continue forever—or at least as long as there were any national parks.

Knowledge and perspectives that emerged from ecology cast the challenge of preserving natural systems, and endangered species such as the Florida panther, in a new light.

Some issues, threats, and problems are never resolved, but reappear again and again to demand response from NPCA and its allies. Threats to park waters have occupied the association since Yard helped organize opposition to Addison Smith's attempted raid on the waters of Yellowstone National Park early in 1920. Decade after decade, defense of Yellowstone waters has been necessary. And the struggle continues. In 1993 NPCA published *Park Waters in Peril*, describing a dozen old and new threats.[4] Yellowstone waters are still at risk, although proposed dams within the park have now been replaced by energy development and mining around it. In the 1920s, Yard could not have imagined that developers drilling for steam, oil, and gas would, in the 1990s, threaten the thermal wonders that led to creation of America's first national park.

The water problems of Everglades National Park have also occupied the association since the 1940s, when Devereux Butcher visited with the Army Corps of Engineers and worked to ensure fresh water for the park's water-dependent ecosystems. In ensuing years the human population of south Florida has grown steadily, as has the Army Corps water-management system. The flow of fresh water into the park is almost totally under human control and comes erratically and often with pollutants that are degrading the natural systems that the park was created to protect. Since the 1930s, for instance, the population of nesting wading birds in the Everglades has decreased 95 percent.[5]

Even the waters of Dinosaur National Monument remain threatened. Dams are still the problem, although no one is proposing to build dams *in* the monument—that much has been achieved. Now the threat involves dams outside the park on the Yampa River, which joins the Green River in the monument. At risk are populations of endangered fish species; the riverine ecosystem is threatened. *Park Waters in Peril* describes twelve of the "most pressing problems confronting efforts to protect park waters," and by no means all such problems in the National Park System.[6] Most problems come from activity on lands and waters outside park boundaries, requiring government action "off the reservation" at a time when, once again, there is a hue and cry—especially in the West—against government and in defense of private property and the right of private enterprise to do business on public land.

This very hue and cry is another threat that returns again and again. "Landgrab," Bernard DeVoto called the uprising of ranchers, loggers, miners, and other public land users in the late 1940s. "Sagebrush Rebellion" it was called in the 1970s, and in the 1990s it is the "Wise Use Movement." This latest incarnation of anti-government and exploitative reaction to conservation and environmentalism aspires to break up the National Park Service and open the national parks to mineral and energy production. Its most ambitious goal is "to eradicate the environmental movement."[7] The Wise Use Movement is a coalition of 250 groups including the American Mining Congress, the American Motorcyclists Association, the National Rifle Association, and the National Cattlemen's Association. It uses the same grassroots organizing and direct-mail tactics that have been effective for environmental groups. Well financed and politically sophisticated, the Wise Use Movement is a modern counterpart to the generations of earlier national park "invaders."

Once again defenders of the national parks must find new ways to counter this old threat. The growing popularity of national parks suggests one way. In 1990 there were more than 252 million recreational visits to national parks. Projections are that visitation will rise to 360 million visits by the year 2000 and perhaps nearly a half-billion by 2010.[8] Although such massive visitations are themselves a threat, they indicate how valuable these places are to people. What, asks NPCA, is more valuable: the oil, gas, and minerals beneath the surface of these popular parks, or the beauty and heritage that is preserved in them? In the pages of *National Parks*, NPCA educates its members to the nature of the Wise Use threat. Complacency among environmentalists is a danger, and the hope is that the educational campaign will help motivate an energetic defense against Wise Use. As Richard Stapleton pointed out in the magazine, "Environmentalists must heed the first rule of politics: learn what people care most about, then answer their needs."[9] Because people care about their national parks, NPCA helps the defense by reminding them that their parks will not be the same if they are mined, logged, and otherwise opened to commodity production. Yard would see in the Wise Use Movement the same "commercial interests" he fought throughout his NPA career.

NPCA has focused on the need to improve the recruitment, training, and support of rangers since Devereux Butcher documented their problems in the 1940s and 1950s.

Other familiar problems continue to occupy the association. Concessions in the parks have always been a concern, especially since the 1940s, and today NPCA has made reforming concession policy a priority. Attracting businesses to provide goods and services in national parks was once difficult; the Park Service provided incentives to lure them such as a monopoly on a closed and captive market, preferential rights to contract renewal, even possessory interest in facilities constructed on federal land. Further, concessioners pay low fees for the privilege of doing business in the parks, and fees they do pay go into the general treasury of the United States rather than to the park or the Park Service.[10] Reform of this situation has been on the NPCA agenda for decades, yet such change proves difficult. A highly organized, special-interest lobby of concessioners successfully fights proposed reforms.

Another persistent problem is National Park Service funding. Despite Mission 66 and other infusions of money over the years, the financial condition of the national parks continues to worsen. In 1992 the General Accounting Office estimated a $2.2-billion-and-growing backlog of park projects. NPCA reports that "Ever-increasing visitor use, dwindling park staff, and limited budgets exacerbate the problem. The number of disintegrating roads and trails, deteriorating and collapsing historic structures, and public safety hazards is increasing with each passing month. Our national parks are, literally, falling apart."[11]

Traveling the parks in the 1940s, Devereux Butcher reported that rangers were overworked, underpaid, frustrated, and often living in miserable conditions. Still, their morale was surprisingly high, for they loved the parks and their work and saw themselves as providing a critically important service. The situation has not improved in the 1990s. In fact, it is worse. Morale is lower today than it was then for many reasons, most of which are related to the inadequate funding of the National Park Service.[12] As it has since Butcher's time, NPCA works to better the situation of the "endangered ranger," lobbying Congress to increase funding and recommending ways the Park Service can improve its personnel practices. Improvement will come slowly. Recession plagues the United States, the national debt grows, and Congress seeks to cut budgets. NPCA recommends improved "intake" programs to recruit better qualified people into the Park Service, an improved salary and promotion system, greater investment in employee development, and increased training in appropriate technical, scientific, interpretive, and administrative fields. All of this will take not only money but organizational change and creativity. Strong leadership is necessary, and even if it appears, the likelihood of achieving all these goals is uncertain. The bureaucratic and political road to improvement is bumpy and steep.

Behind many current problems for the parks and the Park Service are the growing numbers of visitors. When Stephen Mather set out to create a national park bureau and a national park system, one of his first tasks was to increase support for parks. He did this through promotion, alliance with automobile clubs, and development of parks to make them accessible and comfortable for visitors. He succeeded, some might say too well. As new parks joined the system, budgets fell behind need, and the agency struggled for more appropriations. If more money were to go to parks, Congress required that even more people visit them, and the Park Service continued to encourage use with programs such as Mission 66. This growth loop spiraled upward until the visitors themselves became a threat. The National Parks Association worried continuously about this situation. Concerned as it was about protecting the primitive, opposed as it was to mass recreation in "primeval" parks, it fretted and warned time after time that park values were being sacrificed. The refrain continues in the 1990s.

In its 1992 report *Parks in Peril*, NPCA noted that in 1980 visitation to the parks was 210 million, that it grew to 252 million in 1990 and nearly 268 million in 1991. The consequence "mirrors the ills of large urban areas: traffic congestion and gridlock, the degradation of both cultural and natural resources, and human conflict. As the number of visitors increases, so does the damage to our national parks."[13] What should be done? Robert Sterling Yard would approve of current NPCA recommendations: study and identify visitor impacts and plan for their reduction; use education to alert visitors to the problems they create. Anthony Wayne Smith would like the recommendation to use mass transit to reduce vehicle impacts. The 1992 recommendations, in fact, are similar to those he made repeatedly in the 1960s and 1970s—plan, disperse visitation, remove private automobiles from many of the parks.

THE PERENNIAL PROBLEM OF STANDARDS

Has the issue of standards been resolved? Is there widespread agreement on what should be a national park and what a national park should be? National Park Service Bureau Historian Barry Mackintosh has written that "today's system, it is fair to say, is both more and less than it might be.... That its quality has sometimes been compromised was surely inevitable, given the public and political involvement in its evolution befitting a democratic society. All things considered, the wonder is not that the System has fallen short of the ideals set for it, but that it has come so close."[14] The National Park System in the 1990s is different from that envisioned by Yard and his colleagues of the 1920s. Two-thirds of its units are cultural and historical, and it includes urban units.

The system's purposes have changed, and most of this change is accepted by NPCA. Natural areas provide scenery, inspiration, and education, but they are also considered reserves of biological and genetic diversity. Cultural and historic areas associated with minorities offer new opportunities and approaches to interpretation, visitor services, and resource management as demanded by a public that seeks increasingly to be involved in park management.[15] Some parts of the system, particularly the urban units, exist to provide outdoor recreation. NPCA continues to argue that standards should be high and clearly understood, but it accepts the need for diversity of purpose in the system. In its 1988 *National Park System Plan,* it stated that the park system should continue to preserve natural and cultural areas of *national significance.* The association accepts the approach of the 1972 Park Service *National Park System Plan,* which set a goal of identifying and protecting a system of representative natural and cultural areas. Using the 1972 approach, the association concluded in 1987 that "the national park system lacks potential representation of 42 percent of all ecosystems defined by the method employed in the 1972 *NPS Plan."*[16] The Park Service criteria for selecting new areas were judged adequate except that they should be revised "to reflect the possibility of restoring nationally significant but damaged ecosystems."[17] Representative marine and estuarine systems should be included. Many new units should be added to protect deserving areas and keep pace with growing visitor demand.

As its 1988 plan indicates, NPCA remains very involved in the struggle to define the system and hold it to a high standard, but no longer defines either the system or the standards as narrowly as it did in its early days. Recreation is accepted as a legitimate purpose in parts of the system. The association has recommended that "The National Park Service, through its external historic preservation programs, the National Historic Landmark Program and direct management, should invigorate its historical additions, especially in the areas of industrial, labor, architectural/art and ethnic history."[18] Cultural and ethnic diversity in the United States is increasing, and NPCA states that the park system should reflect this change. "The total character of the nation should be commemorated through the preservation and interpretation of significant sites."[19]

Even though the standards have broadened and changed, there must still be standards. NPCA will support "adding *worthy areas* [author's emphasis] to

NPCA has long supported the efforts of National Park Service rangers, interpreters, and environmental educators to reach out to young people and help them appreciate and understand the heritage of the National Park System.

the system and supporting alternative methods of protecting natural and cultural resources that do not qualify for national park status."[19] Its current *Five-Year Plan* (1992-1996) states its mission as "to maintain the integrity of the National Park System and to promote the establishment of new, high-quality areas."[19] Its partners in this will be Congress, the National Park Service, and its allies among environmental groups.

Several people emerge from the NPCA story as pivotal figures. Yard, Wharton, Butcher, Olson, Smith, and Pritchard powerfully influenced the course of association history. All were markedly different personalities and had different styles of leadership, but their impact was great. Yard dominated the early years with his idealistic, sometimes cantankerous, always hard-working presence. Wharton quietly and patiently provided continuity and perspective for four decades, occasionally enduring attacks on his character but persisting when a less dedicated person might have given up. Butcher brought youth, energy, writing and editorial skill, and deep commitment to national park ideals at a time when all of these qualities were greatly needed. Olson thought deeply, wrote eloquently, and continued Wharton's approach. The ambitious, autocratic Smith tried to make the association his instrument for changing the world. And Pritchard brought the organization back from near disaster with his bureaucratic and managerial skill, savvy politics, and pragmatic approach to serious problems.

All of these men profoundly influenced the course of the association. Sometimes they controlled that course, and sometimes not, but they responded to challenge and opportunity as they encountered them and used their particular combinations of abilities and skills to important effect. They did not act alone, but always with a board of trustees. At times the greatest contribution of the board has been the prestige its members have lent the association. Influential scientists, educators, and businessmen have stocked the board. Although their counsel has often been significant, they have sometimes played a supporting role to the prominent leaders. Policy has not always been made by the board but by the executive leadership, most notably during the Smith years. This combination of leadership styles and organizational structure has been inefficient and sometimes problematical, but the organization survives.

When founders launched NPCA into the world of conservation action, they did so because they did not think government could or would do all that was necessary to ensure that national parks served their highest purpose. They knew what those purposes should be and organized to promote and protect them. Their principal tool would be education. Their faith and confidence in the rightness of their mission and their method was high. No statement in their discussions seventy-five years ago suggests how long they thought it might take to achieve their goals. Perhaps they never considered the question. They would start the work, and it would go on until it was done.

The founders were of a privileged class of Americans. They were an elite—men who had achieved success in their professions, who were civic minded, and who sought to improve their society and the world. They came together to advance an idea and a social institution. The idea was that people should understand nature so that they could understand who and what they were. An emerging science and its evolutionary perspective was their inspiration. They believed that people needed places where they could encounter and study nature, where they could be inspired and educated. They saw the national parks as an institution dedicated to this high purpose.

Others identified different purposes for the parks, and throughout its history NPCA has engaged in a debate over the parks' nature and purpose. The association's fortunes have waxed and waned during seven decades, but it has returned time and again to the question of the purpose of national parks. It has specialized in this question more than any other conservation organization. A central theme winding through its story is its struggle to maintain its identity around the issues of national park purpose and standards. The more focused the association has been on these issues, the more successful it has been in terms of organizational health and stability and effect in public debates.

The association has tried to be the conscience of the National Park Service. Yard cast himself in that role with Stephen Mather, and Mather was not pleased. Park Service directors have found themselves explaining and justify-

NPCA's long effort to add new areas to the park system continued in the 1990s with its work for the California Desert Protection Act.

ing to the NPCA leadership their decisions and the agency's policies. They have sometimes been criticized and at other times supported by the association; the NPCA view mattered to them at times, and on other occasions it did not. NPCA has struggled to define its relationship with the Park Service, and that struggle has created serious stress on the association. Despite its conservation idealism and purism, NPCA has never staked out a narrow and inflexible ideological position in its relations with the Park Service. Attempts have been made to test each case against a broad litmus of park values. As a consequence, sometimes the two organizations have agreed; at other times they have not.

The context of the National Park System, the agency managing that system, and the nongovernmental voluntary association watching both of them has changed over the seven-decade span of NPCA history. The park system and its purpose have diversified. A wilderness system has appeared. Environmentalism has redefined some of the issues to be addressed by the association. The politics of conservation has changed. NPCA has been forced to adapt or be rendered useless. The future for the organization is as uncertain now as it was in 1919, but now its leaders know that its work will never be completed. The only certainty is that the national parks will need defenders, because the threats to the National Park System will continue.

Bibliography

Chapter 1
The Stage Is Set

1. George Catlin, *Illustrations of the Manners, Customs and Conditions of the North American Indians*, 2 vols. (London: H.G. Bohn, 1851), 1:262.
2. John Ise, *Our National Park Policy: A Critical History* (Baltimore: Johns Hopkins Press, 1961), 53.
3. Alfred Runte, *Yosemite: The Embattled Wilderness* (Lincoln: University of Nebraska Press, 1990), 21.
4. Ibid., 26.
5. United States, Forty-second Congress, Second Session, February 1872, Stat. 17,32.
6. Aubrey Haines, *Mountain Fever* (Portland: Oregon Historical Society, 1962), 199-200.
7. Ise, *Our National Park Policy*, 122.
8. Ibid., 137.
9. Ibid., 141.
10. Runte, Alfred. *National Parks: The American Experience* (Lincoln: University of Nebraska Press, 1979), 7.
11. American Academy of Political and Social Science, *Annals*, 25 (March 1905), 370.
12. Stephen Fox, *John Muir and His Legacy: The American Conservation Movement* (Boston: Little, Brown and Company, 1981), 132.
13. See Douglas H. Strong, "The Rise of American Esthetic Conservation," *National Parks Magazine*, 44 (February 1970), 5-7; Samuel P. Hays, *Conservation and the Gospel of Efficiency: The Progressive Conservation Movement, 1890-1920* (Cambridge, Mass.: Harvard University Press, 1959), 189-98; and Fox, *John Muir and His Legacy*, 103-47.
14. Fox, *John Muir and His Legacy*, 108.

15. Robert Shankland, *Steve Mather of the National Parks*, 3d ed. (New York: Alfred A. Knopf, 1970), 104.

16. Ise, *Our National Park Policy*, 186.

17. Runte, *National Parks*, 95.

18. Ibid., 94.

19. Ibid., 102.

20. Ibid., 99.

21. Ibid., 103.

22. Ibid., 104.

23. J. Horace McFarland, "Shall We Have Ugly Conservation?" *The Outlook*, 91 (March 13, 1909), 595.

24. Shankland, *Steve Mather*, 7-8.

25. Donald C. Swain, *Wilderness Defender: Horace M. Albright and Conservation* (Chicago: University of Chicago Press, 1970), 44.

26. Ibid., 55.

27. United States, Sixty-fourth Congress, First Session, August 25, 1916. *U.S. Statutes at Large*, 39:535.

28. Shankland, *Steve Mather*, 106.

29. Swain, *Wilderness Defender*, 63.

30. Ibid., 66.

31. Ibid., 67.

32. Minutes of the Board of Trustees, National Parks Association, December 5, 1930. NPCA Papers.

33. Minutes of the Executive Committee, National Parks Educational Committee, April 9, 1919. NPCA Papers.

34. Shankland, *Steve Mather*, 167.

35. Susan R. Schrepfer, *The Fight to Save the Redwoods: A History of Environmental Reform, 1917-1978* (Madison: University of Wisconsin Press, 1983), 234.

36. Robert Sterling Yard, "The Need of [sic] a National Parks Association and its [sic] Scope and Purposes," Minutes, National Parks Educational Committee, June 26, 1918. NPCA Papers.

Chapter 2
The Founding of the National Parks Association

1. Minutes of the Board of Trustees, National Parks Association, May 29, 1919. NPCA Papers.

2. Yard, "The Need of a National Parks Association," Minutes, National Parks Educational Committee, June 26, 1918. NPCA Papers.

3. Minutes of the Executive Committee, National Parks Association, October 13, 1919. NPCA Papers.

4. Minutes of the Board of Trustees, National Parks Association, October 30, 1919. NPCA Papers.

5. Richard A. Bartlett, *Yellowstone: A Wilderness Besieged* (Tucson: University of Arizona Press, 1985), 384.

6. National Parks Association, "The Disaster to the Yellowstone Elk Herd," *News Bulletin*, 4 (December 30, 1919), 2.

7. National Parks Association, "The Complicated Problem of the Yellowstone Elk," *News Bulletin*, 7 (April 30, 1920), 2.

8. Minutes of the Board of Trustees, National Parks Association, November 18, 1920. NPCA Papers.

9. Ise, *Our National Park Policy*, 288.

10. National Parks Association, *The Nation's Parks*, 1 (Summer 1920), 11.

11. Minutes of the Executive Committee, National Parks Association, January 10, 1921. NPCA Papers.

12. Minutes of the Executive Committee, National Parks Association, March 23, 1921. NPCA Papers.

13. Ise, *Our National Park Policy*, 313.

14. National Parks Association, "Essential Facts Concerning the War on Our National Parks: Sixty-Eighth Congress Edition," 10 (October 1923).

15. Minutes of the Board of Trustees, National Parks Association, May 20, 1921. NPCA Papers.

16. Minutes of the Board of Trustees, National Parks Association, January 22, 1920. NPCA Papers.

17. Shankland, *Steve Mather*, 97.

Chapter 3
Coming of Age in the 1920s

1. Lary M. Dilsaver and William C. Tweed, *Challenge of the Big Trees: A Resource History of Sequoia and Kings Canyon National Parks* (Three Rivers, Calif.: Sequoia Natural History Association, 1990), 117.

2. Minutes of the Executive Committee, National Parks Association, April 29, 1922. NPCA Papers.

3. Robert Sterling Yard, "The Association's Position," *National Parks Bulletin*, 24 (January 30, 1922).

4. Horace M. Albright as told to Robert Cahn, *The Birth of the National Park Service: The Founding Years, 1913-33* (Salt Lake City: Howe Brothers, 1985), 125.

5. Shankland, *Steve Mather*, 219.

6. Ibid., 220.

7. Minutes of the Board of Trustees, National Parks Association, June 8, 1922. NPCA Papers.

8. Minutes of the Executive Committee, National Parks Association, June 1, 1922. NPCA Papers.

9. National Parks Association, *National Parks Bulletin*, 30 (November 8, 1922).

10. National Parks Association, "Essential Facts," 68th Congress Edition, 10 (October 1923).

11. National Parks Association, *National Parks Bulletin*, 31 (November 28, 1922).

12. Will P. Lapoint. "Our National Park Unfairly Attacked," *Las Cruces Citizen*, September 1922. Scrapbook, NPCA Papers.

13. Adela C. Holmquist, *Albuquerque Herald*, December 9, 1922. Scrapbook, NPCA Papers.

14. Samuel Eliot Morrison, Henry Steele Commager, and William E. Leuchtenburg, *The Growth of the American Republic*, Vol. 2 (New York: Oxford University Press, 1980), 417.

15. Herbert Hoover to Charles Walcott, Minutes of the Board of Trustees, National Parks Association, February 26, 1924. NPCA Papers.

16. Minutes of the Executive Committee, National Parks Association, November 23, 1925. NPCA Papers.

17. National Parks Association, "Secretary Work's Open-door Policy," *National Parks Bulletin*, 33 (March 8, 1923).

18. Minutes of the Board of Trustees, National Parks Association, January 23, 1923. NPCA Papers.

19. Minutes of the Executive Committee, National Parks Association, June 22, 1923. NPCA Papers.

20. Robert Sterling Yard, *National Parks Bulletin*, 46 (November 1925).

21. National Parks Association, *National Parks Bulletin*, 25 (January 30, 1922).

22. National Conference on Outdoor Recreation, *Proceedings*, U.S. Senate 68th Congress, 1st Session, Senate Document No. 151, June 6, 1924. Government Printing Office, Washington, D.C., 45-48.

23. Ibid., 190.

24. Ibid., 34-35.

25. Ibid., 63-64.

26. National Conference on Outdoor Recreation, *Recreation Resources of Federal Lands*, U.S. Senate, 70th Congress, 1st Session, Senate Documents, Vol. 20, 1928. Government Printing Office, Washington, D.C., 62.

27. Ibid., 64.

28. Ibid., 71.

29. Ibid., 70.

30. James P. Gilligan, "The Development of Policy and Administration of Forest Service Primitive and Wilderness Areas in the Western United States (Ph.D. dissertation, University of Michigan, 1953).

31. Minutes of the Board of Trustees, National Parks Association, May 31, 1928. NPCA Papers.

32. Schrepfer, *The Fight to Save the Redwoods*, 54.

33. Minutes of the Board of Trustees, National Parks Association, May 21, 1925. NPCA Papers.

34. National Parks Association, *National Parks Bulletin*, 37 (January 21, 1924).

36. National Parks Association, *National Parks Bulletin*, 38 (March 31, 1924).

Chapter 4
What Is–or Should Be–a National Park?

1. Minutes of the Executive Committee, National Parks Association, June 22, 1923. NPCA Papers.
2. Minutes of the Board of Trustees, National Parks Association, May 29, 1919. NPCA Papers.
3. Robert Sterling Yard, "Historical Basis of National Park Standards," *National Parks Bulletin*, 10, 57 (November 1929).
4. Ibid., 5.
5. United States, Sixty-fourth Congress, First Session, August 25, 1916. *U.S. Statutes at Large*, 39:535.
6. Albright and Cahn, *The Birth of the National Park Service*, 69.
7. National Park Service, *Annual Report*, 1918.
8. National Parks Association, *National Parks Bulletin*, 37, (January 21, 1924).
9. National Parks Association, *National Parks Bulletin*, 38, (March 31, 1924).
10. Robert Sterling Yard, "Politics in Our National Parks," *American Forests*, (August 1926), 486.
11. Ibid., 488.
12. Ibid., 489.
13. Minutes of the Board of Trustees, National Parks Association, May 19, 1926. NPCA Papers.
14. Robert Sterling Yard, "Standards of Our National Park System," *National Parks Bulletin*, 8, 51 (April 1927), 4.
15. Ise, *Our National Park Policy*, 298.
16. National Parks Association, Information Circular No. 16, December 19, 1928. NPCA Papers.
17. George H. Harvey, Jr., to Nicholas J. Sinnott, April 4, 1928, published in *National Parks News Service*, Release No. 8 (April 12, 1928). NPCA Papers.
18. Statements from organizations are in *National Parks News Service*, Release No. 31 (January 6, 1930).
19. Minutes of the Board of Trustees, National Parks Association, December 5, 1930. NPCA Papers.
20. Minutes of the Board of Trustees, National Parks Association, May 19, 1931. NPCA Papers.
21. Schrepfer, *The Fight to Save the Redwoods*, 45.
22. National Parks Association, "Report of the Committee on Study of Educational Problems in the National Parks," *National Parks Bulletin*, 9, 56 (April 1929), 2.
23. Swain, *Wilderness Defender*, 247.
24. Frederick Law Olmsted, Jr. to Robert Sterling Yard, February 11, 1930. National Archives, Records of Horace Albright, Record Group 79, Entry 17, Box 4.
25. Albright and Cahn, *The Birth of the National Park Service*, 295.
26. National Parks Association, *National Parks Bulletin*, 13, 61 (February 1936).
27. National Parks Association, "Losing Our Primeval System in Vast Expansion," *National Parks Bulletin*, 13, 61 (February 1936), 2.

28. William P. Wharton, "The National Primeval Parks," *National Parks Bulletin*, 13, 62 (February 1937), 4.

29. Minutes of the Board of Trustees, National Parks Association, May 14, 1936. NPCA Papers.

30. William P. Wharton, "Park Service Leader Abandons National Park Standards," *National Parks Bulletin*, 14, 65 (June 1938).

31. Ibid., 5.

32. Minutes of the Board of Trustees, National Parks Association, May 10, 1938. NPCA Papers.

33. Carsten Lien, *Olympic Battleground: The Power Politics of Timber Preservation* (San Francisco: Sierra Club Books, 1991), 185.

34. Cited in Lien, 131.

35. Arno Cammerer, "Maintenance of the Primeval in National Parks," *Appalachia*, 22 (December 1938), 9.

36. Rosalie Edge, "Conservation—Come and Get it," Emergency Conservation Committee Report for 1938, pp. 3-4. Cited in Lien, *Olympic Battleground*, 198.

37. Minutes of the Executive Committee, National Parks Association, October 14, 1937. NPCA Papers.

38. Minutes of the Board of Trustees, National Parks Association, May 18, 1939. NPCA Papers.

39. Swain, *Wilderness Defender*, 252.

40. Ibid., 247.

41. Ise, *Our National Park Policy*, 439.

42. Ibid., 648.

43. National Parks Association, "National Primeval Park Standards: A Declaration of Policy," *National Parks Magazine*, 83 (October-December 1945), 11.

44. National Parks Association, "A National Policy for the Establishment and Protection of National Parks and Monuments," *National Parks Magazine*, 31, 128 (January-March 1957), 7.

Chapter 5
Surviving the 1930s

1. Minutes of the Executive Committee, National Parks Association, November 25, 1929. NPCA Papers.

2. Minutes of the Board of Trustees, National Parks Association, May 26, 1930. NPCA Papers.

3. National Parks Association, *Constitution and By-Laws, as amended December 5, 1930*. NPCA Papers.

4. Robert Sterling Yard to John C. Merriam, March 29, 1934. Yard Collection, #5934, Box 1, American Heritage Center, University of Wyoming.

5. Robert Marshall to Yard, October 25, 1934, Yard Collection, Wyoming.

6. James M. Glover, *A Wilderness Original: The Life of Bob Marshall* (Seattle: The Mountaineers, 1986), 183.

7. Barry Mackintosh, *Interpretation in the National Park Service: A Historical Perspective*. Publication of the National Park Service, History Division, Department of the Interior, Washington, D.C., 1986, 15.

8. C. Frank Brockman, "Park Naturalists and the Evolution of National Park Service Interpretation Through World War II," *Journal of Forest History*, (January 1978), 37-38.

9. See Frank Graham, Jr., *The Audubon Ark: A History of the National Audubon Society* (New York: Alfred A. Knopf, 1990), 41-59.

10. Ise, *Our National Park Policy*, 373.

11. Frederick Law Olmsted, Jr., and William P. Wharton, "Report to the National Parks Association of Its Committee on Study of the Proposed Everglades National Park in the State of Florida," National Parks Association, January 18, 1932. NPCA Papers.

12. The Olmsted/Wharton report was published as Senate Document No. 54, 1934.

13. Minutes of the Executive Committee, National Parks Association, April 5, 1934. NPCA Papers.

14. Darwin Lambert, *The Undying Past of Shenandoah National Park* (Niwot, Colo.: Roberts Rinehart, Inc., 1990), 194.

15. Ibid., 203.

16. Minutes of the Board of Trustees, National Parks Association, May 21, 1925. NPCA Papers.

17. Minutes of the Executive Committee, National Parks Association, November 23, 1925. NPCA Papers.

18. Minutes, NPA, October 14, 1937.

19. Robert W. Righter, *Crucible for Conservation: The Creation of Grand Teton National Park* (Boulder: Colorado Associated University Press, 1982), 89.

20. Robert Sterling Yard, "A Call to Service—National Parks Standards Again Endangered," *National Parks News Service*, Bulletin No. 40 (January 1, 1935), 2.

21. Arno Cammerer to Henry B. Ward, June 4, 1935. File 602-1, Box 1055, GT, NPS, RG79, National Archives.

22. William P. Wharton, "The National Primeval Parks," *National Parks Bulletin*, 13, 62 (February 1937), 3.

23. Righter, *Crucible for Conservation*, 91.

24. Ibid., 92.

25. Dilsaver and Tweed, *Challenge of the Big Trees*, 204.

26. T. H. Watkins, *Righteous Pilgrim: The Life and Times of Harold Ickes* (New York: Henry Holt and Company, 1990), 550.

27. Ibid., 570.

28. Minutes of the Executive Committee, National Parks Association, March 15, 1939. NPCA Papers.

29. U.S. Department of the Interior, Memorandum for the Press, n.d. NPCA Papers.

30. William P. Wharton, "Open Letter to Secretary Ickes," April 10, 1939. NPCA Papers.

31. Robert Sterling Yard to Anne Newman, June 2, 1939. The Wilderness Society Papers, Denver Public Library.

32. National Parks Association, "Kings Canyon National Park Established," *National Parks Bulletin*, 15, 68 (July 1940), 23.

33. See Elmer Louis Kayser, *Bricks Without Mortar: The Evolution of George Washington University* (New York: Appleton-Century-Crofts, 1970), 249-96.

34. Minutes of the Board of Trustees, National Parks Association, May 14, 1934. NPCA Papers.

35. Minutes of the Executive Committee, National Parks Association, October 14, 1937. NPCA Papers.

36. Benton McKaye to Robert Marshall, 12 December 1935. The Wilderness Society Papers, Box 11, File 15, Denver Public Library.

37. National Parks Association, "Definition of Technical Terms," *National Parks Bulletin*, 11, 59 (August 1933), 6.

38. National Park Service, *The National Parks: Shaping the System* (Washington, D.C.: U.S. Government Printing Office, 1991).

39. Ise, *Our National Park Policy*, 368.

40. Minutes of the Board of Trustees, National Parks Association, May 14, 1937. NPCA Papers.

Chapter 6
Protecting Parks in Wartime

1. Minutes of the Executive Committee, National Parks Association, October 18, 1939. NPCA Papers.

2. Minutes of the Executive Committee, National Parks Association, May 7, 1941. NPCA Papers.

3. Minutes of the Executive Committee, National Parks Association, April 22, 1940. NPCA Papers.

4. See Donald C. Swain, "Harold Ickes, Horace Albright and the Hundred Days: A Study in Conservation Administration." *Pacific Historical Review*, 34, 65 (1965), 455-65, and Barry Mackintosh, "Harold L. Ickes and the National Park Service," *Journal of Forest History*, 29, 2 (April 1985), 78-84.

5. Waldo G. Leland, "Newton Bishop Drury," *National Parks Magazine*, 25, 105 (April-June 1951), 62.

6. Donald C. Swain, "The National Park Service and the New Deal," *Pacific Historical Review*, 41 (August 1972), 329.

7. Leland, "Newton Bishop Drury," 44.

8. Swain, "The National Park Service and the New Deal," 331.

9.National Parks Association, "Newton B. Drury is New Director of the National Park Service," *National Parks Bulletin*, 15, 68 (July 1940), 16.

10. Minutes of the Executive Committee, National Parks Association, May 9, 1940. NPCA Papers.

11. William P. Wharton to Harold Ickes, November 9, 1940. NPCA Papers.

12. Minutes of the Board of Trustees, National Parks Association, May 8, 1941. NPCA Papers.

13. Edward G. Ballard, "Integrity of National Monument System Threatened by Current Legislation, " *National Parks News Service*, Release No. 45 (October 11, 1940).

14. "Remarks of the President," Minutes of the Board of Trustees, National Parks Association, May 14, 1942. NPCA Papers.

15. Ise, *Our National Park Policy*, 447-48.

16. William P. Wharton, "The National Parks in Wartime," *National Parks Magazine*, 70 (July-September 1942), 3-4.

17. *Seattle Post-Intelligencer*, June 9, 1943, quoted in Lien, *Olympic Battleground*, 213.

18. Lien, *Olympic Battleground*, 215.

19. Ibid., 218-19.

20. National Parks Association, "Sitka Spruce and the War," *National Parks Magazine*, 73 (April-June 1943), 18.

21. Minutes of the Executive Committee, National Parks Association, January 20, 1943. NPCA Papers.

22. Lien, *Olympic Battleground*, 224.

23. Ise, *Our National Park Policy*, 450.

24. Dilsaver and Tweed, *Challenge of the Big Trees*, 190. See also Newton B. Drury, "The National Parks in Wartime," *American Forests*, (August 1943), 375-78, 411.

25. Minutes of the Board of Trustees, National Parks Association, May 27, 1943. NPCA Papers.

26. Ise, *Our National Park Policy*, 452.

27. Righter, *Crucible for Conservation*, 104.

28. Ibid., 108.

29. Minutes of the Executive Committee, National Parks Association, May 26, 1943. NPCA Papers.

30. National Parks Association, "Jackson Hole National Monument," *National Parks News Service*, Release No. 50, (September 12, 1943).

31. Olaus J. Murie, "The Jackson Hole National Monument," *National Parks Magazine*, 75 (October-December 1943), 37.

32. Ibid., 9.

33. Robert Sterling Yark, "Jackson Hole National Monument Borrows Its Grandeur from Surrounding Mountains," *The Living Wilderness*, 8 (October 1943), 3-5.

34. Righter, *Crucible for Conservation*, 127.

35. Minutes of the Board of Trustees, National Parks Association, May 8, 1941. NPCA Papers.

36. Minutes of the Board of Trustees, National Parks Association, May 5, 1944. NPCA Papers.

37. Minutes of the Special Committee to Study National Park Standards and Classification, National Parks Association, February 17, 1945. NPCA Papers.

38. National Parks Association, "National Primeval Park Standards: A Declaration of Policy," *National Parks Magazine*, 83 (October-December 1945), 6.

39. Minutes of the Board of Trustees, National Parks Association, May 24, 1945. NPCA Papers.

40. Russell Butcher, interview with author, Washington, D.C., March 3, 1993.

41. Horace M. Albright, "Making the Parks Known to the People," *The Living Wilderness*, 10, 14 (December 1945), 6.

Chapter 7
Battles in the Post-War Period

1. Ise, *Our National Park Policy*, 448.

2. Richardson, *Dams, Parks and Politics*, 22-23.

3. Minutes of the Executive Committee, National Parks Association, April 18, 1946. NPCA Papers.

4. Richardson, *Dams, Parks and Politics*, 39.

5. Ibid., 41.

6. Quoted in Wallace Stegner, *The Uneasy Chair: A Biography of Bernard DeVoto* (New York: Doubleday and Company, Inc., 1974), 303.

7. Minutes of the Board of Trustees, National Parks Association, May 9, 1946. NPCA Papers.

8. Fred Packard, Statement on H.R. 4053, in Minutes of the Executive Committee, National Parks Association, September 25, 1947. NPCA Papers.

9. National Parks Association, "Hold the Olympic Park Intact," *National Parks Magazine*, 21,90 (July-September, 1947), 4.

10. See Lien, *Olympic Battleground*, 232-253.

11. Olaus J. Murie, "The Olympic Attack," *National Parks Magazine*, 21, 90 (July-September, 1947), 7.

12. Hearings before the Committee on Public Lands Pursuant to H.R. 93, September 16-17, 1947, Washington, D.C., Government Printing Office, 1948, 14.

13. ElmoRichardson, *Dams, Parks and Politics* (Lexington: The University Press of Kentucky, 1973), 43.

14. Minutes of the Board of Trustees, National Parks Association, May 22, 1947. NPCA Papers.

15. Richardson, *Dams, Parks and Politics*, 44.

16. Fred Packard, interview by S. Herbert Evison, January 7, 1971, for the Oral History Project, National Park Service. Transcript in Conservation History Collection, Western History Department, Denver Public Library.

17. Devereux Butcher, *Exploring Our National Parks and Monuments*, 2d ed. (New York: Oxford University Press, 1949).

18. Ise, *Our National Park Policy*, 484.

19. Sigurd F. Olson, "Swift as the Wild Goose Flies," *National Parks Magazine*, 23, 99 (October-December 1949), 3-9.

20. Sigurd F. Olson, "Wilderness Victory, " *National Parks Magazine*, 24, 101 (April-June 1950), 51.

21. Minutes of the Executive Committee, National Parks Association, October 8, 1946. NPCA Papers.

22. William E. Colby, "Yosemite's Fatal Beauty," *National Parks Magazine*, 21, 88 (January-March 1947). 10.

23. Ise, *Our National Park Policy*, 508-9.

24. National Parks Association, "Everglades National Primeval Park Established," *National Parks News Service*, Release No. 58, June 26, 1947.

25. Devereaux Butcher, "Your Secretary Visits the Everglades," *National Parks Magazine* 22, 93 (April-June 1948), 35.

26. See Marc Reisner, *Cadillac Desert: The American West and Its Disappearing Water* (New York: Viking, 1986), 5. See also Donald Worster, *Rivers of Empire: Water, Aridity and the Growth of the American West* (New York: Pantheon, 1985).

27. John Widtsoe, *Success on Irrigation Projects* (New York, 1928), 138, quoted in Wallace Stegner, *The American West as Living Space* (Ann Arbor: University of Michigan Press, 1987), 45.

28. Reisner, *Cadillac Desert*, 180.

29. National Parks Association, "Glacier View Dam," *National Parks Magazine*, 22, 95, (October-December 1948), 4.

30. Richardson, *Dams, Parks and Politics*, 46.

31. Quoted in National Parks Association, "Glacier View Dam—A Victory," *National Parks Magazine*, 23, 98 (July-September 1949), 9.

32. See Richardson, *Dams, Parks and Politics*, 39-113; Fox, *John Muir and His Legacy*, 281-86; Russell Martin, *A Story That Stands Like a Dam: Glen Canyon and the Struggle for the Soul of the West* (New York: Henry Holt, 1989).

33. Fox, *John Muir and His Legacy*, 286.

34. Ibid., 285.

35. Ibid., 286.

36. National Parks Association, "Editorial — Dinosaur Monument and San Jacinto," *National Parks Magazine*, 24, 102 (July-September 1950), 111.

37. Fred Packard to Morris L. Cooke, chairman, President's Resources Policy Commission, n.d., attached to Minutes of the Executive Committee, National Parks Association, May 18, 1950. NPCA Papers.

38. All of the following appeared in *National Parks Magazine*: "Grand Canyon Park and Dinosaur Monument in Danger," by Fred Packard, October-December 1949; "Stop the Dinosaur Power Grab," by Devereux Butcher, April-June 1950; editorial, "Dinosaur Monument and Mount San Jacinto," July-September 1950. "This Is Dinosaur," by Devereux Butcher, October-December 1950; "Alternative Sites for Dinosaur Dams," from a report by General U.S. Grant III, October-December 1951; "The Menaced Dinosaur Monument," by Arthur H. Carhart, January-March 1952; "Dinosaur Dams Again," By Fred M. Packard, January-March 1954.

39. Sigurd F. Olson, "Statement on S. 5000, Echo Park Dam and the National Park System." March 5, 1955. Copy in Devereux Butcher Papers, Box 2, Dinosaur National Monument File, American Heritage Center, University of Wyoming.

40. Fox, *John Muir and His Legacy*, 289.

41. Ibid., 281.

42. Michael Cohen, *The History of the Sierra Club* (San Francisco: Sierra Club, 1988), 155.

43. Leland, "Newton Bishop Drury," 65.

44. Ise, *Our National Park Policy*, 478.

44. Richardson, *Dams, Parks and Politics*, 63.

46. Ibid., 66.

47. Ibid., 67.

48. Minutes of the Board of Trustees, National Parks Association, May 10, 1951. NPCA Papers.

49. Minutes of the Executive Committee, National Parks Association, April 8, 1952. NPCA Papers.

Chapter 8
The Sigurd Olson Years

1. Minutes of the Board of Trustees, National Parks Association, May 21, 1953. NPCA Papers.

2. Sigurd F. Olson, "We Need Wilderness," *National Parks Magazine*, 84 (January-March 1946), 19.

3. Robert Keith Olson, "Introduction," in Mike Link, ed., *The Collected Works of Sigurd F. Olson: The Early Writings, 1921-1934* (Stillwater, Minnesota: Voyageur Press, Inc., 1988), xxi.

4. Ise, *Our National Park Policy*, 534.

5. Bernard DeVoto, "Let's Close the National Parks," *Harper's Magazine* 207 (October 1953), 49-57.

6. Ronald A. Foresta, *America's National Parks and Their Keepers* (Washington, D.C.: Resources for the Future, 1984), 52.

7. Ise, *Our National Park Policy*, 547.

8. Minutes of the Board of Trustees, National Parks Association, May 6, 1955. NPCA Papers.

9. Conrad L. Wirth, *Parks, Politics, and the People* (Norman: University of Oklahoma Press, 1980), 256.

10. Ibid., 258-60.

11. Minutes of the Board of Trustees, National Parks Association, May 10, 1956. NPCA Papers.

12. Minutes of the Executive Committee, National Parks Association, September 12, 1957. NPCA Papers.

13. Devereux Butcher, "The Termites," *National Parks Magazine*, 28, 119, (October-December 1954), 148.

14. Devereux Butcher, "A Letter to the Board of Trustees," n.d., in Minutes of the Executive Committee, National Parks Association, May 23, 1957. NPCA Papers.

15. Minutes of the Executive Committee, National Parks Association, May 23, 1957. NPCA Papers.

16. Wirth, *Parks, Politics, and the People*, 258.

17. Michael Frome, *Regreening the National Parks* (Tucson: University of Arizona Press, 1992), 65.

18.National Parks Association, "A National Policy for the Establishment and Protection of National Parks and Monuments," *National Parks Magazine*, 31, 128 (January-March 1957), 7-11.

19. Russell Butcher, interview with author, Washington, D.C., March 3, 1993.

20. Minutes of the Board of Trustees, National Parks Association, May 24, 1957. NPCA Papers.

21. Olaus Murie to Conrad Wirth, January 7, 1958. NPCA Papers.

22. Horace Albright to Conrad Wirth, January 15, 1958, copied from National Archives. NPCA Papers.

23. Minutes of the Board of Trustees, National Parks Association, January 23, 1958. NPCA Papers.

24. Fox, *John Muir and His Legacy*, 241.

25. Minutes of the Executive Committee, National Parks Association, May 10, 1956. NPCA Papers.

26. Senate Interior Committee, Hearings on S. 1176, June 19-20, 1957, pp. 107-11, quoted in Craig W. Allin, *The Politics of Wilderness Preservation* (Westport, Conn.: Greenwood Press, 1982, 110.

27. Michael Frome, *Battle for the Wilderness* (New York: Praeger, 1974), 139.

28. Minutes of the Board of Trustees, National Parks Association, September 23, 1955. NPCA Papers.

29. Cohen, *The History of the Sierra Club*, 178.

30. Elizabeth Cushman and Martha Hayne, "A Worthwhile Summer," *National Parks Magazine*, 32, 133 (April-June 1958), 74.

31. Minutes of the Executive Committee, National Parks Association, December 13, 1956. NPCA Papers.

32. Cushman and Hayne, "A Worthwhile Summer," 76.

33. Minutes of the Executive Committee, National Parks Association, June 23, 1959. NPCA Papers.

34. Minutes of the Board of Trustees, National Parks Association, May 19, 1960. NPCA Papers.

35. Minutes of the Executive Committee, National Parks Association, March 3, 1959. NPCA Papers.

36. Anthony Wayne Smith, Report to the Trustees, March 5, 1959. Bound in Minutes, 1959. NPCA Papers.

37. Emilie Martin, "Student Volunteers in the National Parks," *National Parks and Conservation* 47, 2 (February 1973), 24-27.

38. Fox, *John Muir and His Legacy*, 286.

39. Foresta, *America's National Parks and Their Keepers*, 55.

40. Minutes of the Executive Committee, National Parks Association, September 10, 1958. NPCA Papers.

41. Foresta, *America's National Parks and Their Keepers*, 62.

Chapter 9
Anthony Wayne Smith Takes Charge

1. Donald A. McCormack to Executive Committee, National Parks Association, July 25, 1959, attached to Minutes, same date. NPCA Papers.

2. William P. Wharton to Charles Woodbury, January 3, 1957. The Wilderness Society Papers, Box 66, File 6, Denver Public Library.

3. William P. Wharton to Devereux Butcher, January 25, 1957. Devereux Butcher Papers, Box 1, Misc. Corresp. File, American Heritage Center, University of Wyoming.

4. Wharton to Woodbury, August 23, 1957. The Wilderness Society Papers, Box 66, File 6, Denver Public Library.

5. Minutes of the Executive Committee, National Parks Association, September 19, 1959. NPCA Papers.

6. Minutes of the Board of Trustees, National Parks Association, November 19, 1959. NPCA Papers.

7. Wharton to Woodbury, December 30, 1959. The Wilderness Society Papers, Box 66, File 6, Denver Public Library.

8. Woodbury to Wharton, December 19, 1959. The Wilderness Society Papers, Box 66, File 6, Denver Public Library.

9. Wharton to Woodbury, March 26, 1960. The Wilderness Society Papers, Box 66, File 6, Denver Public Library.

10. Wharton to Woodbury, April 17, 1960. The Wilderness Society Papers, Box 66, File 6, Denver Public Library.

11. Fox, *John Muir and His Legacy*, 333.

12. Cohen, *The History of the Sierra Club*, 163.

13. Minutes of the Executive Committee, National Parks Association, January 29, 1959. NPCA Papers.

14. Anthony Wayne Smith, Memorandum to the Trustees, National Parks Association, November 5, 1959. NPCA Papers.

15. Woodbury to Wharton, September 12, 1959. The Wilderness Society Papers, Box 66, File 6, Denver Public Library.

16. Wharton to Woodbury, January 7, 1964. The Wilderness Society Papers, Box 66, File 6, Denver Public Library.

17. Sigurd F. Olson to Devereux Butcher, November 27, 1972. Devereux Butcher Papers, Box 1, Misc. Corresp. File, American Heritage Center, University of Wyoming.

18. Alfred Runte, *Yosemite: The Embattled Wilderness* (Lincoln: University of Nebraska Press, 1990), 194-97.

19. Anthony Wayne Smith, "The Tioga Road," *National Parks Magazine*, 33, 136 (January 1959), 10-13.

20. Anthony Wayne Smith to Conrad Wirth, September 27, 1959. NPCA Papers.

21. Conrad Wirth to Anthony Wayne Smith, October 6, 1959. NPCA Papers.

22. Foresta, *America's National Parks and Their Keepers*, 63.

23. Roderick Nash, *Wilderness and the American Mind* (New Haven: Yale University Press, 1982), 222.

24. Richardson, *Dams, Parks and Politics*, 201.

25. Foresta, *America's National Parks and Their Keepers*, 67.

26. Minutes of the , National Parks Association, May 19, 1960. NPCA Papers.

27. United States, 64th Congress, First Session, August 25, 1916. *U.S. Statutes at Large*, 39:535.

28. Public Law 787, 81st Congress, September 1950.

29. "A Statement of Policy Concerning Hunting in the National Parks, Approved by the Board of Trustees of the National Parks Association at the Semi-Annual Meeting, November 3, 1960," *National Parks Magazine*, 35, 160 (January 1961), 15.

30. Conrad Wirth to Anthony Wayne Smith, February 20, 1961, in *National Parks Magazine*, 35, 164 (May 1961), 14.

31. Anthony Wayne Smith to Conrad Wirth, March 21, 1961, in *National Parks Magazine*, 35, 164 (May 1961), 14.

32. "A Statement of Policy Concerning Hunting in the Parks," Board of Trustees, National Parks Association, May 25, 1961, in *National Parks Magazine*, 35, 168 (September 1961), 17.

33. National Park Service, "Wildlife Conservation and Management in the National Parks and Monuments," in Memorandum from the Director to Secretary of the Interior, October 25, 1961. In *National Parks Magazine*, 36, 172 (January 1962), 14.

34. A. Starker Leopold et al., "Report of the Advisory Board on Wildlife Management," March 4, 1963. Insert in *National Parks Magazine*, 37, 186 (April 1963).

35. Anthony Wayne Smith, Report to the Trustees, National Parks Association, May 21, 1963. Bound in Minutes, 1963. NPCA Papers.

36. Russell Martin, *A Story That Stands Like a Dam*: Glen Canyon and the Struggle for the Soul of the West (New York: Henry Holt & Company, 1987), 215-29.

37. Public Law 485, 84th Congress, April 11, 1956.

38. Anthony Wayne Smith, "Precedent Is Important," *National Parks Magazine*, 34, 149 (February 1960), 3.

39. N.B. Bennett, Jr., to Anthony Wayne Smith, October 13, 1961, in *National Parks Magazine*, 36, 173 (February 1962), 15, 19.

40. Stewart L. Udall to Anthony Wayne Smith, March 18, 1963, quoted in Anthony Wayne Smith, "Rainbow Bridge: Record and Requiem," *National Parks Magazine*, 37, 187 (May 1963), 19.

41. Ibid, 265.

43. Anthony Wayne Smith, "Report to the General Membership of the National Parks Association," *National Parks Magazine*, 39, 212 (May 1965), insert.

43. Martin, *A Story That Stands Like a Dam*, 272.

44. Anthony Wayne Smith, "Good News on the Grand Canyon," *National Parks Magazine*, 41, 234 (March 1967), 1.

45. Minutes of the Board of Trustees, National Parks Association, November 19, 1963. NPCA Papers.

46. Anthony Wayne Smith, "Campaign for the Grand Canyon," *National Parks Magazine*, 36, 175 (April 1962), 15.

47. Anthony Wayne Smith, Report to the Trustees, National Parks Association, December 4, 1962, 7. Bound in Minutes, 1962. NPCA Papers.

48. Anthony Wayne Smith, Report to the Trustees, National Parks Association, December 10, 1964. Bound in Minutes, 1964. NPCA Papers.

49. Anthony Wayne Smith, "Potomac Prospect," *National Parks and Conservation*, 45, 2 (February 1971), 1.

50. Outdoor Recreation Resources Review Commission, *Outdoor Recreation for America*, Washington, D.C.: Government Printing Office, 1962, 127.

Chapter 10
The Identity Crisis: 1965-1975

1. Fox, *John Muir and His Legacy*, 306.

2. Anthony Wayne Smith, Report to the Trustees, December 20, 1962, 20. Bound in Minutes, 1962. NPCA Papers.

3. Anthony Wayne Smith, Report to the Trustees, November 19, 1963, 13-14. Bound in Minutes, 1963. NPCA Papers.

4. Anthony Wayne Smith, Policy Report to the Trustees, November 14, 1968, 21. Bound in Minutes, 1968. NPCA Papers.

5. Ernest M. Dickerman to James C. Charlesworth, April 30, 1971. NPCA Papers.

6. Anthony Wayne Smith, Policy Report to the Trustees, November 19, 1970, 14-18. Bound in Minutes, 1970. NPCA Papers.

7. Anthony Wayne Smith, "A New Forestry Program," *National Parks and Conservation Magazine*, 45,5 (May 1971), 35-37.

8. Anthony Wayne Smith, "Report to the General Membership by the President, May 22, 1969." Supplement to *National Parks Magazine*, (May 1969), iii.

9. Anthony Wayne Smith, Policy Report to the Trustees, May 30, 1975. Bound in Minutes, 1975. NPCA Papers.

10. Public Law 88-577, 88th Congress, S4, September 3, 1964, Section 3c. In Stewart M. Brandborg, *A Handbook on the Wilderness Act* (Washington, D.C.: Wilderness Society, n.d.), 3.

11. Anthony Wayne Smith, Policy Report to the Trustees, November 28, 1967. Bound in Minutes, 1967. NPCA Papers.

12. Anthony Wayne Smith, Policy Report to the Trustees, October 21, 1969. Bound in Minutes, 1969. NPCA Papers.

13. Anthony Wayne Smith, Policy Report to the Trustees, December 10, 1964. Bound in Minutes, 1964. NPCA Papers.

14. Anthony Wayne Smith, "A Yellowstone Regional Plan," *National Parks Magazine*, 39,208 (January 1965), 4-12.

15. Anthony Wayne Smith, "Why Not Plan Big?" *National Parks Magazine*, 40,225 (June 1966), 23.

16. NPCA, "Supplement on Regional Planning," in *Preserving Wilderness in Our National Parks* (Washington, D.C.: NPCA, 1971), xxi.

17. J.N. Clarke and D. McCool, *Staking Out the Terrain: Power Differentials Among Natural Resource Management Agencies* (Albany: State University of New York Press, 1985).

18. Foresta, *America's National Parks and Their Keepers*, 176.

19. F. Fraser Darling and Noel D. Eichhorn, *Man and Nature in the National Parks* (Washington, D.C.: The Conservation Foundation, 1967), 32.

20. Minutes of the Executive Committee, National Parks Association, March 20, 1970. NPCA Papers.

21. Anthony Wayne Smith, Policy Report to the Trustees, November 19, 1971. Bound in Minutes, 1971. NPCA Papers.

22. Anthony Wayne Smith, Policy Report to the Trustees, Appendix F, October 26, 1973. Bound in Minutes, 1973. NPCA Papers.

23. Anthony Wayne Smith, Policy Report to the Trustees, November 18, 1976, 9. Bound in Minutes, 1976. NPCA Papers.

24. Anthony Wayne Smith, Policy Report to the Trustees, May 21, 1976, 15. Bound in Minutes, 1976. NPCA Papers.

25. Anthony Wayne Smith, "Our Undefended Borders," *National Parks and Conservation Magazine*, May 1978, in NPCA, *Visions of Tomorrow—Work of Today* (Washington, D.C.: NPCA, 1979), 200.

26. Frank Graham, Jr., *The Audubon Ark: A History of the National Audubon Society* (New York: Alfred A. Knopf, 1990), 243.

Chapter 11
A Crisis of Leadership

1. Anthony Wayne Smith, Memorandum on Financing to the Executive Committee, February 20, 1969. NPCA Papers.

2. Anthony Wayne Smith, Policy Report to the Trustees, May 25, 1973. NPCA Papers.

3. Anthony Wayne Smith to the Board of Trustees, October 23, 1974. NPCA Papers.

4. Memorandum from O.J. Neslage to Board of Trustees, November 15, 1974. NPCA Papers.

5. Staff Reports, Minutes of the Board of Trustees, National Parks and Conservation Association, May 19, 1978. NPCA Papers.

6. Anthony Wayne Smith, Policy Report to the Trustees, May 18, 1979. NPCA Papers.

7. Minutes of the Executive Committee, National Parks and Conservation Association, September 28, 1979. NPCA Papers.

8. Minutes of the Executive Committee, National Parks and Conservation Association, November 29, 1979. NPCA Papers.

9. Richard A. Watson, "Notes," May 26, 1972. NPCA Papers.

10. Carl Reidel to Richard A. Watson, February 22, 1974. NPCA Papers.

11. Richard A. Watson, "Notes," May 17, 1974. NPCA Papers.

12. Richard A. Watson to April Young, December 9, 1979. NPCA Papers.

13. Lawrence Merriam to Gilbert Stucker, December 24, 1979. NPCA Papers.
14. Staff Letter to Trustees, National Parks and Conservation Association, February 21, 1980. NPCA Papers.
15. Anthony Wayne Smith, Final Report to the Board of Trustees, National Parks and Conservation Association, March 28, 1980. NPCA Papers.
16. NPCA Staff Briefing Paper for the Executive Committee, National Parks and Conservation Association, April 1, 1980. NPCA Papers.
17. Eugenia Horstman Connally, Report to the Board of Trustees, National Parks and Conservation Association, "New Directions for NPCA's Magazine," May 1, 1980. NPCA Papers.
18. Richard A. Watson to Lawrence Merriam, November 8, 1973. NPCA Papers.

Chapter 12
Reaffirming the Mission

1. National Parks and Conservation Association, "NPCA Adjacent Lands Survey: No Park Is An Island," *National Parks and Conservation Magazine*, 53,3 (March 1979), 4-9; and "NPCA Adjacent Lands Survey: Part II," *National Parks and Conservation Magazine*, 53,4 (April 1979), 4-7.
2. NPCA, "NPCA Adjacent Lands Survey: No Park Is An Island," 4.
3. NPCA, "NPCA Adjacent Lands Survey: Part II," 5.
4. Ibid., 7.
5. U.S. Department of Interior, National Park Service, *State of the Parks 1980: A Report to Congress* (Washington, D.C.: U.S. Government Printing Office, 1980).
6. Robert Cahn, "The Conservation Challenge of the '80's," in Eugenia Horstman Connally, ed., *National Parks in Crisis* (Washington, D.C.: National Parks and Conservation Association, 1982), 11.
7. Paul Pritchard, interview with author, Washington, D.C., July 7, 1993.
8. Minutes of the Board of Trustees, National Parks and Conservation Association, November 18, 1982. NPCA Papers.
9. Connally, *National Parks in Crisis*, 1.
10. Harold R. Kennedy, "The Shameful State of America's National Parks," *U.S. News and World Report*, 90 (May 25, 1981), 51-52.
11. Russell E. Dickenson, "Time to Catch Up," in Connally, *National Parks in Crisis*, 186-87.
12. Michael Frome, "Building a Broader Park Constituency," in Connally, *National Parks in Crisis*, 199, 202.
13. The State of the National Parks: Planning for the Future Conference, "Recommendations," in Connally, *National Parks in Crisis*, 211.
14. Dee Frankfourth, "National Park Ideals," in Connally, *National Parks in Crisis*, 207.
15. T. Destry Jarvis, "Do We Really Need Another National Park System Plan?" in Connally, *National Parks in Crisis*, 175.
16. Ibid., 177.
17. Robert and Patricia Cahn, "Disputed Territory," *National Parks*, 62, 5-6

(May-June 1987), 30.

18. Dave Simon, interview with author, Washington, D.C., July 7, 1993.

19. Laura Loomis, interview with author, Washington, D.C., July 7, 1993.

20. Minutes of the Board of Trustees, National Parks and Conservation Association, November 18, 1982. NPCA Papers.

21. Robert Cahn, "The National Park System: The People, The Parks, The Politics," *Sierra* (May-June 1983), 52.

22. Minutes of the Board of Trustees, National Parks and Conservation Association, May 13, 1983. NPCA Papers.

23. Gilbert F. Stucker, letter to author, July 30, 1993.

24. Gilbert F. Stucker to Stephen McPherson, in Minutes of the Board of Trustees, National Parks and Conservation Association, May 18, 1985. NPCA Papers.

25. NPCA, *Five Year Plan: 1985-1990* (Washington, D.C.: NPCA, 1984) 10-11.

26. Ibid., 1.

27. Paul Pritchard, interview with author, Washington, D.C., July 12, 1993.

28. William Lienesch, interview with author, Washington, D.C., July 13, 1993.

29. Paul Pritchard, interview with author, July 12, 1993; T. Destry Jarvis, interview with author, July 13, 1993, Arlington, Va.

30. See Robert and Patricia Cahn, "Disputed Territory," and John Kenney, "Interior Sub Rosa," *National Parks*, 63, 9-10 (September-October 1989), 12-14.

31. Thomas J. St. Hilaire, interview with author, Washington, D.C., July 13, 1993.

32. See Nicholas Freudenberg and Carol Steinsapir, "Not in Our Backyards: The Grassroots Environmental Movement," in Riley E. Dunlap and Angela G. Mertig, eds., *American Environmentalism: The U.S. Environmental Movement, 1970-1990* (Philadelphia: Taylor and Francis, 1992), 27-37.

33. Robert Cameron Mitchell, "From Conservation to Environmental Movement: The Development of the Modern Environmental Lobbies," in Michael J. Lacey, ed., *Government and Environmental Politics: Essays on Historical Development Since World War Two* (Washington, D.C.: Woodrow Wilson Center Press, 1991), 94.

34. Ibid., 48.

35. Information on membership provided by Terry Vines, interview with author, Washington, D.C., July 14, 1993.

36. National Parks and Conservation Association, *Investing in Park Futures, Vol. 2: Research in the Parks, An Assessment of Needs* (Washington, D.C.: NPCA, 1988).

37. "NPCA Commission Examines Park Science," *National Parks*, 62, 7-8 (July-August 1988), 10.

38. John C. Gordon, "The Scientific Method," *National Parks*, 63, 5-6 (May-June 1989), 16.

39. Commission on Research and Resource Management Policy in the National Park System, *National Parks: From Vignettes to a Global View* (Washington, D.C.: NPCA, 1989), 13.

40. Ibid., 4.

41. Gordon, "The Scientific Method," 17.

42. Alan Graefe, Fred R. Kuss, and Jerry J. Vaske, *Visitor Impact Management: The Planning Framework*, Vol. 2 (Washington, D.C.: NPCA, 1990), 10.

43. Paul Pritchard, "Foreword," in Fred R. Kuss, Alan R. Graefe, and Jerry J. Vaske, *Visitor Impact Management: A Review of Research*, Vol. 1 (Washington, D.C.: NPCA, 1990), iii.

44. Helen B. Byrd, Memorandum to the Executive Committee and the Board of Trustees, National Parks and Conservation Association, April 1, 1980. NPCA Papers.

45. Darwin Lambert, *Great Basin Drama: The Story of a National Park* (Niwot, Colo.: Roberts Rinehart Publishers, 1991), 174.

46. Lambert, *Great Basin Drama*, 188.

47. Ibid., 194.

48. Gilbert F. Stucker, letter to author, July 30, 1993.

Chapter 13
NPCA: Seventy-five and Working

1. Schrepfer, *The Fight to Save the Redwoods*, 27.

2. Riley E. Dunlap and Angela G. Mertig, "The Evolution of the U.S. Environmental Movement From 1970 to 1990: An Overview," *Society and Natural Resources*, 4, (1991), 210.

3. Ibid., 211.

4. National Parks and Conservation Association, *Park Waters in Peril* (Washington, D.C.: NPCA, 1993).

5. Ibid., 54.

6. Ibid., 1.

7. See Richard M. Stapleton, "Greed vs. Green," *National Parks*, 66, 11-12 (November/December 1992), 32-37, and Stapleton, "A Call to Action," *National Parks*, 67, 3-4 (March/April 1993), 37-40.

8. National Parks and Conservation Association, *A Race Against Time* (Washington, D.C.: NPCA, 1991), 25.

9. Stapleton, "A Call to Action," 40.

10. NPCA, *A Race Against Time*, 18.

11. National Parks and Conservation Association, *Parks in Peril: The Race Against Time Continues* (Washington, D.C.: NPCA, 1992), 2.

12. National Parks and Conservation Association, *The National Park Service: Its Organization and Employees* in Vol.9 *Investing in Park Futures: A Blueprint for Tomorrow* (Washington, D.C.: NPCA, 1988).

13. NPCA, *Parks in Peril*, 13.

14. Barry Mackintosh, *The National Parks: Shaping the System* (Washington, D.C.: National Park Service, 1985), 98.

15. NPCA, *The National Park Service: Its Organization and Employees*, 11.

16. National Parks and Conservation Association, *Investing in Park Futures: The National Park System Plan—Executive Summary* (Washington, D.C.: NPCA, 1988), 34.

17. Ibid., 35.

18. National Parks and Conservation Association, *Five-Year Plan: 1992-1996* (Washington, D.C.: NPCA, 1992), 3.

19. Ibid., 2.

Index